Ernest A. Zadig is the author of two other popular Prentice-Hall books, *The Complete Book of Boating* and *The Complete Book of Pleasure Boat Engines*. Widely known for his writings, lectures, and radio and television appearances on the subject of boating, he was trained as an electrical and mechanical engineer at Cornell University and gained his seamanship and navigation skills in the Navy. Now retired as owner and operator of an electronics manufacturing plant, he is an adjunct professor of electrical engineering and author-in-residence at Florida Institute of Technology. He resides nearby on his 85-foot yacht complete with a photographic darkroom, an electronic laboratory, and a workshop.

THE COMPLETE BOOK OF BOAT ELECTRONICS

Ernest A. Zadig

PRENTICE-HALL, INC.
Englewood Cliffs, N.J. 07632

Library of Congress Cataloging in Publication Data

Zadig, Ernest A.
The complete book of boat electronics.

Includes index.
1. Boats and boating--Electronic equipment.
I. Title.
VM325.Z33 1984 623.8'504 84-13963
ISBN 0-13-156050-6

This book is available at a special discount when ordered in bulk quantities. Contact Prentice-Hall, Inc., General Publishing Division, Special Sales, Englewood Cliffs, N.J. 07632.

10 9 8 7 6 5 4 3 2 1

Printed in the United States of America

ISBN: 0-13-156050-6

Prentice-Hall International, Inc., *London*
Prentice-Hall of Australia Pty. Limited, *Sydney*
Prentice-Hall Canada Inc., *Toronto*
Prentice-Hall of India Private Limited, *New Delhi*
Prentice-Hall of Japan, Inc., *Tokyo*
Prentice-Hall of Southeast Asia Pte. Ltd., *Singapore*
Whitehall Books Limited, *Wellington, New Zealand*
Editora Prentice-Hall do Brasil Ltda., *Rio de Janeiro*

To
AUDREY
mate, wife, and above all else, sweetheart
and to
JANE

Contents

Author's Foreword

Electronics has become almost as important aboard a pleasure boat as its motive power, be that sail or engine. What skipper feels comfortable underway without at least a depth sounder to warn him when the water is "so thin" that grounding becomes a danger? Radio telephones are so widely recognized as a safety feature that often even the smallest boats are so equipped. And that is only the introduction to a choice that encompasses direction finders, radars, lorans, auto pilots, and exotic devices that "latch" themselves electronically to satellites for superaccurate bearings.

However, electronics is a new and unfamiliar shipmate to most boatmen—and that is the reason for this book. Although an uninitiated glance into the assembly of a modern instrument can be overwhelming, the following pages will allay the mystery in logical fashion. The how and the why are traced from their inception in the electron and explained in simple language.

My aim is to make you comfortable and familiar with the electronics now on your boat and to help you choose additions. This book tells you how to use the instruments most efficiently, how to maintain them properly, and how to make emergency repairs.

The text is arranged to give you the choice of studying electronics in depth, going directly to the whys and wherefores of a particular instrument, or even doing both these things. Part I is concerned with basic theory and the functioning of basic components. Part II studies the behavior of these components when they are connected together to form the standard modules that make up electronic circuits.

Part III is an introduction to the physical makeup of the electronic navigational equipment that is the subject of this book and that is dealt with individually and at length in Part IV.

Part V explains the ancillary equipment necessary for the operation of the instruments discussed in Part IV. Part VI gives the skipper the know-how for dealing with the troubles that hopefully will never arise but somehow seem to happen on the best of ships.

Turn those switches on for happy cruising!

Ernest A. Zadig
Florida Institute of Technology
Melbourne, Florida

PART ONE
THE WORLD OF ELECTRONICS

1. *Electronics Aboard Boats*

Figure 1-1. The complexity of instrumentation shown above is in the radio shack of the steamer Leviathan circa 1926. This wireless equipment enabled the vessel to remain in communication world-wide. A modern yacht carrying both a VHF transceiver and an SSB transceiver, each about shoebox size, can rival this performance. (RCA)

A tiny piece of pointed spring wire, called a cat whisker, rests delicately on a pea-sized piece of galena crystal in an ultrasimple detector. An operator with phones clapped tightly to his ears moves the cat whisker haphazardly over the crystal surface. Suddenly he finds a sensitive spot. In his earphones he hears dots and dashes that convey a message originating in a wireless station perhaps halfway around the globe. That was radio communication within the memory of men still living.

In the space of that one long lifespan, the radio phenomenon has become indispensable at sea, whether in the form of a conveyor of messages, a locator of geographical position, or as a detector of objects beyond the power of human vision to see. Not only indispensable to large vessels, it is equally so to pleasure boats of all sizes. Even a small outboard boat without some form of radio aboard is difficult to find.

The infancy of radio coincided with the early days of pleasure boating, and both progressed along parallel lines to their present state of maturity. Radio became more reliable, less expensive, smaller, and able to function on greatly reduced amounts of electric current. Pleasure boats developed larger internal space, installed adequate sources of electricity, and ranged over greater distances so that communication became a needed safety measure. That the two should wed was a natural consequence.

The electronics aboard the well-equipped pleasure boat have far wider scope than the mere transmission and reception of radio frequencies. Depth sounders employ sonic vibrations and measure echos against time. Autopilots orient themselves from a compass and use electronics to maintain a desired course. Hailers have amplifiers that pick up distant voices and holler back. Electronic gadgetry sniffs flammable fumes and warns of danger. Even gasoline engines have become more reliable, thanks to electronic ignition.

The sinking of the Titanic brought shipboard radio to world attention, although the desperate flashing of SOS did not bring help quickly enough to avert the huge loss of life. The tragedy caused radio apparatus aboard all vessels to assume greater importance and to be manned for greater periods of time. Today, passenger-carrying ships must maintain a continuous watch; this does not apply to pleasure boats, however.

Modern technology has reduced the large rack of heavy equipment that characterized early radio installations into an attractive cabinet often smaller than a shoebox. With the reduction in size came a cut in the electric current requirement to an amount that pleasure boat sources could supply. A modern transceiver, adequate for small craft cruising coastal waters, can operate and make hardly a dent in the charge of an average marine storage battery. (The photograph in Figure 1–1 dramatically contrasts the past with the present.)

The earth is surrounded by a medium that makes radio transmission possible over a tremendous range of frequencies or wavelengths. This medium, in turn, is enclosed by ionized shells like immense hollow balls

Figure 1-2. Electromagnetic waves range from low frequency radio up through radar, visible light, and into cosmic rays.

Band	Abbreviation	Range of frequency	Range of wave length
Audio frequency	AF	20 to 20,000 cps	15,000,000 to 15,000 m
Radio frequency	RF	10 kc to 300,000 mc	30,000 m to 0.1 cm
Very low frequency	VLF	10 to 30 kc	30,000 to 10,000 m
Low frequency	LF	30 to 300 kc	10,000 to 1,000 m
Medium frequency	MF	300 to 3,000 kc	1,000 to 10 m
High frequency	HF	3 to 30 mc	100 to 10 m
Very high frequency	VHF	30 to 300 mc	10 to 1 m
Ultra high frequency	UHF	300 to 3,000 mc	100 to 10 cm
Super high frequency	SHF	3,000 to 30,000 mc	10 to 1 cm
Extremely high frequency	EHF	30,000 to 300,000 mc	1 to 0.1 cm
Heat and infrared*		10^6 to 3.9×10^6mc	0.03 to 7.6×10^{-5} cm
Visible spectrum*		3.9×10^6 to 7.9×10^6 mc	7.6×10^{-5} to 3.8×10^{-5} cm
Ultraviolet*		7.9×10^6 to 2.3×10^{10} mc	3.8×10^{-5} to 1.3×10^{-6} cm
X-rays*		2.0×10^6 to 3.0×10^{13} mc	1.5×10^{-5} to 1.0×10^{-9} cm
Gamma rays*		2.3×10^{12} to 3.0×10^{14} mc	1.3×10^{-8} to 1.0×10^{-10} cm
Cosmic rays*		$> 4.8\times10^{15}$ mc	$< 6.2\times10^{-12}$ cm

*Values approximate.

that prevent man's radio activities from being lost into space. Some radio transmissions of extremely high frequency, however, are able to penetrate the shells and travel, perhaps forever, into the great beyond. The total system of frequencies is the so-called radio spectrum.

The early use of the spectrum was willy-nilly, and the comparatively few users chose radio frequencies pretty much at random. Increasing congestion naturally brought government regulation. Today the spectrum is sharply divided and a federal agency assigns channels and frequencies that must be adhered to under penalty for violation. International agreement reserves certain frequencies for worldwide emergency use. The chart in Figure 1–2 has a complete rundown.

Agreement among nations also mandates call signs

Figure 1-3. Radio transmissions travel to their destination either as ground waves, sky waves, or line of sight. In the drawing, (a) is the direct ground wave, (b) is line of sight, and (c) is the sky wave or "skip" that is bounced back by the ionosphere. Local conditions and radio frequency determine which mode takes place.

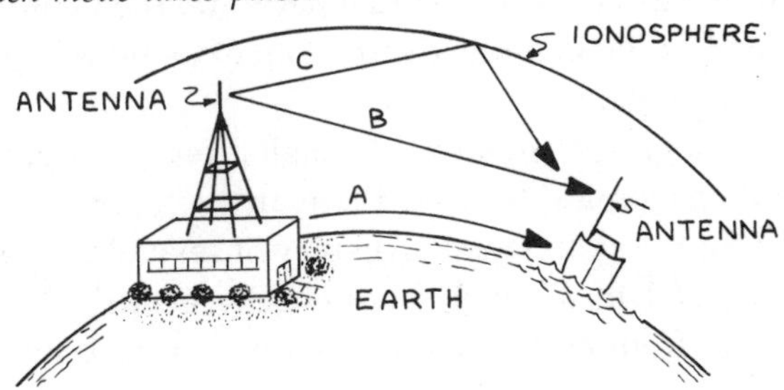

that make it possible instantly to recognize the country of origin of a radio transmitter. For instance, a station bearing a call sign beginning with either the letters K, N, or W is under the jurisdiction of the United States.

Radio waves have certain unique characteristics that affect their usage, and these are linked to frequency or wavelength. (Frequency and wavelength relate in inverse order; the higher the frequency, the shorter the wavelength.) Very low frequencies travel through or along the ground by what is known as a ground wave. Medium frequencies may alternate with ground and sky waves. Ultra-high frequencies go forth like light, are known as line of sight, and theoretically cannot follow the curvature of the earth. Abnormalities occur—a sky wave in its travel may bounce back and forth many times between the earth and the ionized shell surrounding it; a high frequency transmission that should be valid for only 15 or 20 miles suddenly "skips" and is heard 500 miles away. (See Figure 1–3.)

Heinrich Hertz demonstrated the wireless phenomenon with two split wire rings, one the transmitter and the other the receiver, but practical radio requires antennas. Antennas have ranged progressively from very long wires to short whips. As working frequencies went up into the millions of cycles per second with modern technology, the whips became shorter. The multi-wired antenna more than 100 feet long between the masts of an early ocean liner and the standard whip on today's pleasure boat show how the art has advanced.

It was soon discovered that antennas functioned more effectively in certain directions than in others, and that this directivity could be maximized by scientific design. The result is the loop antenna that makes radio direction finders possible. Other specialized antenna configurations that confine the radio energy to a narrow beam are found on marine radar sets and enable a skipper to "see" through fog and the darkness of night.

The electronics aboard a pleasure boat is not at all confined to radio communication, however. The rapid evolution from the crystal detector through the vacuum tube to the ubiquitous transistor has affected almost all the operational gear of the well-equipped small craft. Compasses, engines, steering systems, navigational procedures, even entertainment—all have felt the touch of electronics and have been the better for it.

The vacuum tube enabled the first giant step forward. Instead of the feeble whispering in an earphone, sound now could blast forth from a loudspeaker. A skipper could hail a marina over a distance too great for his unaided voice. Compass courses could be presented in digital form that eliminated misreading. A series of numbers appearing on an instrument face could be referred to a chart for an exact geographical position. A small box could steer a boat more accurately than a human helmsman. And of course television, that superb waster of time, could while away the hours with the boat "on the hook" or tied to a pier.

The wonder of the vacuum tube was overshadowed by its gluttonous appetite for electric current. The tube's standby requirements are a continuously glowing filament and high voltage for its plates and screens. Supplying these wants depletes the electricity normally in short supply on a pleasure boat. In addition, the vacuum tube is fragile, bulky, and generates heat. The transistor took over and spawned another revolution. Electronic equipment shrank to a fraction of its former size, and current drains became almost minuscule.

The transistor is an anomalous device. It depends for its functioning on some of the most esoteric concepts in physics and yet it also leans back slightly to the crystal experience of early radio. The transistor's juxtaposition of miniature size and large capability is fantastic. Within the space of a tiny bead, it can amplify, control, or switch, and it can perform these miracles with the power available from a flashlight battery. Heat is generated only minimally even by the husky transistors at the end of a circuit that delivers appreciable output. Theoretically, a transistor can last "forever."

The miniaturization made possible by the transistor has brought aboard a host of electronic instruments that formerly were excluded by the cramped space available for navigation on a small pleasure boat. A transceiver that can send and receive on more than 55 VHF channels is less than three inches high and occupies less than a square foot of space on a console. Radars that were really at home only on a battleship bridge now in smaller versions do their magic from an easily accommodated cabinet. And the end is not yet because everything is being designed smaller.

Even the individual transistor itself, which seemed the epitome of size reduction, now looms like a giant next to an integrated circuit component, smaller than a dime, that contains hundreds of transistors. One integrated circuit chip comprises an entire system and can function as a major part of a marine instrument or even as a complete calculator. The integrated circuit, or IC, succeeded the printed circuit board on which discrete components were mounted and connected together. The absence of these individual parts and their connections makes the integrated circuit an extremely rugged and reliable item.

The integrated circuit made digital readouts practical and economical because the many components required for the simultaneous display of several numbers are enclosed in a small, hermetically sealed plug-in unit, safe from environmental harm. Loran receivers, depth sounders, remote compass readouts—all tell their findings in everyday numbers. Deluxe transceivers announce with a digital display the channels to which they are tuned.

The gasoline engine would seem an unlikely participant in the electronic revolution, but it has been a major beneficiary. Electronics solved the engine's inherent ignition problem. The breaker points and condenser that have plagued motorist and boatman alike no longer appear on modern engines. The new electronic components show no wear because they are not in physical contact, and the spark they produce is hotter and faster. Not only is reliability better, but response and power have been improved.

Of the many goodies that electronics has brought aboard the pleasure boat, communication is probably the ace. Nothing so engenders a feeling of despair as being in trouble on an apparently limitless sea. That is when a call to the Coast Guard becomes a morale restorer as well as an actual lifeline. Frequencies especially assigned for emergency use are under constant watch. Channels are available in the VHF (very high frequency) system over which a skipper can talk to bridge tenders and to pilots of commercial vessels in the interest of safe navigation.

As an ancillary bonus, electronics is slowly doing away with the tedium of sight reduction calculations in celestial navigation. Pocket calculators are taking over. Some are even programmed to eliminate the need for the bulky sight reduction tables that have been the hallmark of the navigator throughout history.

The rapid march of electronics into modern boat operation may give the impression that the skipper is being replaced, but such a conclusion would be entirely wrong. Electrons, en masse, can accomplish wonders, yet they do not have experienced judgment. Instrumentation can reduce the skipper's tedium, can sharpen his senses, and can quicken his responses, but it cannot replace his brain. What the dials and numbers report has little value until it has been subjected to human thinking.

A tragic example of the need for intelligent human supervision happened off the coast not many years ago. A liner and a freighter collided, resulting in much loss of life. Both vessels were equipped with every modern navigational aid; the radars on both vessels were turned on and functioning. However, no brain on either ship interpreted the radar picture and reacted with correct commands.

The prevalence of flybridges on pleasure craft has prompted manufacturers to make many instruments reasonably weather- and waterproof. Such devices can survive the hazards encountered on these open control stations but, even here, common sense advises protection when not in use. One outstanding unit has O rings to seal the holes around shafts that go through the panel.

As contrary as it may seem, tightly sealing the cabinet of an electronic instrument may not always be the best answer, because of the insidious moist air at marine locations. Somehow or other, this air nearly always manages to get in, despite barriers. Once inside, condensation produces the water that is the killer. Some large interior volumes are provided with "weep" holes so that this water can escape.

The proliferation of electronic equipment has brought an entirely new type of serviceman to the marina, the electronic technician, trained in the use of the necessary sophisticated test instruments such as the oscilloscope, the counter, the signal generator, the standing wave ratio meter, and others. Federal Communications Commission rules mandate that he have either a first class or a second class radio license, preferably endorsed for radar. The high cost of this test equipment has tended to make this group of technicians financially more substantial and more reliable than most service people.

The small size and easy portability of modern electronic equipment for pleasure boats permits the offending unit to be taken to the repair shop instead of having the repair man come to it. Naturally this reduces costs. Radars and radio transceivers, with a few exceptions, do require the technician to come aboard for a personal check. He is required to sign the log and give his license number.

To paraphrase the Bard, what to buy, or what not to buy—that is the question. It is easy enough, figuratively at least, to sink the boat with equipment beyond the reasonable need. Catalogues are enticing, and manufacturers' paeans of praise lull one into a dream of desire. What should a skipper buy?

Obviously, the skipper of an outboard and the skipper of a goldplater would have different shopping lists, but the list could be different even from one goldplater to another. It depends on what use is intended for the boat. Cruising on an inland river, coasting along on the ocean, or passagemaking across the sea, each form of boating has a unique basic requirement.

A deluxe VHF transceiver equipped to operate on all the channels used around the world is an unnecessary expense for a pleasure boat that will run the intracoastal or fish a few miles off the shore. A radar powerful enough to have a 32-mile range will employ only a small portion of its capability on a winding river. Auto pilots are great but only on open water. Radio direction finders are not needed when numbered buoys tell you where you are. Surely a small craft on inland waters will not need Loran. In all cases, common sense, the pocketbook, and a modicum of experience must make the decision.

One basic need for all boats is communication in case of trouble. An outboard cruising local waters may be satisfied with a Citizen's Band radio, although CB is marginal for marine use. The Coast Guard has decided to place a cursory watch on the citizen's band emergency channel, but the probable result of a call for aid on CB will be an individual listener who will use the land line telephone to summon help. The main problems with CB, aside from its limited range, are the interference and the complete lack of legal behavior on the channels.

Very high frequency (VHF) transceivers have taken the place of the outlawed 2-3 megahertz AM radios that were standard on pleasure boats. Whereas the old AM's were available with output powers of 150 watts, the new VHF's are limited by law to 25 watts with the required capability of emitting only 1 watt for local calls. This, plus the line-of-sight character of the high frequency, limits the effective range, although VHF is closely monitored by the Coast Guard from a number of strategically placed high antennas.

A VHF transceiver is considered the basic radio installation for a pleasure boat of any size. Well-developed, commercially available antennas make the most of what power there is and, in effect, concentrate and thereby "amplify" it (see Chapter 24). The VHF sets progress in complexity and cost from 14 channel models to 78 channel models. Access to the channels is made by a dial on the lower-priced units and by numbered buttons, a la touch-tone telephones, on high-priced sets. The legally required frequency stability is achieved with separate quartz crystals for each channel in the older circuits, but the newer development employs one master crystal whose output is synthesized for all channels (see Chapter 7).

Communication for the passagemaker would add a single sideband (SSB) radio transmitter-receiver. This is a world-wide system and with sufficient power would keep the skipper in touch regardless of his position on any sea.

From a practical operating standpoint, a depth sounder ranks with the radio in importance. The choice here is between those that simply indicate bottom depth and those that add a recorded graph of the bottom contour. Dyed-in-the-wool fishermen like the recorder because it yields clues to likely fish locations. The instruments announce the sounded depth either by a scale on the graph, a flash on a dial, or by actual numbers.

A radio direction finder is great for quick and easy determination of position when offshore. This instrument gives an azimuth direction to a selected radio station onshore; this may be transferred to a chart as a line. Two or more such lines fix the boat's location by their intersection. Radio direction finders are nothing more than a radio receiver with a loop antenna and a curser marked in degrees; consequently, they are moderate in price.

A radar is useful—if your type of cruising creates a need for one. In many installations on small boats, the suspicion arises that the radar is more for status than for service.

The original Loran sets used by the Navy and the Air Force in WWII required a genius to operate and a mystic to read the pips. Modern Loran-C receivers are fantastically complicated and yet almost unbelievably easy for even an unskilled person to use. Once the desired station chain is selected, the internal electronics do everything else. Numbers appear that are correlated with correspondingly numbered lines on a chart. The accuracy of fix is so great that fishermen can return unerringly to a productive hole.

Many items remain on the list of goodies, such as autopilots, hailers, and protective devices and alarms. As stated earlier, the man who runs the boat and pays the bills uses his experience to decide what he really needs and how much extra he can afford. The photo in Figure 1–4 shows some representative electronic installations.

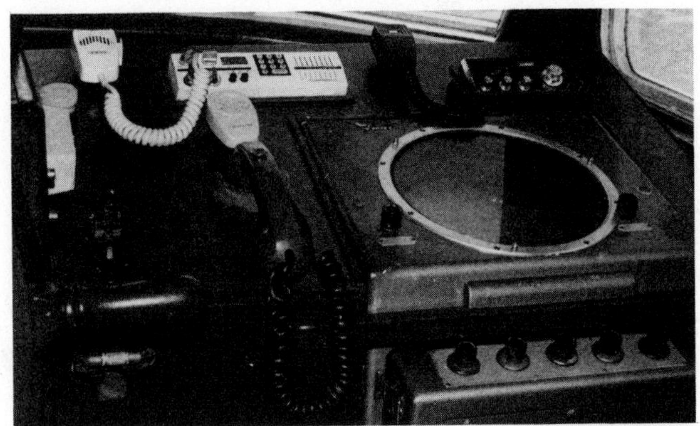

Figure 1-4. This radio console provides VHF communication in addition to Citizen's Band, intercom, and radar.

2. *The Ubiquitous Electron*

The ubiquitous, invisible electron is nature's most basic building block. By its position and number the electron determines the identity of a substance, as well as its physical and chemical characteristics. Whether a material will conduct electricity or act as an insulator and block the flow of current is also decided by its electrons. The most powerful microscope cannot reveal the electron, yet it has been measured, weighed, and clocked by fantastic feats of scientific research and deduction.

A close examination of the electron's life and habits is fascinating in itself, but there is a far more important reason for understanding how and why this submicroscopic bit of matter functions. The electron's unique capability is the basis for modern solid state electronics—the world of the transistor. The electron, and the "hole" it creates by its absence, are the current carriers in these marvels of germanium and silicon that have superseded the vacuum tube and have made miniaturization possible. (See Figure 2–1.)

Figure 2-1. The diode (a two-terminal device) and the transistor (a three-terminal device) are close relatives, both derived from semiconductor material.

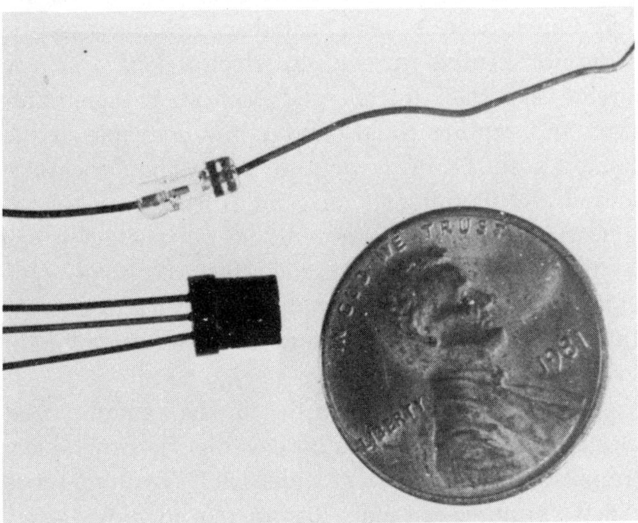

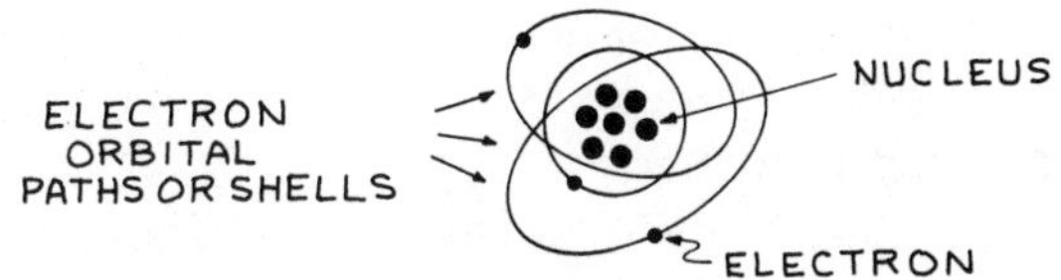

Figure 2-2. The atom contains a nucleus of protons with a positive charge about which the electrons rotate in orbital paths or shells.

The world of the electron is so minute that the imagination hardly can grasp it. Measured by the smallest units and expressed numerically, the diameter of an electron requires thirteen zeros to the right of the decimal point before a significant digit appears! The electron behaves in its sphere very much as the earth does in our solar system—it spins and it also revolves in orbit about a central, sun-like nucleus that is the heart of every atom. Whether this atom is hydrogen, or gold, or whatever is fixed by the number of electrons spinning in these orbits.

The rotation of the electrons of an atom about their nucleus takes place within several concentric shells that have the nucleus as their center. There may be as many as seven shells with sub-shells. Electrons are assigned irrevocably to these shells by a geometric law. Thus, the innermost shell may never have more than two electrons, the second shell never more than eight, the third shell a maximum of eighteen, and so on in orderly progression. Sub-laws also govern—a shell must be filled before the next outer shell may have a tenant. Furthermore, any outermost shell is considered filled when it has eight electrons regardless of its theoretical capacity and then the next inner shell is limited to eighteen. While all this may seem involved, Figure 2–2 shows that the concept is actually straightforward and simple.

Atoms combine into molecules; this, too, is the province of electrons. Whether or not an atom will combine with another is determined by how easily one of its outer electrons can be attracted away, or else by how able it is to snatch an electron from a nearby atom.

This intermingled loss and acquisition of what the chemists call "valence electrons" is the basis of the composition of matter.

An excellent, yet simple illustration of how combination takes place is afforded by a molecule of common table salt, sodium chloride, that is composed of one atom of sodium and one atom of chlorine. The outer shell of the sodium atom contains one loosely held electron. The outer shell of the chlorine atom has seven electrons with room for one more. The chemical marriage is consummated when the loose sodium electron is enticed into the open spot in the chlorine shell and the two atoms join to form a molecule of salt. (See Figure 2–3.)

Figure 2-3. As shown, the sodium atom and the chlorine atom are both neutral. However, when the chlorine atom captures the single electron from the sodium atom, the chlorine becomes negative and the sodium positive. Opposites attract and the two join to become table salt, NaCl.

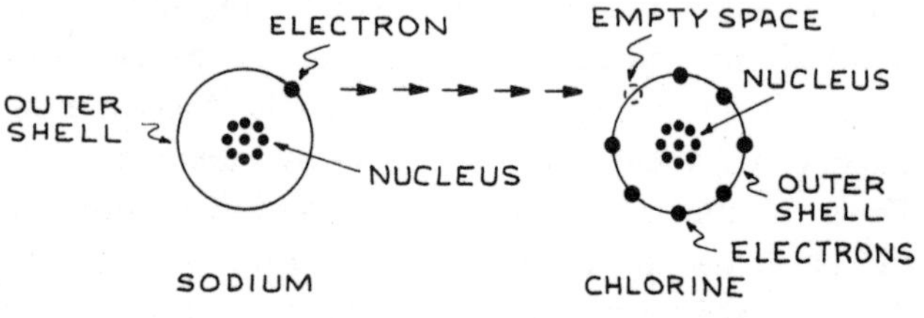

Much may be deduced about the physical and chemical properties of a substance by examining the arrangement of the electrons of its atom. If the outermost shell be filled with the full allowable eight electrons, then that atom has no means of combining with others. In other words, it is chemically inert. When that outer shell has four or less electrons, the chances are good that these electrons can be wooed away easily, thereby pulling that atom into combination with others. An outer shell with only a few empty spaces has great attracting power for the electrons of other atoms and consequently accelerates combinations.

One particular type of electron placement and number is especially important because of its total bearing on modern solid state electronics—the case in which all atoms have only four electrons in their outer shells. Here adjacent atoms share electrons with each other and use them jointly to fill all outer shells to the limit of eight. The resulting substance is the semiconductor material that makes the transistor and the diode possible. (The theoretical matrix of this important semi-conductor is shown in Figure 2–4.) Germanium and silicon, the semi-conductors that dominate today's solid state electronic devices, are in this category.

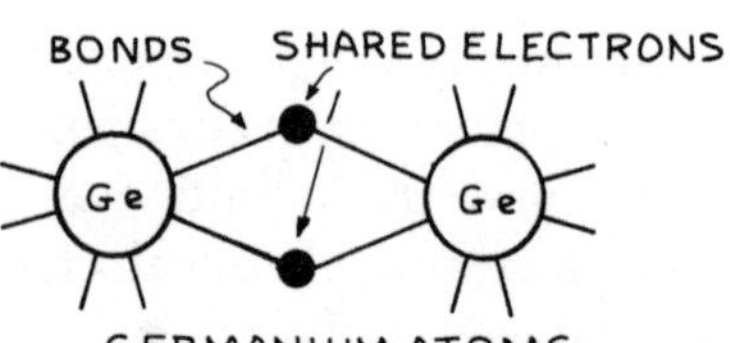

Figure 2-4. Germanium atoms share their electrons with neighboring germanium atoms in the form of "covalent bonds," an important feature from the standpoint of semiconductor operation.

Electrons can terminate their allegiance to a particular atom if they are given sufficient energy to make the leap. Several methods are able to accomplish this result. One is heat. A metal heated above a characteristic minimum temperature ejects electrons in a steady stream, especially if a charged electrode is there to attract and capture them. This is the principle of the heated cathode in a vacuum tube. The process is known as "thermionic emission." (See Figure 2–5.)

Certain rare substances emit electrons when struck by photons or quanta of light. This is "photoelectric emission." Many photographic light meters employ this phenomenon to measure the intensity of light available to a camera.

Many metals respond to the impact of high speed electrons on their surface by ejecting their own electrons. This is "secondary emission." Vacuum tubes called "multipliers" make use of this to provide tremendous amplification. The X-ray tube owes its performance to secondary emission. A small tungsten

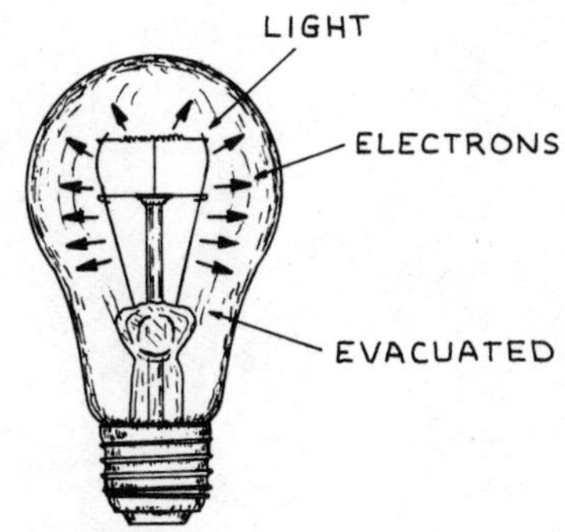

Figure 2-5. When a metal is heated beyond a critical temperature, electrons are thrown off. This is the "Edison effect" and explains the purpose of the cathode in a vacuum tube.

electrode in the tube is the target for ultra high speed electrons that knock out the secondary X-rays used in medicine. (Figure 2–6 explains this.)

Although the methods of causing electron emission differ, the underlying principle always is the same: Energy is added to the electron until it has attained the potential of breaking its bonds to a nucleus. The amount of energy required varies proportionately with the shell in which the electron normally resides. Whether it be heat, light, or impact, all are forms of energy transmission. Even a strong electrostatic field, for instance a nearby plate maintained at a high positive voltage, can draw electrons without touching the donor substance.

An electric current is a flow of electrons. Substances that permit their electrons to wander easily and join such a flow therefore are conductors of electricity. The metals are the most common examples. Conversely, if

Figure 2-6. The electrons from the heated cathode, accelerated to high velocity by the positive voltage, strike the target and generate X rays.

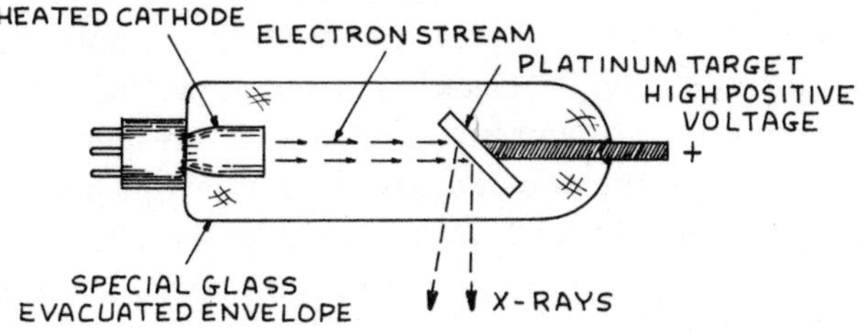

the electrons are held tightly and are not allowed to roam, the material cannot convey electricity and is an insulator. Then there exists a gray area between these two extremes—this is the realm of the semiconductors. The name is self-descriptive. The two most prominent semiconductors are germanium and silicon, and they, as already stated, are the basis of modern solid state electronics.

No one has ever actually seen an atom, but scientists are agreed on its internal makeup. At the center is the nucleus of protons and neutrons with a positive electric charge. Orbiting about the nucleus are the electrons with negative electric charges. An atom in its normal state is electrically neutral because the positive and negative charges are equal and balance each other out. If an electron be removed from the atom, the positive charge is now in the majority and the atom becomes a positive ion. Were an electron to be added, the result would be a negative ion. Once either of these changes takes place, the atom becomes subject to the attractions and repulsions that characterize electrically charged bodies.

All of the foregoing is in the realm of the invisible, but a simple experiment can prove that the movement of electrons actually occurs and changes the electric charge. Take a glass rod and rub it with silk. The rubbing removes electrons from the previously neutral rod and makes it positive. Watching the hair on your arm stand up when the electrified rod is brought near proves the existence of the charge.

Chemists and physicists arrange the elements in the order of their "atomic numbers." These numbers are determined by, and are identical with, the number of electrons possessed by the atom. Germanium, for instance, has 32 electrons and bears the atomic number 32.

Earlier it was said that the mind boggles at any attempt to visualize the infinitely small sizes involved in an atom. Perhaps looking at our solar system may help. Most of this system is empty space between the sun and the planets. So too in the atom—empty space

Figure 2-8. Lee de Forest inspects the triode "Audion" that he invented. He added a third element, a "grid," to the two-element vacuum tube of Fleming. The Audion not only could detect wireless signals but it could also amplify them. (RCA)

pervades between the nucleus and its electrons. Remembering this empty space makes it easier to realize that electrons can move about freely and thus fulfill their chore of carrying electric charges under the right conditions. If we could magnify it sufficiently, we would find that a "solid" copper wire is not solid at all, but has plenty of open space for electron movement.

With electrons orbiting dizzily, what keeps them in the atom instead of flying off? What keeps the earth in its orbit about the sun? The answer to both questions is the same: Centrifugal force is balanced by the attractive force from the nucleus or the sun.

Both an excess and a dearth of electrons create electric charges; these manifest themselves as fields of force. One of nature's basic laws specifies how these fields shall react to each other. The rule states that like polarities shall repel and unlike shall attract. Thus two negatives or two positives will push apart while a negative and a positive will come together. The forces in both cases are considered to extend in straight lines, as shown in Figure 2–7. The force of these attractions and repulsions varies inversely as the square of the distance between them. (Twice as far away, one-fourth the force.)

All of electronics is basically a study of the electron and its life style. Familiarity with the electron therefore becomes the key to understanding electronic instrumentation.

The importance of electrons to electronics dawned in 1883 when Edison discovered that "something" was being given off by the heated filament in his newly invented incandescent lamp. Remarkably, this something could carry an electric current. Furthermore, experiments revealed that the current could be made to flow in only one direction. Of course the "something" was the ubiquitous electron. The interesting "Edison effect" was to wait years before it would be put to practical work.

Figure 2-7. The fields of electric charges of the same polarity (a) repel each other. Unlike polarities (b) attract each other.

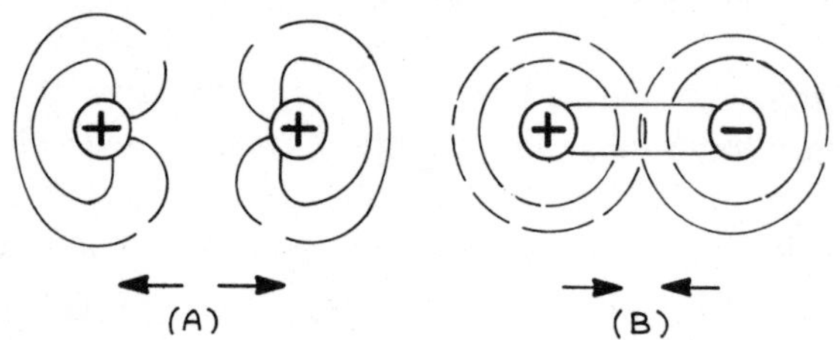

At the turn of the century, an Englishman named Fleming reasoned that this one-way habit of the electron would rectify alternating current into direct current and could be used as a detector of wireless signals. The Fleming valve was born and replaced the much cruder mineral detector. It consisted of two electrodes, one heated, in an evacuated bulb (in other words, a diode).

The real breakthrough came in 1906 when deForest added another electrode inside the Fleming valve and created the triode. Whereas the diode could only control the direction of electric current flow, the triode had the power to amplify, to change a whisper into a shout. Signals previously too weak for intelligibility now could be brought forth from a loudspeaker. This was the real turning point in wireless communication. (See Figure 2–8.)

3. *The Transistor and the Diode*

The transistor arrived in the 1940s and made most of the then existing electronic technology obsolete. The vacuum tube had delivered a similar knockout blow to its own predecessor about forty years before. Now the vacuum tube had to go to make room for miniaturization, greater reliability, and extreme economy of electric current.

The vacuum tube and the transistor are as different from each other as black is from white, yet both perform the same functions in an electronic circuit. The glass bulb of the vacuum tube puts it high on the list of fragile items; it is difficult to find a way to damage physically the small plastic or metal enclosure of a transistor. A radio with a number of vacuum tubes requires a husky power supply, whereas its counterpart built with transistors operates on a couple of flashlight batteries. As to miniaturization, devices that formerly were housed in large cabinets now fit into the palm of the hand.

The transistor and the solid state diode depend upon the same phenomena of molecular physics for their functioning. Very loosely, the diode may be considered as half a transistor, and the transistor, in turn, as two diodes placed together back to back.

A diagrammatic representation of the internal construction of a transistor and of a diode is shown in Figure 3–1. Lest this highly enlarged sketch lead to a misconception of the true physical dimensions, the actual size is indicated by the black dot in the drawing. The complete transistor element, exclusive of connecting leads and surrounding case, occupies the volume of that black dot!

The transistor element is a three part sandwich of two different materials called "p-type" and "n-type." Two possible arrangements are used. There are either two n-type and one p-type elements arranged in n-p-n order or else two p-type and one n-type arranged as p-n-p. The first results in an "n-type transistor" while the second produces a "p-type transistor." A little

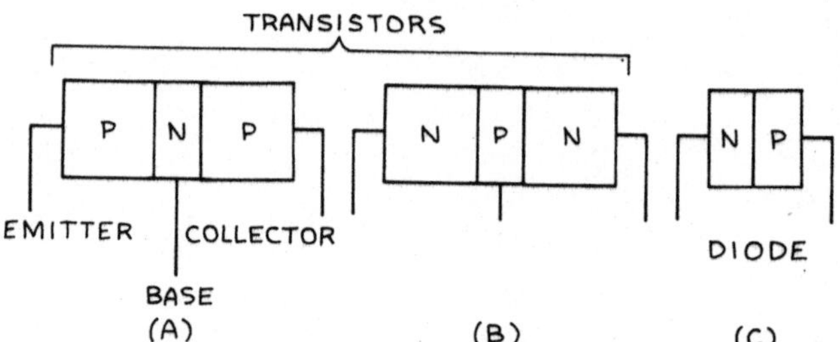

Figure 3-1. The semiconductor arrangements that make up transistors and diodes are shown above. At (a) a PNP transistor, at (b) a NPN transistor, and at (c) a diode.

backtracking explanation should remove whatever mystery surrounds the n and p designations.

Semiconductor materials, such as germanium and silicon, are composed of atoms that have only four electrons in their outer shells and thus are very amenable to bonding with others. These atoms combine by sharing electrons with neighbors until each has a full outer shell of eight electrons, four of its own and four shared. This is easily visualized from Figure 3–2; physicists call it "co-valent bonding."

The bonding forms the atoms into a perfect lattice

Figure 3-2. Although adjacent Germanium atoms are held together by eight bonds, each atom supplies only four. The four valence electrons from each atom work together to form the eight covalent bonds as shown.

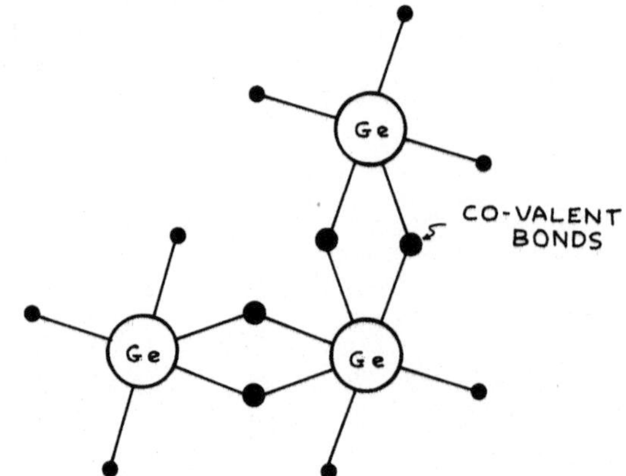

structure. Unfortunately, such a structure has minimal ability to carry an electric current; it lacks the free electrons necessary to do this, since all are in bonds. The inventors dug deep into theoretical physics to overcome this problem. The solution turned out to be the addition of an impurity to the supremely pure semiconductor material. This was the stroke of genius that enabled semiconductors to become the basis for modern solid state electronics.

The impurity usually chosen is arsenic or indium. The addition is extremely sparse, perhaps one atom of impurity to several million atoms of semiconductor. This inordinate dilution is amazingly effective; it produces the desired reaction without interfering with the necessary lattice structure. The impurity chosen determines the nature of the ensuing semiconductor; arsenic yields n-type material, while indium results in p-type.

The choice of using arsenic and indium is logical and clever. Arsenic has five electrons in its outer shell. When it fits itself into the four-prong lattice structure, one electron is left over and free. This action, multiplied many times, provides the stream of free electrons needed for conducting electricity.

Indium maintains only three electrons in its outer shell. When the indium atom places itself into the lattice, it is shy one electron and leaves a "hole" at the fourth position. Electrons play musical chairs with this empty spot and, as positions are vacated, the hole in effect migrates through the semiconductor. Migrating holes also are able to carry electric current.

Quite logically, arsenic is called a "donor" impurity because it donates its fifth electron to the lattice to become a current carrier. With equal logic, indium is an "acceptor" impurity; it accepts electrons and creates the holes that form the other method of electric conduction through a semiconductor. (The relationship between semiconductor material and impurity is shown clearly in Figure 3–3.)

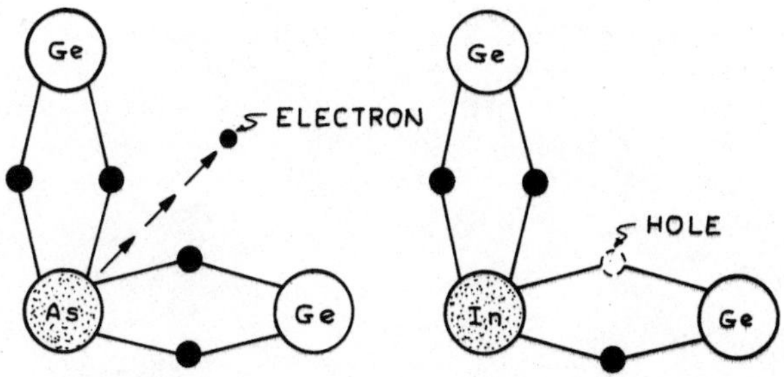

Figure 3-3. Impurities in the Germanium are the secret of its semiconductor qualities. The impurities chosen are either five electron Arsenic or three electron Indium. As shown above, the fifth electron of Arsenic goes into the lattice as a free electron and the lack by Indium of a fourth electron causes a "hole" in the lattice.

Granted all this acceptance and donation of electrons, how does this tiny semiconductor sandwich function? What makes alternate layers of n and p material able to perform the wonders of which modern transistors are capable? A closer look is in order, and greater familiarity will follow.

Familiarity begins when components are known by their proper names so that references to them can be made without ambiguity. The standard, common garden variety transistor, known as a "bipolar junction type" (BJT), consists internally of three semiconductor parts: The "collector," the "base," and the "emitter." These are shown in greatly enlarged form as they appear on an actual device in Figure 3–4. By happy coincidence, collector, base, and emitter perform almost identically with the plate, grid, and cathode, respectively, of the bygone vacuum tube. This was a boon to the oldtimers who had to transfer their thinking from the old to the new.

In transistors and diodes at rest, the holes and electrons are in random motion, due solely to ambient temperature, and the resultant cancellations leave everything in equilibrium without current flow. Attach a battery and the ensuing electrostatic field imposes

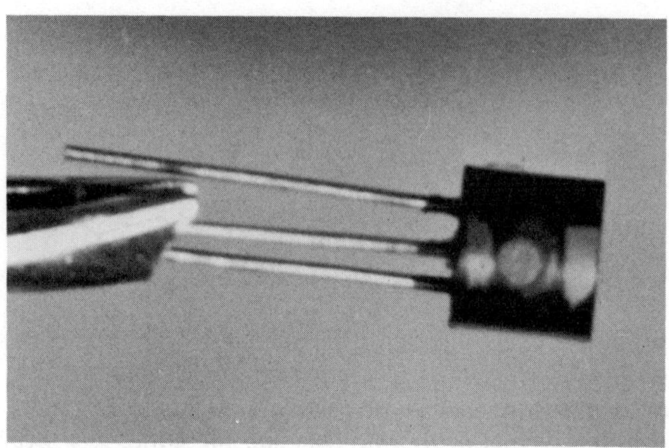

Figure 3-4. The three terminals of a simple transistor are the emitter, the base, and the collector. (AT&T Photo Center)

order and direction, turning the random motions into the coherent streams of electrons and holes necessary for the transmission of electricity. The polarity of the field decides the direction of motion—negative fields repel electrons and attract holes, while positive fields have the opposite effect.

The important development resulting from all this is the ability to control these current-carrying streams. The mere capability to transmit electricity to a greater or lesser degree is shared by many materials. But add control and you have a device that can amplify and perform wonders. That is what the transistor is all about. A tiny expenditure of energy controls a large amount of energy; in other words, amplification is achieved.

The secret of the transistor lies in what takes place at the junctions of the p and n material. The imaginative sketch in Figure 3–5 allows a close examination. Two

Figure 3-5. How the three areas of a transistor are oriented to the circuit is shown above.

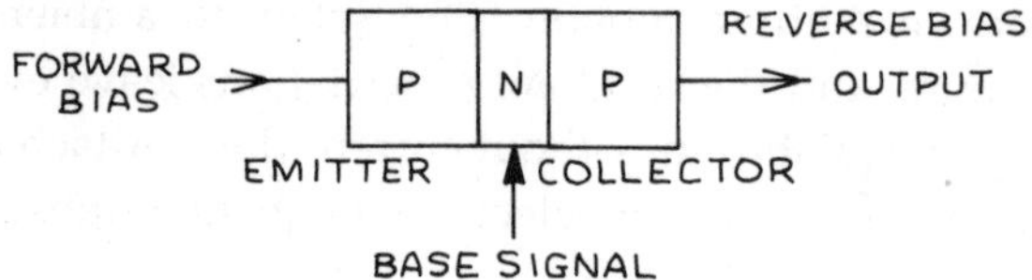

junction areas between the p and the n materials are where the action is. Early transistor circuits required two separate batteries, but practical transistor circuits operate with only one battery.

The terms "forward biased" and "reverse biased" will be encountered and must be understood. When the positive pole of a battery is connected to p material, it is considered forward biased because this is the condition of maximum current flow. An n material is forward biased when it is connected to the negative pole. Reverse these polarities and reverse bias results, a condition of minimum current flow. It is important to remember that p and n transistors are always connected with opposite battery polarities.

Referring to the drawing, a small voltage is applied between emitter and base, with the former positive and the latter negative. This, of course, is forward biasing for p type material, and holes from the emitter are repelled into the base. The base is so narrow that the holes traverse it before they have a chance to recombine with electrons. The holes now enter the collector.

A voltage also is applied between base and collector, but in reverse bias for p type material as shown. Conditions in the collector now are attractive to the electron-hungry holes. The holes recombine into a neutral state with the electrons supplied by the battery. This stream of electrons constitutes the collector current.

The significant fact that emerges from all these goings-on is that a small current in the base circuit controls a large current in the collector circuit. By definition, this is amplification. An illustration of its practical application is the whisper at the microphone that emerges as a roar from the loudspeaker. The vacuum tube also is capable of doing this, but at the cost of much greater energy expenditure and with the evolution of useless heat.

Since the base current of the transistor controls the collector current, it can easily be seen that a desired variation may be introduced at the input to reappear greatly strengthened at the output. That variation could be a very weak radio signal applied at the terminals marked "signal" in the drawing.

The reason for the name "bipolar junction transistor" is revealed by the actions that take place at the p and n junctions. Current transmission in the device is carried on by both positive holes and negative electrons; that is to say, in bipolar fashion. The illustration uses a p type transistor as the example; the action in an n type unit is identical except that polarities are reversed and holes and electrons are interchanged. The important parameters in both cases are the forward and reverse biased junctions.

The solid state diode has only one p element and one n element and therefore only one junction; as stated earlier, it may be considered as half of a transistor. In this single circuit, current can flow in only one direction, the forward biased direction. This is equivalent to saying that a diode can rectify alternating current into direct current; this is one of the principal uses for the device, as later text will show. (See Figure 3–6.)

In the shorthand of electronics found in technical literature, the emitter, base, and collector are referred to by the capital letters E, B, and C, respectively. These are combined with the well-known V for voltage and I for current. For instance, V_{EB} refers to the voltage between emitter and base, while I_{EC} is the current in the emitter-collector circuit. Many other combinations, equally clear, are used. They are of peripheral interest except to those boatmen who are going to delve deeply into theory.

Figure 3-6. What distinguishes the diode is that current will flow in one direction and not in the other. The sketch shows the polarities that determine each condition.

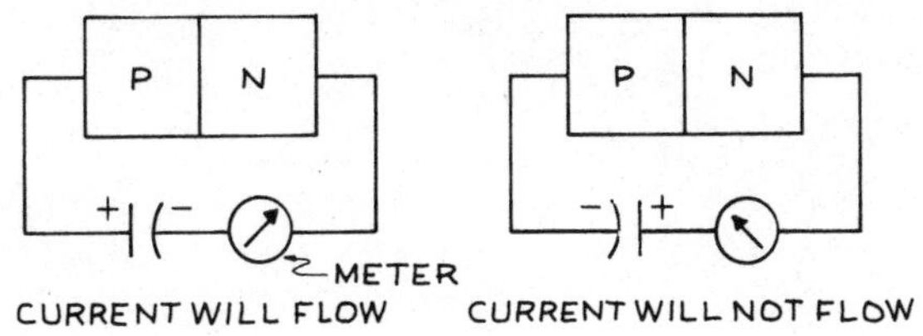

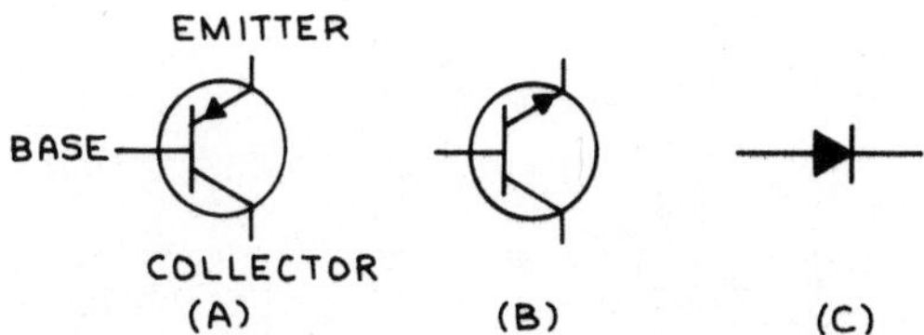

Figure 3-7. The symbols for the PNP transistor (a) for the NPN transistor (b) and for the diode (c) are all shown above.

Technical shorthand also encompasses symbols, and an understanding of these is necessary for deciphering wiring diagrams. The two most pertinent symbols denote the npn transistor and the pnp transistor. As shown in Figure 3–7, the difference between these two consists only in the direction in which the emitter arrow points. An emitter arrow directed at the base denotes a pnp device; the arrow pointed away changes this to an npn unit. The symbol for the diode is an arrow meeting a bar, with arrow the anode and bar the cathode.

Just as more complex forms of vacuum tubes succeeded the original three-element triode, so also have more sophisticated transistors been developed from the simple one just described. One such is the "field effect transistor" (FET) that has proved highly effective in handling very small signals. The drawing in Figure 3–8 shows the construction of an FET diagrammatically.

The field effect transistor brings with it some new names for components not found in the simpler device. The source (S), the drain (D), and the gate (G) are identified in the diagram. Their functions are similar to the emitter, base, and collector already discussed. Note that the symbol for the FET differs from that for the bipolar junction transistor, but that the direction of the arrow again denotes whether the unit is n type or p type. Incidentally, in contrast to the bipolar transistor discussed earlier that makes use of both holes and electrons in its operation, the field effect transistor uses either holes or electrons and accordingly is called "unipolar."

Figure 3-9. These photos show some of the many stages in the manufacture of semiconductor components. All operations are carried out in so-called "clean rooms" that vie with hospital operating rooms for spotlessness. Air and temperature are controlled and workers wear dust-free coveralls. (RCA Solid State Division, Somerville, NJ)

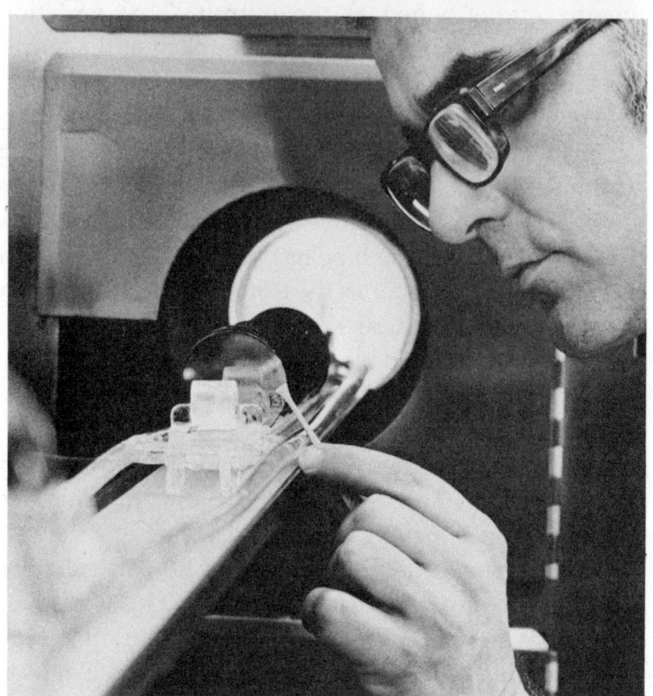

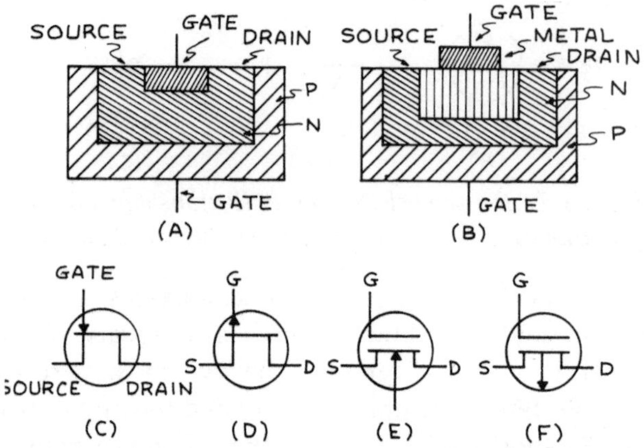

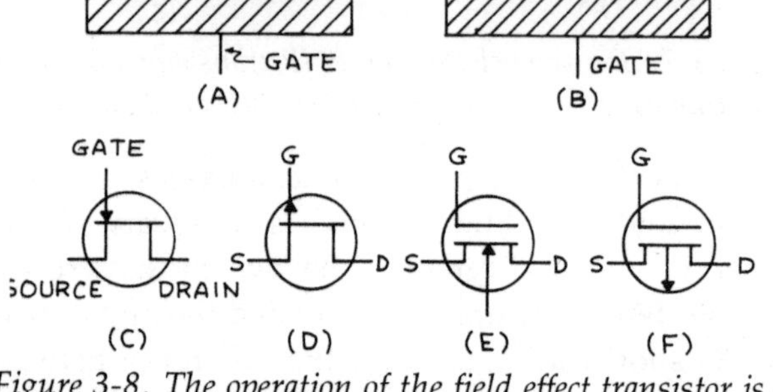

Figure 3-8. The operation of the field effect transistor is described in the text. At (a) the construction of the FET is described schematically. At (b) is the construction of the metal oxide field effect transistor (MOSFET). At (c) is the symbol for the N channel FET, at (d) for the P channel and at (e) for the N channel MOSFET while (f) is the symbol for the P channel MOSFET. (At (b) an insulating film is under the gate.)

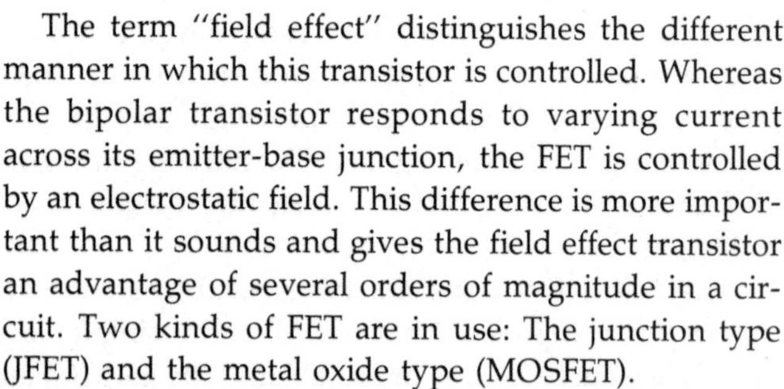

The term "field effect" distinguishes the different manner in which this transistor is controlled. Whereas the bipolar transistor responds to varying current across its emitter-base junction, the FET is controlled by an electrostatic field. This difference is more important than it sounds and gives the field effect transistor an advantage of several orders of magnitude in a circuit. Two kinds of FET are in use: The junction type (JFET) and the metal oxide type (MOSFET).

The manufacture of transistors is a superbly accurate melding of the arts of photography and chemistry. The configuration of the transistor elements is first drawn on a very large scale. This is photographed, optically reduced to a microscopic size, and imprinted on the prepared semiconductor material. Chemical etching now takes away unwanted intervening sections to provide the desired electrical paths. Stages in the manufacture of semiconductors are shown in Figure 3–9.

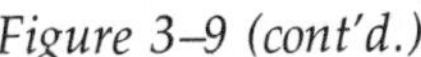

Figure 3–9 (cont'd.)

What began as an arcane laboratory process is now a thriving industry that employs large numbers of people and constantly advances its technology. Several distinct production methods have evolved—diffusion, epitaxial, planar, MOS, masking, and junction. Intimate knowledge of these is not required outside the laboratory, but a few words of explanation of each will enable the skipper to recognize them from catalog descriptions and to make his choice. (See Figures 3–10.)

Diffusion is affected by heating the semiconductor in a vapor of the desired impurity and then hastening penetration by raising the temperature.

The word "epitaxy" denotes arranging something upon a surface. Its use in transistor production means forming semiconductor layers upon each other to achieve the desired combination.

In planar construction all components begin or terminate on a single plane surface where electrical connections are made.

The MOS is somewhat similar to planar in its use of a silicon dioxide layer through which holes are etched for component placement.

The masking technique actually is part of all methods. Sections of p and n material are fused together to form junctions, as for diodes. None of the methods is completely original unto itself.

The simple diode also has branched out. One form that the man at the helm increasingly has before his eyes is the light emitting diode. This is the little gizmo that forms the digital readouts on modern navigational instruments by illuminating seven short bars into arabic numbers. Such readouts now are common on everything from depth sounders to Lorans and beyond.

The light emitting diode (LED) is an example of energy being transformed from one state into another without loss—just as the laws of physics proclaim. Under proper conditions of semiconductor and voltage, the energy released when the current carriers of a

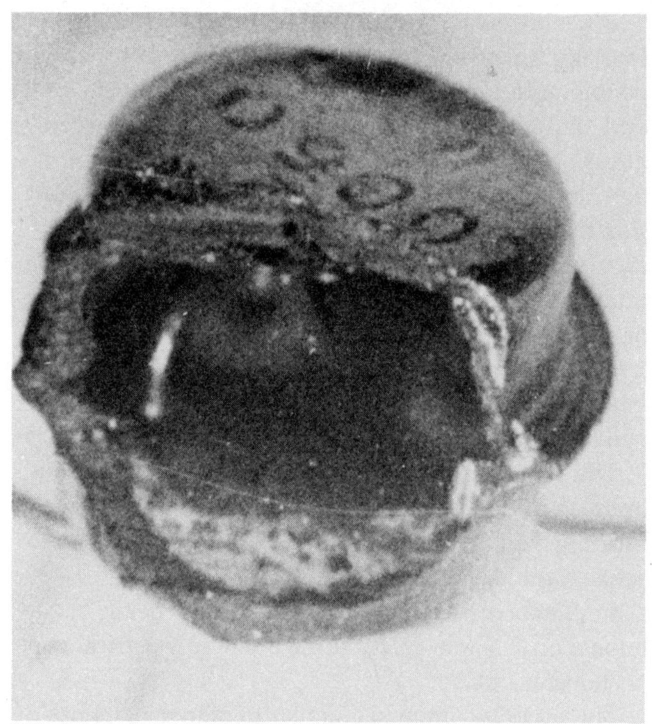

Figure 3-10. This greatly enlarged cutaway view of a metal-housed transistor shows the tiny chip of semiconductor material within that performs the wonders.

diode recombine is emitted as light. The color of the emission, whether infrared or of visible hue, is determined by the semiconductor material. Yet another form of diode employed in digital readouts is the "liquid crystal diode" (LCD).

Transistors are strongly affected by temperature and can self-destruct if permitted to become overheated. This is especially true of power transistors that handle heavy current, a part of which is turned into heat because of internal resistance. This heat must be dissipated harmlessly and "heat sinks" are called upon to do this. A heat sink spreads and disperses the heat over a large surface in contact with the atmosphere. (See Figure 3-11.)

A caution that always must be borne in mind when working with transistors is that they are crucially sensitive to polarity. Make a mistake in polarity when connecting a transistor to a source of voltage and 99 times out of 100 you will have a dead transistor. Some electronic devices guard against the possibility of such an error by including protective diodes that prevent the flow of wrong polarity current.

The variety of diodes and transistors available today is so great, and their electrical characteristics so diverse, that intelligent selection is almost impossible without an up-to-date manual from a leading manufacturer. The containers that house the units and the contact pins that project may be equally confusing without such a technical guide. The photographs in Figure 3–12a portray those diodes and transistors more likely to come within the ken of the electronically equipped skipper. The diagrams (Figure 3–12b) explain the pin connections of the units shown.

Figure 3-11. Heat sinks provide a large radiating surface through which the transistor can shed its heat to the surrounding air.

Figure 3-12. Some common transistors and diodes are shown in this photo at (a). The diagrams at (b) identify the connecting pins. (ARRL Radio Amateur's Handbook)

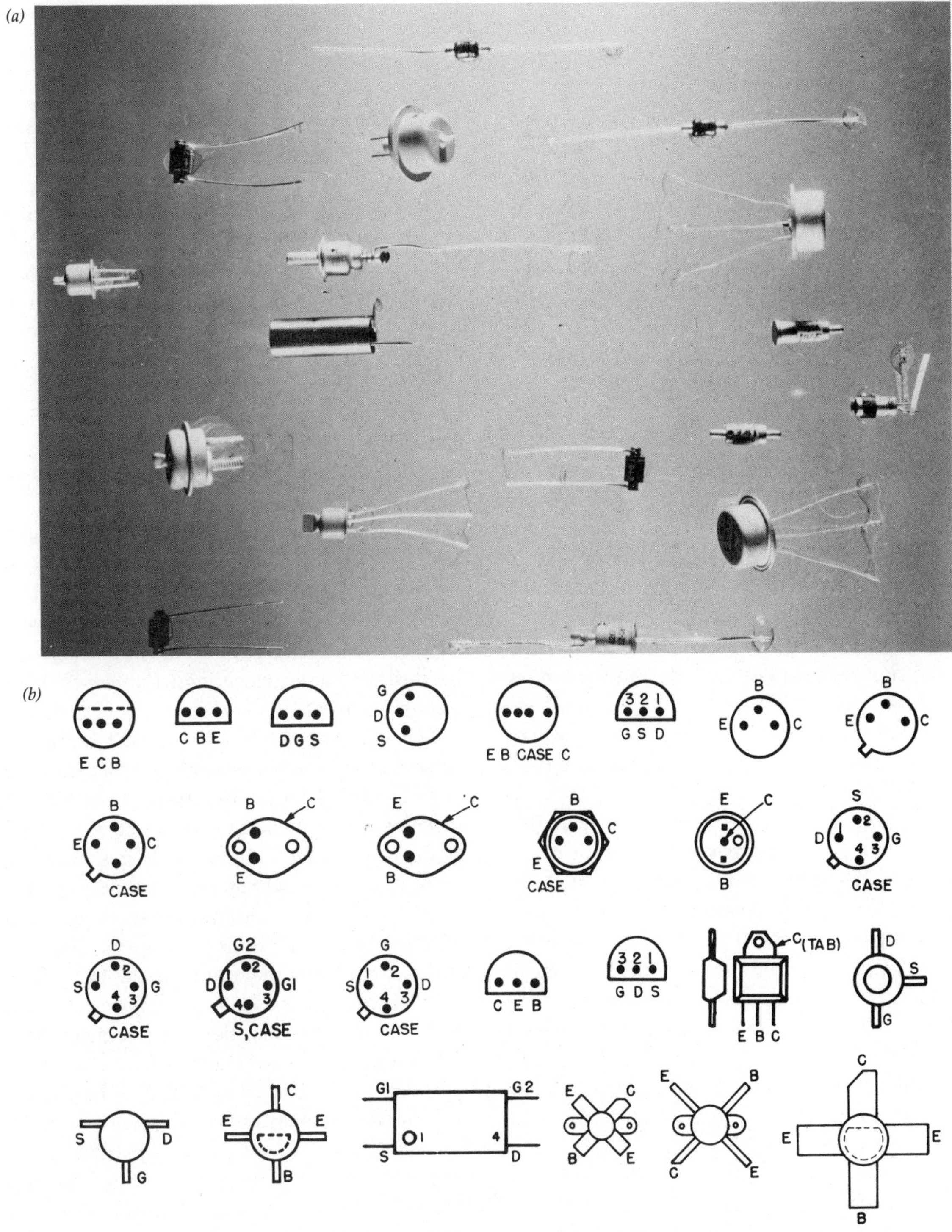

The leads are marked C - collector, B - base, E - emitter, G - gate, D - drain, and S - source.

The earliest transistors, including the original model made by the inventors, utilized a spring wire point contact with the semiconductor material instead of the modern junction method. Some diodes intended for extreme high frequency usage still are manufactured in that manner. The fine wire point, the cat whisker, is a nostalgic reminder of the infant days of radio when crystal detectors were in use.

Everything about transistor and semiconductor diode manufacture depends upon microscopic techniques and, in fact, the final assembly actually is accomplished under microscopes. The semiconductor pellet itself may be no more than a few hundredths of one inch square, and connections must be made accurately to emitter, base, and collector! In turn, these must be led to the proper pins and the whole must then be encapsulated for protection—a remarkable feat.

The manufacturer's manual lists the electrical ratings for each transistor and each diode and these must be observed for successful operation in a circuit. With current-carrying connections of such microscopic size, it is easy to understand that overloading could be ruinous.

The ratings in the manual include many highly technical parameters that are beyond the need of the casual worker in electronics. A few, however, are basic and should be understood. Among these are the voltage and current ratings; the type of service, whether for audio, radio, or extreme high frequency circuits; the amplifying ability; the peak inverse voltage (PIV) that may be applied to a diode used as a rectifier of alternating current.

Heading the list of warnings is the maximum dissipation that may be allowed at the collector of the transistor in question. Current flowing through the collector encounters an internal resistance, and when any current passes through any resistance, heat is generated. The transistor is vulnerable to this heat to the point of self-destruction if the rating be exceeded. Ancillary to this rating is the listing of the maximum current and the maximum voltage that the transistor can handle without damage. (A typical data sheet is reproduced in Figure 3-13.)

The amplifying abilities of a transistor under varying conditions are listed as the Greek letter alpha and the Greek letter beta. Alpha always is less than unity and the choice lies with the value closest to one, as for instance 0.99. Beta may reach any value and here the choice is the highest number for a given type. Both of these parameters reflect how effectively a small input can control a large output—in other words, amplifying action.

As already stated, the diode is a one way device, comparable to a check valve in plumbing, that permits current flow in only one direction. The ideal, theoretically perfect diode would act as an open circuit in the reverse condition and as a path with almost zero resistance in the forward condition. Naturally, such a perfect device does not exist in this imperfect world. But a good diode comes close, and the closer it comes, the greater its merit.

The PIV or peak inverse voltage rating of a diode prescribes the maximum voltage that may be applied to its terminals without damage. In rectifier duty this specifies the highest alternating current potential the diode can handle. Higher voltages than this cause breakdown, short circuit, and total destruction.

Two very special forms of the diode may be encountered in marine electronic equipment. One is the Zener diode, the other is the tunnel diode. Both differ little in internal construction from standard diodes and rely for their action on peculiarities in the semiconductor material of which they are made.

The Zener diode has the ability to hold a voltage at a predetermined level and consequently is much used in regulators. The tunnel diode is extremely efficient in high frequency circuits. Incidentally, the description "tunnel" does not refer to its form of mechanical construction, but rather to the paths taken by its electrons when in operation.

One highly technical aspect of one type of the

SMALL-SIGNAL TYPES

No.	Type	Maximum Ratings				Characteristics			Other Data				Application
		Material[1]	Diss. (Watts)	V_{CEO} (Volts)	I_C (dc)	h_{FE} (Min.)	f_T (Typ.)	Noise Fig. (dB)	Use (Typ.)	Case Style	Base Conn.	Manufacturer[2]	
2N406	PNP	G	0.15	–18	—	34	0.65 MHz	—	Gen. Purpose	TO-1	7	R	Gen. Purpose
2N706A	NPN	S	0.3*	20	50 mA	20	400 MHz	—	rf	TO-18	8	M	rf, Switching
2N718A	NPN	S	0.5	50	150 mA	40	60 MHz	—	—	TO-18	8	R	Switching
2N1179	PNP	G	0.080*	–30	–10 mA	100	—	—	hf Amp.	TO-45	5	R	rf Mixer
2N1302	NPN	G	0.15	25	0.3A	20	—	—	Computer	TO-5	8	R	Osc., Amp.
2N1306	NPN	G	0.15	25	0.3A	60	—	—	Computer	TO-5	8	R	Osc., Amp.
2N2222	NPN	S	1.8	30	800 mA	35	250 MHz	—	Gen. Purpose	TO-18	8	M	vhf Amp., Osc.
2N2925	NPN	S	0.2*	25	100 mA	170	160 MHz	2.8	Gen. Purpose	—	1	GE	Osc., rf, i-f, af
2N3391A	NPN	S	0.2*	25	100 mA	250	160 MHz	1.9	Audio	—	1	GE	Low-noise Preamps.
2N3394	NPN	S	0.31	25	100 mA	55	—	—	Gen. Purpose	TO-92	2	M	Audio Amp.
2N3565	NPN	S	0.2	25	50 mA	150	—	—	—	TO-106	7	—	—
2N3568	NPN	S	0.3	60	500 mA	120	60 MHz	—	—	TO-105	—	—	—
2N3638	PNP	S	0.3	–25	–500 mA	100	150 MHz	—	—	TO-105	—	—	Switching
2N3663	NPN	S	0.12*	12	25 mA	20	900 MHz	4	rf	—	1	GE	vhf/uhf Osc., Amp., Mix.
2N3702	PNP	S	0.31	–25	–200 mA	60	100 MHz	—	Gen. Purpose	TO-92	2	M	vhf Osc., Amp.
2N3866	NPN	S	5	3	400 mA	5	800 MHz	—	Gen. Purpose	TO-39	8	M	uhf Amp., Osc.
2N3904	NPN	S	0.21	40	200 mA	40	300 MHz	—	Gen. Purpose	TO-92	2	M	vhf Amp., Osc.
2N3906	NPN	G	0.15	25	300 mA	60	—	—	Computer	TO-5	8	R	Osc., Amp.
2N4123	NPN	S	0.21	30	200 mA	50	250 MHz	—	Gen. Purpose	TO-92	2	M	vhf Amp., Osc.
2N4124	NPN	S	0.3	25	200 mA	120	250 MHz	5	Audio-rf	—	2	M	—
2N4126	PNP	S	0.3	–25	200 mA	120	250 MHz	4	Audio-rf	—	2	M	—
2N4275	NPN	S	0.28	15	—	18	—	—	—	—	—	—	Switching
2N4401	NPN	S	0.31*	40	600 mA	20	250 MHz	—	Gen. Purpose	TO-92	2	M	Osc., rf, i-f, af
2N4410	NPN	S	0.31*	80	250 mA	60	250 MHz	—	Gen. Purpose	TO-92	2	M	Osc., rf, i-f, af
2N4957	PNP	S	.2	30	30 mA	20	1600 MHz	2.6	rf Amp.	TO-72	9	M	rf Amp., Mix., Osc.
2N4959	PNP	S	.2	30	30 mA	20	1500 MHz	3.2	rf Amp.	TO-72	9	M	rf Amp., Mix., Osc.
2N5032	NPN	S	.2	10	20 mA	25	2000 MHz	3.0	rf Amp.	TO-72	9	M	Low-noise rf Amp.
2N5087	PNP	S	0.310*	–50	–50 mA	200	150 MHz	1	rf Amp.	TO-92	2	M	Low-noise rf Amp.
2N5089	PNP	S	0.310*	–25	–50 mA	450	175 MHz	2	rf Amp.	TO-92	2	M	Low-noise rf Amp.
2N5109	NPN	S	3.5*	40	0.4 A	70	—	3	vhf Amp.	TO-39	8	R	Wide-band Amp.
2N5179	NPN	S	0.200*	12	50 mA	25	900 MHz	4.5	rf Amp.	TO-72	9	M	uhf Amp., Osc., Mix.
2N5183	NPN	S	0.5	18	1 A	120	200 MHz	—	Gen. Purpose	TO-104	8	R	vhf Osc., Amp.
2N5222	PNP	S	0.310*	–15	–50 mA	20	450 MHz	—	rf Amp.	TO-92	18	M	rf Amp., Mix., Video i-f
2N5829	PNP	S	.2	30	30 mA	20	1600 MHz	2.3	rf Amp.	TO-72	9	M	rf Amp., Mix., Osc.
40231	NPN	S	0.5*	18	100 mA	55	60 MHz	2.8	Audio	TO-104	7	R	Preamps. and Drivers
40235	NPN	S	0.18*	35	50 mA	40	1200 MHz	3.3	rf	TO-104	9	R	vhf/uhf Amp., Osc., Mix.
HEP51	PNP	S	0.6	–25	–600 mA	80	150 MHz	—	—	TO-5	8	M	rf Amp.
HEP53	NPN	S	0.6	30	600 mA	85	200 MH	—	—	TO-5	8	M	rf Amp.
HEP56	NPN	S	0.31	20	100 mA	70	750 MHz	—	—	TO-92	18	M	uhf Osc.
MPS918	NPN	S	0.310*	15	—	20	200 MHz	6	Amp. Osc.	TO-92	2	M	uhf Amp., Osc.
MPS2926	NPN	S	0.31	18	100 mA	35	300 MHz	—	Gen. Purpose	TO-92	2	M	vhf Osc., Amp.
MPS3394	NPN	S	0.31	25	100 mA	55	—	—	Gen. Purpose	TO-92	2	M	Audio Amp.
MPS3563	NPN	S	0.310*	12	—	20	200 MHz	—	Amp. Osc.	TO-92	2	M	uhf Amp., Osc.
MPS3693	NPN	S	0.310*	45	—	40	200 MHz	4	rf Amp.	TO-92	2	M	50 MHz Amp.
MPS3694	NPN	S	0.310*	45	—	100	200 MHz	4	rf Amp.	TO-92	2	M	50 MHz Amp.
MPS3702	PNP	S	0.31	–25	–200 mA	60	100 MHz	—	Gen. Purpose	TO-92	2	M	vhf Osc., Amp.
MPS3706	NPN	S	0.310*	20	600 mA	600	100 MHz	—	af Amp.	TO-92	2	M	Audio Amp.

Figure 3-13. Charts of transistor specifications, such as the one reproduced above, are used by technicians in selecting the proper transistor for any given circuit. (ARRL Radio Amateur's Handbook)

ever-useful diode is that the capacitance of its junction varies when the voltage applied is varied. This remarkable phenomenon is the basis for some so-called "electronic tuning" circuits. (See Chapter 6.) The type of diode that can accomplish this tuning function is called a "varactor."

The basic transistor circuit can take one of three possible forms. These are the "common base," the "common emitter," and the "common collector." The designation "common" indicates that the electrode that it precedes is grounded and connected to the other two electrodes—in other words, it is common to all. (See Figure 3–14.)

Since germanium and silicon play such an important role in solid state electronics, it is interesting to know the sources of these minerals. Germanium occurs in many ores, such as zinc, and is refined from them. Its existence was predicted theoretically many years before it actually was discovered. Silicon is an abundant constituent of most rocks and is the second most common element in the crust of the earth. It occurs in nature as an oxide and is refined from that.

Vacuum tubes and transistors perform identical tasks in very similar electronic circuits. However, there is one great technical difference between them that is important to technicians: The vacuum tube is voltage operated; the transistor is current operated.

Figure 3-14. A simple transistor may be connected into a circuit in any of the configurations shown above. These are named (a) common base, (b) common emitter, and (c) common collector.

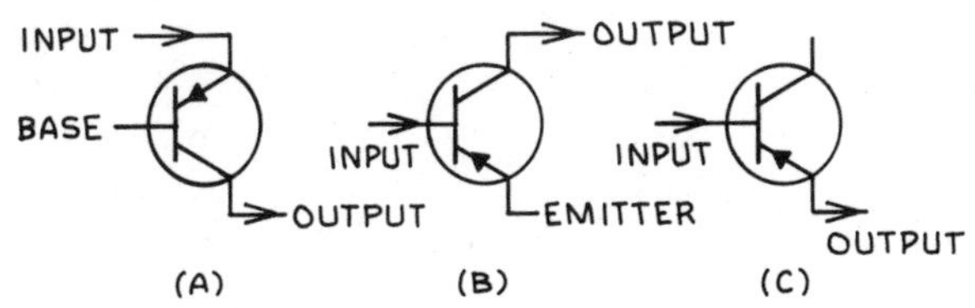

4. *Circuit Components*

Electronic circuits are composed of combinations of five basic components that may be present singly or in profusion, depending upon the complexity of the circuit. Certain combinations, like building blocks, may be repeated many times within the circuit in order to accomplish the amplification or oscillation or whatever is intended. Sometimes these components are designed to function passively; at other times a source of energy, such as a battery, may be added to enable a boost in output.

The five basic components are conductors, resistors, insulators, inductors, and capacitors. Each plays a distinctive role; in correct combinations they perform the wonders of modern electronics. Ancillary to these five basics are vacuum tubes and semiconductors, discussed in Chapter 5 and Chapter 3, respectively.

All metals are conductors, some better than others. Previous text explained that the conduction of electric current is carried on by free electrons. It follows logically that the metal whose electrons have the most freedom should transmit electricity best. Copper is a good conductor and is in wide general use. But silver has a looser hold on its outer electron and is therefore better. Although gold retains its valence electron with slightly greater force than silver, it is the best conductor of all from the standpoint of resistance to corrosion. (All contacts in high quality electronic equipment are plated with gold.)

The amount of current that a copper wire can carry without undue heating and loss is determined by its cross-section; the larger the wire, the greater the current. It is most convenient to express this size as a diameter, since diameter and cross-section in a round wire bear a known relationship. In turn, the diameter is listed in "mils," each mil being one-thousandth of one inch, and the diameter multiplied by itself gives "circular mills," a designation for cross-section.

At this point the American wire gauge steps in to provide a simple means of specifying wire size for commercial use. In this system #0000 is the largest cross-section and a heavy cable while #60 is approximately the thickness of a human hair. The gauge intervals are so arranged that wire size doubles every three numbers toward zero and halves every three numbers toward 60. Pocket-size measuring gauges, with numbered slots of varying width, enable any wire to be checked. Figure 4–1 correlates mils and gauge numbers. It also shows one form of wire measuring gauge.

The wire or metal path that connects one component with another in a circuit is purposely chosen to introduce as little resistance as possible. This is analogous to the piping from a tank to a faucet that is made large to avoid a reduction in flow. But sometimes the function of a circuit demands that the flow of electricity be resisted to some degree. In such a case, the high conductivity wire is replaced by a so-called "resistance wire."

Pure metals are better conductors than their alloys. This apparent detriment is brought to good use in the manufacture of resistance materials. As an example, an alloy of nickel and chromium, known commercially as Nichrome, is almost seventy times more resistant to the passage of electric current than copper. Where the resistance must be great and need not be in wire form, carbon may be used. The resistance of carbon is several *thousand* times greater than that of copper. (All of these resistive conditions relate directly to the paucity of electrons.) Figure 4–2 compares resistivity of various materials.

The "resistor" is one of the most repetitively used components in electrons. Even a small radio may have dozens of them. Resistors take many forms and sizes, as shown in Figure 4–3. A commom form is a small capsule containing the resistance material with a wire pigtail at each end for connections. Resistors are rated in watts and fractions of a watt for the amount of power they can safely dissipate. The capsule of high power resistors generally contains resistance wire while carbon paste or metal oxide is the material for low power units.

CONDUCTOR AMPACITY

ALLOWABLE AMPACITIES OF INSULATED COPPER CONDUCTORS 40°C (104°F) AMBIENT (4)

CONDUCTOR SIZE AWG	NOMINAL CM AREA See Note (1)	SINGLE CONDUCTOR IN AIR MAXIMUM INSULATION TEMPERATURE RATING						NOT MORE THAN 3 CONDUCTORS—See Note (2) MAXIMUM INSULATION TEMPERATURE RATING					
		60°C (140°F)	75°C (167°F)	85°C (185°F)	90°C (194°F)	105°C (221°F)	125°C (257°F)	60°C (140°F)	75°C (167°F)	85°C (185°F)	90°C (194°F)	105°C (221°F)	125°C (257°F)
16	2,580	12	15	25	25	25	30	8	12	20	20	20	25
14	4,110	15	20	25	25	35	40	10	15	25	25	25	30
12	6,530	20	20	35	35	45	50	15	20	25	25	30	40
10	10,380	35	35	50	50	60	65	25	25	35	35	40	50
8	16,510	45	55	65	65	75	85	30	40	45	45	55	60
6	26,240	65	85	90	90	115	120	45	55	65	65	75	80
4	41,740	85	110	125	125	150	160	55	75	80	80	100	110
2	66,360	115	150	165	165	200	215	80	100	110	110	130	140
1	83,690	135	170	190	190	230	250	90	115	125	125	150	160
1/0	105,600	160	200	225	225	270	290	105	130	140	140	180	190
2/0	133,100	185	235	260	260	310	340	120	155	170	170	205	220
3/0	167,800	215	275	300	300	365	390	135	175	190	190	230	250
4/0	211,600	245	320	350	350	420	450	160	200	215	215	260	295

NOTES:

(1) To recognize stranded conductors made of AWG size elements, the actual nominal cm area may differ from the specified nominal cm area, but by no more than 7 per cent.

(2) Current ratings are for not more than 3 current-carrying conductors in a raceway or a cable. Reduce the current rating to 80 per cent of values shown for 4 to 6 current-carrying conductors. Reduce the current rating to 70 per cent of values shown for 7 to 24 current-carrying conductors. See Note (3).

(3) A conductor used for equipment grounding and a neutral conductor which carries only the unbalanced current from other conductors, as in the case of normally balanced circuits of three or more conductors, are not considered to be current-carrying conductors.

(4) The correction factors listed below are to de-rate the ampacity for 60°C (140°F) ambient

Max. Insulation Temperature Rating	*60°C(140°F)*	*75°C(167°F)*	*85°C(185°F)*	*90°C(194°F)*	*105°C(221°F)*	*125°C(257°F)*
Correction Factor	– –	*.66*	*.74*	*.78*	*.84*	*.87*

Figure 4–1 (above). A conductor carrying more current than its normal rating will overheat and may cause a dangerous situation. The table above shows allowable currents for various wire sizes and conditions. (American Boat & Yacht Council)

Wire Size A.W.G. (B&S)	*Ohms per 1000 ft. 25° C.*
1	.1264
2	.1593
3	.2009
4	.2533
5	.3195
6	.4028
7	.5080
8	.6405
9	.8077
10	1.018
11	1.284
12	1.619
13	2.042
14	2.575
15	3.247
16	4.094
17	5.163
18	6.510
19	8.210
20	10.35
21	13.05
22	16.46
23	20.76
24	26.17
25	33.00
26	41.62
27	52.48
28	66.17
29	83.44
30	105.2
31	132.7
32	167.3
33	211.0
34	266.0
35	335
36	423
37	533
38	673
39	848
40	1070

Figure 4-2. This table of resistance versus length of copper wire makes it possible to calculate voltage drop for any given run. (AARL Radio Amateur's Handbook)

Figure 4–3 (below). Resistors come in all sizes, styles, and power-handling capabilities.

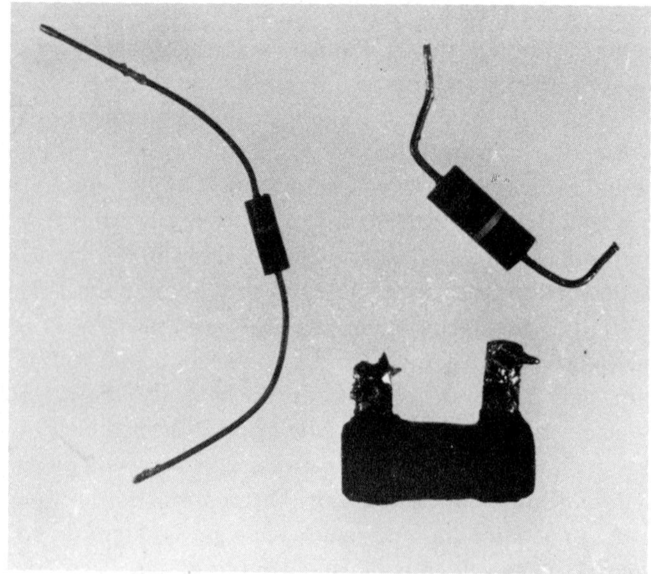

For convenience in handling, resistors are marked with their resistance value. This marking is in a simple code and consists of colored bands. (The table in Figure 4–4 explains this code.)

The foregoing refers to what are known as "fixed" resistors. Some methods of control of electronic circuits require that the resistors be continuously and easily variable from some low value to some maximum. This is accomplished by making the resistance material into circular form over which a radial contact slides. Such a device is a "potentiometer" if both ends of the resistance material are connected to the circuit and as a "rheostat" if only one end is connected. The illustration in Figure 4–5 has photographs in addition to the conventional symbols used in wiring diagrams.

The precision resistors in use today are aeons away from their crude counterparts that served the early days of radio. It was common practice to draw a heavy line on a small piece of cardboard with a soft pencil and then to employ this carbon/graphite track as a resistance. Exact values did not bother anyone then.

The ability of metals to carry electricity would have little practical value without some means of preventing current leakage to adjoining surfaces. Such means must be impervious to electric current, and must be unable to conduct it. According to theory, they would have no free electrons. Such materials are called "insulators."

Air is a good insulator. So also are paper, mica, porcelain, and glass—all found routinely in electronic assemblies. A good insulator cuts off all current flow up to a breakdown point known as its "dielectric strength." This is expressed as the voltage placed across one millimeter of the material that will puncture it. For purposes of comparison, the dielectric strength of air is approximately 3,000, of paper about ten times as much, while for mica it could soar to as much as a quarter of a million. (The voltages in modern transistorized radio receivers are too low to be a challenge to any of these limits; some voltages in transmitters may rise high enough to be.)

(a)

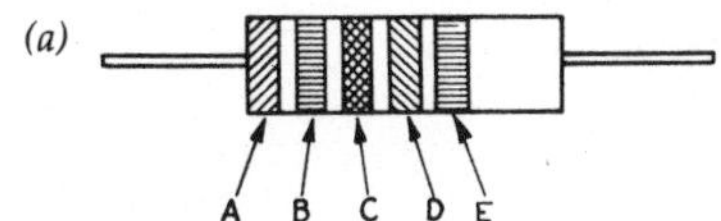

(b)

Resistor-Capacitor Color Code

Color	*Significant Figure*	*Decimal Multiplier*	*Tolerance (%)*	*Voltage Rating**
Black	0	1	–	–
Brown	1	10	1*	100
Red	2	100	2*	200
Orange	3	1,000	3*	300
Yellow	4	10,000	4*	400
Green	5	100,000	5*	500
Blue	6	1,000,000	6*	600
Violet	7	10,000,000	7*	700
Gray	8	100,000,000	8*	800
White	9	1,000,000,000	9*	900
Gold	–	0.1	5	1000
Silver	–	0.01	10	2000
No color	–		20	500

* Applies to capacitors only.

The colored areas have the following significance:
A – First significant figure of resistance in ohms.
B – Second significant figure.
C – Decimal multiplier.
D – Resistance tolerance in percent. If no color is shown the tolerance is ±20 percent.
E – Relative percent change in value per 1000 hours of operation; Brown, 1 percent; Red, 0.1 percent; Orange, .01 percent; Yellow, .001 percent.

Figure 4-4. The colored bands on fixed composition resistors (and on their look-alike capacitors) have the designations given in the above table. (ARRL Radio Amateur's Handbook)

Figure 4-5. Potentiometers (with three terminals) and rheostats (with two terminals) find use in many parts of electronic circuits.

The insulation for copper wire used in coil windings generally is a paint-like enamel coating applied during manufacture. For exposed usage, the insulation on wire is a close-fitting vinyl sheath easily removed when making connections. Where wires must cross, or where any other possibility of unwanted contact exists, the wires in electronic assemblies are encased in insulating tubes that look like spaghetti. All insulation, of whatever form, is there solely to maintain the integrity of an electric circuit, to isolate it from others.

The electrons in insulators are very tightly held. The breakdown voltage is so excessive for any given material that its electrons are violently wrenched from their orbits. The resultant mechanical stresses are great enough to split, puncture or otherwise destroy the material. A wooden mast struck by lightning is split open.

Figure 4–6 depicts two metal plates separated by a sheet of glass. This is a simple "capacitor" and has the ability to store an electric charge. The glass is the "dielectric." In more conventional capacitors, the plates are replaced by two long ribbons of metal foil separated by waxed paper, rolled up into a small space and sealed in a can for protection. Two movable plates separated by air as the dielectric form the common "variable capacitor." See Figure 6–3.

The capacitor is an example of electrons at work. Connect a battery to the two plates and the capacitor will charge. Remove the battery and the capacitor will retain the charge until poor insulation allows it to leak away. The charge and the discharge are the manifestations of electron displacement and return.

Figure 4-6. Two metal plates separated by glass or other dielectric form the basic capacitor.

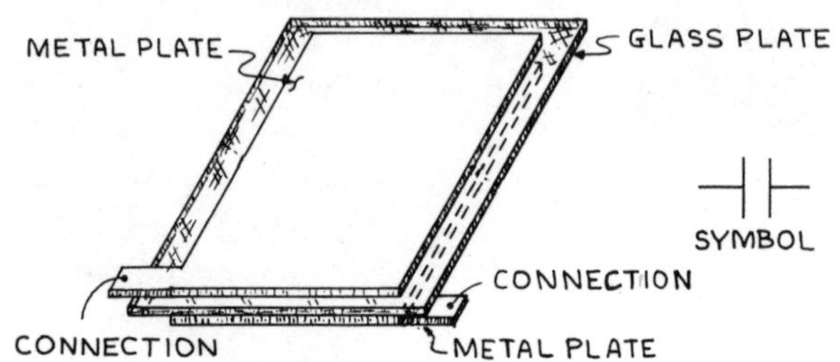

When the battery is connected, its negative terminal forces electrons into one plate while its positive terminal draws electrons from the other plate. A point of equilibrium is reached when the voltage of that battery can cause no further electron movement. The capacity of that capacitor has been reached and it is fully charged.

If the plates were larger in area, more electrons could be displaced and the capacity of this capacitor would be increased. If the glass were thinner, the plates would be closer with magnified attractive and repellant forces on the electrons; again the capacity would be greater. It follows that the capacity of a capacitor is directly proportional to the area of its plates and inversely proportional to the separation of the plates.

The nature of the dielectric material that separates the plates also affects capacity. The degree to which this takes place is called the "dielectric constant" of that material. For instance, the dielectric constant of glass is 10; the dielectric constant of air is 1. If air were substituted for the glass in the illustration, the capacity of the unit would be only one-tenth as much. Some representative values of dielectric constants are given in Figure 4–7. (Dielectric constant and dielectric strength should not be confused.)

The electrolytic capacitor, commonly found in electronic instruments, differs from those so far described in that its dielectric is a conductive chemical solution generally in paste form. One plate is a large area of aluminum foil. The solution acts as the other plate and also as the dielectric. In actuality, the dielectric is a microscopically thin film of oxide created when the solution attacks the aluminum. Since the dielectric is so thin, the capacity is extraordinarily high. This capacitor is housed in an aluminum can as shown in Figure 4–8.

Dielectric Constants and Breakdown Voltages

Material	*Dielectric Constant**	*Puncture Voltage***
Air	1.0	
Alsimag 196	5.7	240
Bakelite	4.4–5.4	300
Bakelite, mica-filled	4.7	325–375
Cellulose acetate	3.3–3.9	250–600
Fiber	5–7.5	150–180
Formica	4.6–4.9	450
Glass, window	7.6–8	200–250
Glass, Pyrex	4.8	335
Mica, ruby	5.4	3800–5600
Mycalex	7.4	250
Paper, Royalgrey	3.0	200
Plexiglass	2.8	990
Polyethylene	2.3	1200
Polystyrene	2.6	500–700
Porcelain	5.1–5.9	40–100
Quartz, fuxed	3.8	1000
Steatite, low-loss	5.8	150–315
Teflon	2.1	1000–2000

* At 1 MHz ** In volts per mil (0.001 inch)

Figure 4-7. The higher the dielectric constant of an insulator, the greater its effectiveness in forming a capacitor. (ARRL Radio Amateur's Handbook)

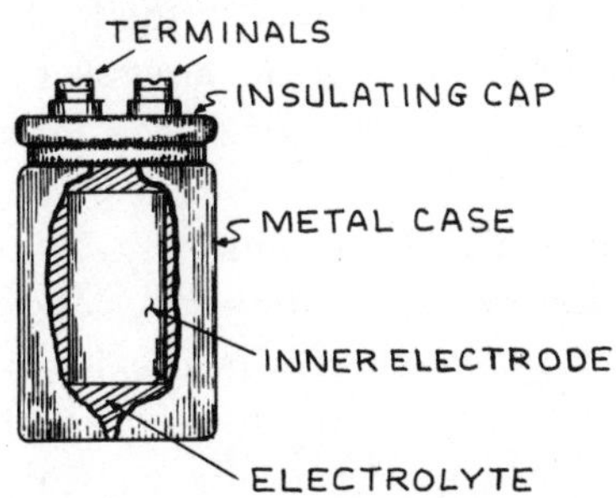

Figure 4-8. The dielectric in an electrolytic capacitor is the oxide formed on the inner electrode by the electrolyte. Capacities are high because the dielectric film of oxide is thin.

Electrolytic capacitors, the name obviously derived from the electrolyte solution contained within them, have their pluses and their minuses. They pack great capacity into very small space, much more than could be obtained from solid dielectric capacitors. But they are sensitive to polarity to the point of destruction, and they always allow a small current leakage. However, at certain places in electronic circuits they are indispensable, even though they have a short life when compared to the theoretical "forever" of solid dielectric types.

The color code marking described for resistors has been carried over for capacitors. The numbers denoted

Color Code for Ceramic Capacitors

Color	Significant Figure	Decimal Multiplier	Capacitance Tolerance: More than 10 pF (in %)	Capacitance Tolerance: Less than 10 pF (in pF)	Temp. Coeff. ppm /deg. C.
Black	0	1	±20	2.0	0
Brown	1	10	± 1		– 30
Red	2	100	± 2		– 80
Orange	3	1000			–150
Yellow	4				–220
Green	5				–330
Blue	6		± 5	0.5	–470
Violet	7				–750
Gray	8	0.01		0.25	30
White	9	0.1	±10	1.0	500

Figure 4-9. The table above explains the color code for ceramic capacitors. Capacities are in pico farads. (ARRL Radio Amateur's Handbook)

by the colors are the same but the reference now, naturally, is to capacity. (This code is explained by Figure 4-9.)

Electric current flowing through a conductor creates an invisible magnetic field around it as shown at (a) in Figure 4-10. If the conductor be wound into a coil, as shown at (b), the magnetic fields add together to form a powerful magnetic force. This force has many uses, depending upon the size of the coil and the magnitude of the current involved. It can do the delicate job of closing the contacts in a small relay or it can develop the tremendous thrust needed to mesh a starting motor pinion with the flywheel gear of an engine. Or the magnetic field, without moving parts, can induce current in an adjoining coil as in a transformer.

Figure 4-10. Current passing through a wire or a coil causes a magnetic field as shown.

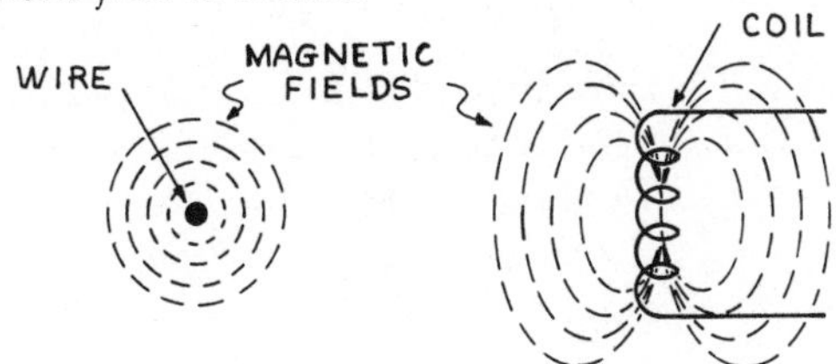

If an iron core be placed within the coil, the magnetic field is greatly multiplied because iron is a better path for magnetism than air. Iron or steel, as such, forms the core for only those magnetic devices operating on direct current or low frequency alternating current. For the higher frequencies found in marine electronic equipment, the iron is used in powdered form, moulded into a suitable core with inert binders.

Any coil of wire through which current may be passed has an electrical property called "inductance" or "self-inductance" and the general term for these units is "inductors." Inductors are an important component in most electronic circuits.

The perfect conductor exists only in theory. In actuality, every conductor offers some resistance to the passage of electric current. This resistance is measured and expressed in "ohms." The algebraic relationship between ohms and volts (potential) and amperes (current) is known as Ohm's law and is stated as "ohms equals volts divided by amperes." By simple algebraic inversion, this also becomes "amperes equals volts divided by ohms" and "volts equals amperes multiplied by ohms." The three versions are shown mathematically in Figure 4–11 together with examples. Incidentally, a resistance of one ohm exists in a current or path when a pressure of one volt can force only one ampere through it.

The ability of a capacitor to hold a charge of electricity is measured in "farads." However, the farad is an impractically large unit and most specifications in electronic circuitry will be in "microfarads" (millionths of one farad) and "picofarads" (billionths of one farad).

Figure 4-11. The three forms of Ohm's Law for simple direct current circuits are shown above. I is current in amperes. E is potential in volts. R is the resistance in ohms.

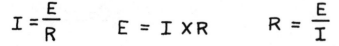

$$I = \frac{E}{R} \qquad E = I \times R \qquad R = \frac{E}{I}$$

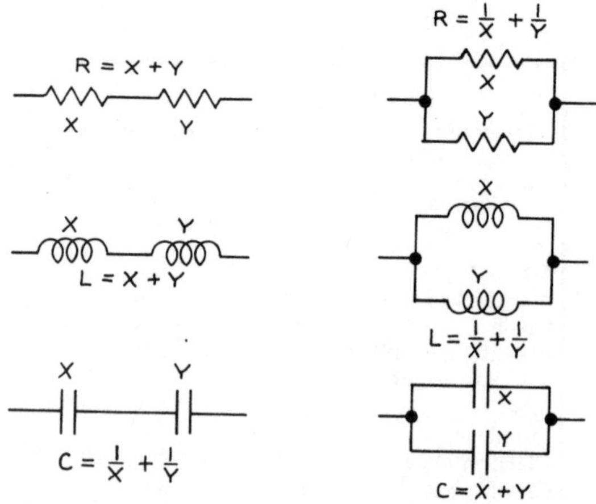

Figure 4-12. The results of placing resistors, inductors, and capacitors in series and in parallel are illustrated above. R is resistance, L is inductance, and C is capacitance.

The farad may be compared loosely to the volume measurement of a liquid container.

The inductance of an inductor is rated in "henries." This is a big unit and only large coils with heavy iron cores will have values of one henry or more. The inductors found in electronic circuits will be marked in "millihenries" (thousandths of one henry) or "microhenries" (millionths of one henry).

Resistors dissipate energy in the form of heat while both capacitors and inductors are able to store energy. The capacitor stores energy in the form of dielectric stress in its insulation and only leakage prevents the charge from remaining indefinitely. The energy in an inductor when current is applied resides in the magnetic field that is created. This field exists only while the current is connected. Upon disconnection, the field collapses and the energy is either returned to the circuit or appears as a spark at the switch. (The condenser at the ignition breaker points of a gasoline engine is there to absorb this spark from the coil, an inductor.)

The many units in an electronic circuit are hooked together in "series" or in "parallel" or in a combination of the two called "series-parallel." The series connection is like the elephant parade in a circus—single file with each trunk holding the preceding tail. In the parallel connection, each unit is a rung in a long ladder. (See Figure 4–12.)

To find the total resistance created by resistors in series it is necessary only to add the individual resistances. Figuring out the final resistance of resistors in parallel becomes a little more complicated and is stated as "the reciprocal of the sum of the reciprocals." This apparent double talk becomes clear in the drawing. The basic fact to remember is that the series connection *increases* the total value while the parallel connection *decreases* it.

The computation for inductors is the same as for resistors—series increases, parallel decreases. The values of all the inductors in series are added to find the total inductance. The reciprocal of the sum of the reciprocals gives the total inductance for inductors in parallel. (See the drawing.)

The computations for capacitors merely interchange the words "series" and "parallel" in the formulas. Thus the total capacitance of capacitors in parallel is the sum of the individual capacitances, while for capacitors in series it is the reciprocal of the sum of the reciprocals. (Again, see Figure 4–12.)

Resistance in ohms is the easiest of the three parameters to measure. The common VOM (volt-ohm-meter) is the instrument used and a simple reading of the scale gives the value. Capacitance in farads is more difficult to determine and requires either a so-called "bridge" or a source of alternating current with a specially calibrated meter. The measurement of inductance is still more complex; and an inductance bridge is normally used.

Of the three parameters, the only one the skipper will have occasion to evaluate exactly in routine maintenance or repair is the resistance in ohms. The VOM he will need is widely available at nominal cost. Actually, the VOM is such a versatile instrument that its place should be assured on every boat with any electrical installation. Voltage, resistance, and amperage are all read on the dial. The most modern VOMs have even discarded the dial in favor of a digital readout that gives its answers in arabic numerals.

A few words of caution are in order to alert the skipper against the two kinds of electricity he will meet, whether afloat or ashore. They are direct current (DC) and alternating current (AC). One of the two wires in a direct current circuit will always be positive and the other always negative. In an alternating current circuit the two wires will constantly alternate polarity. This deeply affects the kind of devices that may be connected to either.

A lamp bulb does not care whether its current is direct or alternating as long as the voltage is correct. A transformer will burn out quickly if connected to a direct current source. Some small motors are "universal" and will perform almost equally on both types of electricity, but motors that have permanent magnet fields must have direct current. The power supply in electronic circuits is always direct current although this may be provided from an alternating current line through a rectifier.

The usual source of direct current on a pleasure boat is a storage battery or perhaps a bank of batteries supplying a specified voltage. Whatever is connected must be of this same voltage. The alternating current aboard is furnished either by a line from the pier or from an engine room generating plant—either one at a time but never both together.

The first item to inspect when connecting any device to any circuit is the nameplate; the answers to the following questions are there: Direct current or alternating current? What voltage? What current requirement? Any cautions?

The type of current supplied and required must agree. So must the voltage. The stated current requirement permits a decision on whether the circuit is able to sustain the device. The amount of current the circuit can deliver depends upon the size of the source, the diameter or gauge number of the wires, and the rating of the fuses or circuit breakers. Motors take a large gulp of current to start and then simmer down when running.

The terms "discrete" and "integrated" are often seen in manufacturers' descriptions of their electronic products. A discrete chassis has all the actual capacitors, resistors, and transistors needed by the circuit mounted "discretely" upon it. In an integrated circuit, these components are simulated and contained within a miniature, sealed module. Most of the interconnections required by the discrete chassis in the form of wires or printed paths are eliminated by the integrated method. As a consequence, reliability of the integrated form is greater because there is less to cause trouble.

Much is heard about the high fidelity (or hi-fi) of electronic equipment that reproduces music and speech. No great importance is attached to the high fidelity of marine voice components. The only criterion is intelligibility and this can be achieved without the extreme high and low frequencies necessary for good entertainment listening. The telephone company proved long ago that transmitting only a narrow band of middle frequencies of the voice created clarity.

Metric Multiplier Prefixes
Multiples and submultiples of fundamental units (e.g., ampere, farad, gram, meter, watt) may be indicated by the following prefixes.

Prefix	*Abbreviation*	*Multiplier*
tera	T	10^{12}
giga	G	10^{9}
mega	M	10^{6}
kilo	k	10^{3}
hecto	h	10^{2}
deci	d	10^{-1}
centi	c	10^{-2}
milli	m	10^{-3}
micro	μ	10^{-6}
nano	n	10^{-9}
pico	p	10^{-12}

Figure 4-13. The prefixes used in stating the values of electronic components are explained in this table. (ARRL Radio Amateur's Handbook)

5. *The Vanishing Vacuum Tube*

The vacuum tube dominated electronics for a generation. Its commercial life began as an oversized, spherical glass bulb screwed into a fixture on the front of a radio cabinet that made it look like a miniature street lamp. It quickly became the beneficiary of tremendous advancements that grew out of the new technology. Its constantly improved size and shape varied from tiny pea-like beads of glass intended for hearing aids to huge "bottles" that generated the radio power for major broadcasting stations. It became the heart of every active electronic device. And then the transistor cut it down! Today it is the vanishing vacuum tube.

As noted earlier, the vacuum tube had its indirect birth when Edison discovered a stream of electrons being thrown off by the heated filament in his incandescent lamp. Fleming added a positive electrode inside the evacuated bulb and created the diode. DeForest placed a grid between these two elements and the triode was born. The triode, with its ability to amplify, was the real beginning of the vacuum tube age. From here on, subsequent developments were really refinements of what had been done originally.

The vacuum tube is a superb example of the ubiquitous electron at work in its primal form. The tube's internal functioning is easier to understand than the action within a transistor because it is so straightforward. An actual stream of electrons is the motivating force in a vacuum tube, in contrast to the mystic movement of holes and electrons in a transistor. Even the more complex forms of vacuum tubes depend upon this electron stream and merely add electrodes for more precise control and greater amplification.

Another grid was added to the triode, and because the tube now contained four electrodes, it became a tetrode. Yet another grid inside the bulb produced the even more versatile pentode. The additions went on through hexode, heptode, and octode, although these names were rarely used commercially. The internal population of the bulb continued to grow as improvement followed improvement and approached twenty in the sophisticated multiplier tubes that are able to amplify a signal a billion-fold. Yet the principle of operation remains the same dependency upon the ubiquitous electron.

The drawing in Figure 5-1 shows at (a) what actually exists inside the bulb of a triode vacuum tube. The three electrodes, cathode, grid, and plate, are arranged axially and supported by the glass stem of the bulb much like the filament in a light bulb. The bulb is evacuated to a high degree to avoid collisions between electrons and air molecules. At (b) is a diagrammatic sketch to help explain what takes place when the tube is in operation and electrons flow from cathode to plate.

The grid is imagined to be a venetian blind interposed between cathode and plate and directly across the electron stream. When the blind is open, electrons pass freely to the attraction of the plate; when the blind is closed, the electron stream is cut off. If the blind opening were varied (modulated) in conformance with an applied signal, it is obvious that the electron stream would assume the modulation of that signal. Earlier text shows that moving electrons constitute an electric current. Thus the varying quantities of electrons that impinge upon the plate cause a current that is an amplified version of the signal applied to the grid.

Of course there are no moving parts within the bulb of the triode, no actual opening and closing of a vene-

Figure 5-1. The three-element vacuum tube, the triode, contains a plate, a grid, and a cathode arranged as shown. The schematic symbol for the triode also is shown.

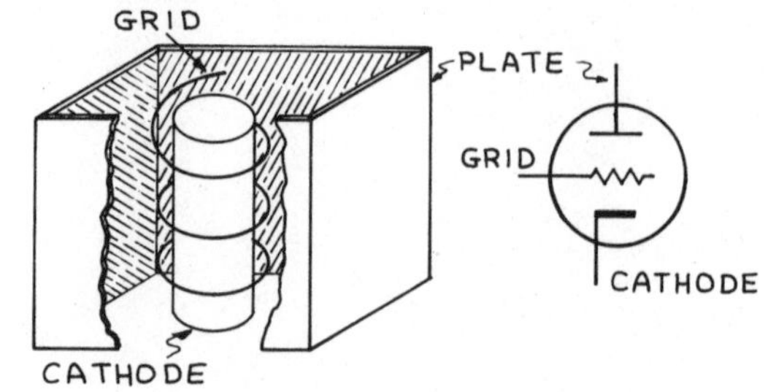

tian blind, although the electrical effect is similar. As a matter of fact, the grid is a loosely wound coil that offers little resistance to electron passage. The grid functions by varying an invisible negative cloud near the cathode that inhibits the escape of the expelled electrons. When the grid is made fully negative, the cloud is too dense for electron traverse and none reaches the plate. Make the grid positive and the cloud virtually disappears to allow electrons free passage. In other words, the potential on the grid determines plate current from full cutoff to complete saturation. This potential in an operating tube is maintained between upper and lower limits unique to the tube type and the circuit in which it is connected.

The source of the electron stream is the cathode heated to a high temperature. Although the material and the configuration are different, the phenomenon is the same Edison found in his crude light bulb. The sketch in Figure 5–2 shows one form of cathode that is common to many vacuum tubes. An electrically heated central wire brings the cathode material up to its emitting temperature. Incidentally, the need for this wasteful heating current is one of the drawbacks the vacuum tube suffers when compared to the transistor.

Many materials will spew out electrons when heated to high enough temperatures, but only a few are sufficiently efficient to be useful in vacuum tubes. Of these, tungsten stands out. But even tungsten's efficiency, the number of electrons thrown out per watt of expended heating power, is comparatively low. It was found that the oxides of certain metals perform better.

Figure 5-2. The source of electrons for the vacuum tube is the cathode. The emissive material is brought to high temperature by an internal heating coil.

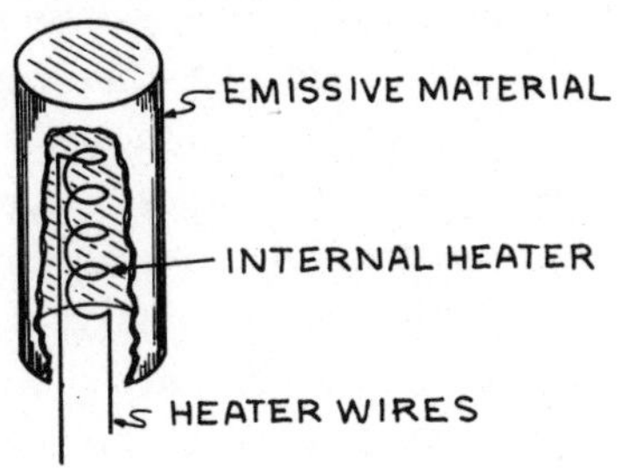

The oxides in use today are barium oxide and strontium oxide. These are applied as surface coatings on nickel sleeves. This combination is approximately fifty times as efficient as tungsten alone. In this design, tungsten is used merely as the central heating wire, a service for which it is well suited because of its extremely high melting point and its resistance to evaporation. Some large power tubes utilize the tungsten heating wire as a combined filament and cathode and coat it with thorium to increase its efficiency.

As circuits became more complex, certain shortcomings of the triode became apparent. One of these is the capacitative effect between the grid and the plate because, as with any two adjacent conductors, they formed a capacitor (see Chapter 4). This capacitor interfered with tube operation at very high signal frequencies, and its neutralization would prove a step forward. An additional grid, making the triode a tetrode, solved the problem.

The new grid, placed between the control or main grid and the plate, is known as the "screen grid." It reduces the objectionable capacitative effect without interfering too much with electron travel. A bonus that came with the tetrode is higher amplification ability.

An electron travelling at high speed and hitting the plate dislodges a secondary electron, causing an unwanted condition. Again, the problem is solved by the addition of still another grid to the four electrodes in the tetrode, thereby creating the pentode. This new grid is called the "suppressor grid," an apt name considering its function. It is close to the plate, as shown in Figure 5-3. The pentode is one of the most versatile of all vacuum tubes and it found wide use in radio receivers of the superheterodyne type.

With electrons travelling at the tremendous speeds of light, it is hard to believe that, under certain condi-

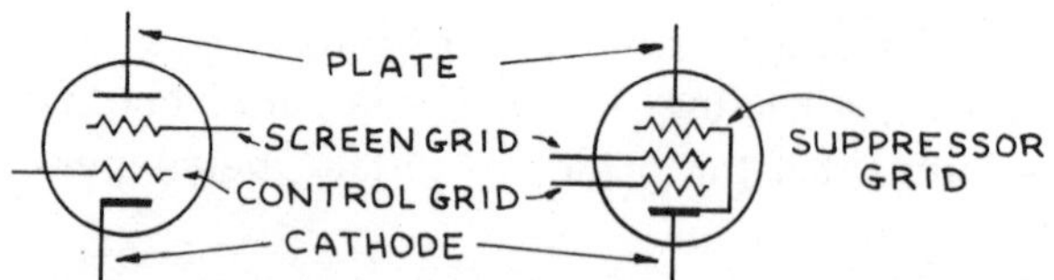

Figure 5-3. Adding a screen grid to the triode creates the tetrode. The addition of a suppressor grid to the tetrode results in the pentode. The ability to amplify increases from triode to tetrode to pentode.

tions, their passage from grid to plate could be too slow. Yet, at extremely high radio frequencies, running into many millions of cycles per second, the distance from grid to plate may be too great for the electron to cover in time. This has led to designs of vacuum tubes in which the electrodes are infinitely close together in terms of transit time. One such is called the acorn because of its shape.

One recurrent problem in the heyday of vacuum tubes was posed by bulbs that had lost part of their vacuum and were designated "soft." Often these could be spotted visually through the purplish glow that enfolded their electrodes when in use. A soft tube was a candidate for rejection in many circuits because its operating characteristics deteriorated below the required level. Soft tubes also ran hotter in service.

The creation of heat is one of the bugaboos of vacuum tube circuits, and when many tubes are enclosed in limited space, this heat may become an acute problem. A usual feature in such cases is a blower to circulate cooling air or, at the very least, louvers for natural circulation. By contrast, circuits with an equivalent number of transistors function practically without noticeable heat generation except perhaps for large power transistors—and even these are able to dissipate their heat through fins or heat sinks.

The need for maintaining the grid between fixed negative potential limits, mentioned earlier, brings what is called the "biasing" circuit. In the earliest vacuum tube receivers the biasing or maintaining of grid potential was accomplished by a separate battery known as the C battery. In this era the voltage for the plate came from a B battery and the voltage for the heaters from an A battery. Later circuits eliminated these troublesome batteries and substituted a "power supply" that derived all needed voltages from the house main alternating current. The voltages required for vacuum tube operation are generally high and the power supply therefore is heftier and more complex than the equivalent supply for transistors.

The ability of a vacuum tube to do a desired job is determined by a perusal of the technical ratings given it by its manufacturer and listed on available data sheets. The more important of these ratings are the "amplification factor" denoted by the Greek letter mu (μ), the "transconductance" listed as g_m, and the "plate resistance," r_p. A skipper need not go into the intricacies of derivation of these parameters, but a cursory acquaintance with them will help choose the correct tube for a given spot if the need arises.

The amplification factor is an overall indication of the tube's ability to amplify a signal under optimum conditions. The larger this number, the more able the tube. However, the maximum value is seldom reached in practical, commercial circuits.

The transconductance is a good overall figure of the merit of the tube, and this is the characteristic measured by good tube testers. Some testers will give this as a meter reading that can be compared with the manufacturer's data sheet. Most testers, especially those provided in stores for layman use, will have a dial marked in red, yellow, and green sectors. Once such a tester is set for a particular tube, a meter needle in the green indicates a good tube, yellow means poor, and red shows defective or dead. (See Figure 5-4.)

Electric current passing through any medium encounters some resistance and the path from cathode to plate via the electron stream is no exception. The plate resistance data evaluates this and makes it possible for technicians to calculate the effects in a circuit. The plate resistance is the genesis of some of the heat developed by a vacuum tube in operation.

Figure 5-4. The tester for vacuum tubes was one of the important pieces of shop equipment before the age of the transistor. (Eico Electronic Instruments Co.)

There is one spot from which the transistor has not been able to oust the vacuum tube—its function as a cathode ray tube, the familiar picture tube of television. Aboard ship, the cathode ray tube is the image screen of the radar receiver.

The cathode ray tube depends upon a stream of electrons for its operation, just as do all other vacuum tubes. The electrons are ejected by a heated cathode, again as in all others. Focusing electrodes sharpen the stream into a pencil-like beam that is aimed at any point on the screen to form the desired image. The screen fluoresces where the beam hits, thereby making the image visible. These large image tubes are unique in the extremely high potentials they require, often into the thousands of volts. The cathode ray tube from a radar receiver is shown in Figure 5–5.

A transistor is in business the moment its current is turned on; a vacuum tube is not. The tube's cathode must be brought up to its emitting temperature before action can begin, and this takes finite time (although some tubes have been designed to heat so quickly as to be almost "at once"). To overcome this problem, electronic equipment designed for vacuum tubes usually has a "stand-by" switch. Only the cathode heaters are in circuit during the stand-by condition; subsequent application of plate voltage makes the unit operative immediately. The advantage of this scheme is the saving of plate power during stand-by periods.

To save space and heater current, two or more vacuum tube assemblies are often placed in one bulb with only one cathode. The types of tubes that are thus sealed together are varied; for instance two triodes, or a diode and a triode, or two pentodes, or whatever. Each unit is able to function without interfering with the other because all electrodes are connected to separate pins.

The long array of vacuum tubes includes some types designed for special purposes. Among these are the thyratron, the voltage regulator, and the beam power tube.

The thyratron is a control tube often found in the older high power vacuum tube transmitters. In essence, this tube is a diode plus a control grid sealed within a gas-filled bulb. Unlike an ordinary diode, this one does not conduct the moment a positive voltage is connected to its plate or anode; a proper pulse also

Figure 5-5. The indicating tube, on whose face the radar picture is presented, is a specialized form of cathode ray tube. The electron beam is accelerated to high velocity by the anodes and forms bright spots where it hits the fluorescent coating on the inside of the tube face. A rotating deflection coil on the neck of the tube keeps the beam in synchronism with the rotation of the antenna.

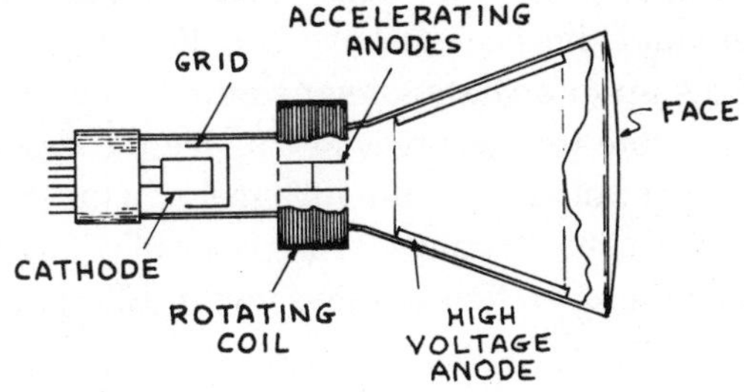

must be applied to its grid. Thus a small pulse can turn a large current on and off, making this an efficient, non-mechanical switch.

The voltage regulator is not truly a vacuum tube, although outwardly it resembles one. A gaseous discharge inside the bulb when the regulator is in operation maintains a fixed voltage across the terminals of this tube. Accordingly, it finds its use in regulated power supplies where it keeps the output voltage constant regardless of the load imposed.

The beam power tube is a four element tube to which small, flat metal electrodes have been added to concentrate the electron beam on its passage to the plate. The electrical action of this tube is similar to that of the pentode, but it can handle larger power with less loss. The drawing in Figure 5–6 shows the beam-forming flat plates in relation to the four basic electrodes.

Figure 5-6. The beam power tube is a special form of the pentode. The normal suppressor grid of the pentode (see Figure 5-3) is replaced by beam-forming electrodes that concentrate the electrons flowing to the plate into powerful beams.

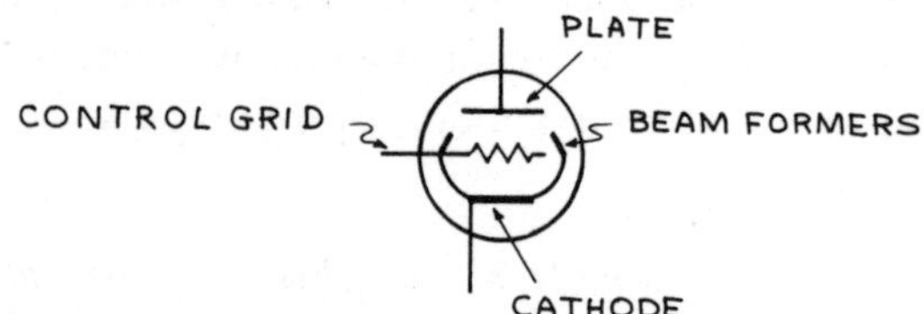

One device whose outward appearance may fool the observer into thinking it a vacuum tube is the thermal time delay tube. Actually, it is only a thermally timed switch sealed into a vacuum tube format. In a vacuum tube transmitter, this delay tube withholds plate power until the heaters have had time to bring the cathodes up to operating temperature. To apply plate power before this occurs is to damage the tubes.

The literature of vacuum tube manufacturers shows what are called "families of curves" that graphically describe the operation of tubes under different conditions. One family, for instance, enables plotting the current drawn by the plate for varying combinations of plate voltage and grid voltage. This is basic information for engineers and designers and helps technicians substitute tube types when necessary.

Vacuum tubes are made with base pins that fit into appropriate sockets for connecting them into circuits. The earliest commercial tubes, triodes, had large phenolic bases cemented to the glass bulbs with four projecting pins. As tube internals became more complex, the number of pins increased, and bulbs and bases became smaller. On many miniature tubes the pins project directly from the glass and forego the plastic base. Some means such as a slot or uneven pin spacing always is provided to prevent incorrect tube insertion.

The photo in Figure 5–7 is a striking summation of the transistor-versus-vacuum-tube situation. The transistor occupies a fraction of the hulk of the tube, has an infinitely longer life, requires no wasted heater current, and yet stands shoulder to shoulder with the vacuum tube in its output of power.

Figure 5-7. The photo dramatizes the striking difference between vacuum tubes and transistors for approximately equal service.

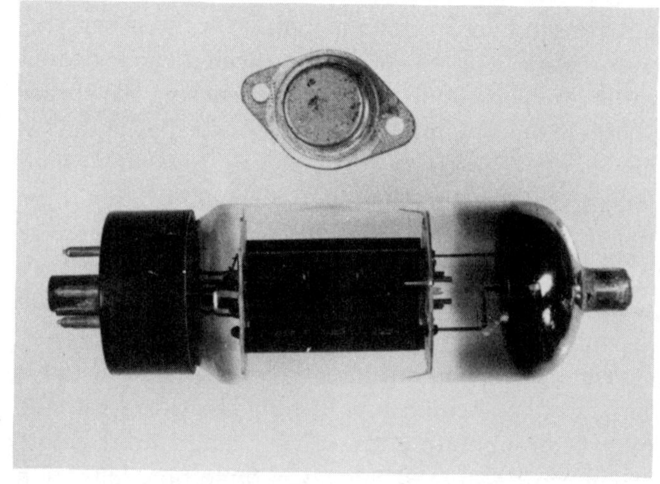

PART TWO

BASIC ELECTRONIC CIRCUITS

6. *Tuners*

Tuning, in the electronic sense, is the ability to receive a desired signal while simultaneously rejecting undesired signals of higher and lower frequencies. This tuning is accomplished by various methods that involve circuits from the very simple to the highly complex. In essence, tuning establishes resonance between the desired signal and the receiving equipment.

The meaning of the term "resonance" may be illustrated with two identical tuning forks. If one be struck into vibration, the other, if nearby, will vibrate sympathetically. Since they are identical physically, the frequency at which they vibrate is the same. The sound waves given off by one will actuate the other. They are in resonance. They are tuned to each other. A tuning fork of a different pitch will not affect them.

Tuning forks will vibrate up to frequencies of a few thousand cycles per second. They are tuned by varying their length and their mass. The frequencies usually involved in electronics are many thousand (and even million and billion) cycles per second, and the tuning is accomplished with various combinations of resistors, inductors, and capacitors, the electronic components discussed in Chapter 4.

The term "tuners" often is found in advertisements of high fidelity radio receivers; this may cause a misconception. The tuners mentioned there are complete radio receivers minus only the audio amplifiers needed for loudspeaker operation. The tuners discussed in this chapter encompass only the basic circuits that change the frequency of electrical resonance. In turn, these circuits are incorporated in many electronic devices.

A radio wave may be described by its frequency or by its wavelength; the two terms are interdependent. The wavelength is the distance between two successive upper crests; the frequency is the number of times the wave cycle is repeated within one second. Radio waves travel at a speed of 300,000,000 meters per second and the interdependency relates this to the frequency. Thus, a radio broadcaster transmitting at a frequency of 1,500 kilocycles per second (or 1,500,000 cycles) is propagating a wave 200 meters long—300 million divided by 1½ million.

Amateur radio operators seem to prefer to describe their transmissions by the length of the waves in meters. In the marine electronics world, the specifications are all in frequency of cycles per second. The new unit is the Hertz, one cycle in one second. This is amplified into kilohertz for 1,000 hertz, abbreviated KHz, and megahertz for 1,000,000 hertz, or MHz. (The broadcaster in the example above is on a frequency of 1,500 kilohertz.) Prefixes for even higher orders of frequency begin with "giga" for billion.

The idea of resonance in an electronic circuit may be understood without delving into the depths of alternating current theory. It all depends upon the unique characteristics of capacitors and inductors. Capacitors offer a barrier to the passage of alternating current, called their "reactance." This reactance varies inversely with the frequency of the electricity applied. In other words, for any given capacitor, as the frequency goes up, the reactance decreases.

Inductors also offer reactance to alternating currents, but in a manner opposite to that of capacitors. The reactance of an inductor increases when the applied frequency goes up. Since the inductive and capacitive reactances are directly opposite, they are said to be 180 degrees out of phase. (Remember that a 180 degree change of course turns the ship directly back.)

The combination of these opposite-going reactances is the secret of electrical resonance. In the circuit of Figure 6–1 a capacitor (C) and an inductor (L) have been placed in series with an applied variable frequency alternating current AC. When the frequency is

Figure 6-1. This combination of an inductor L and a capacitor C constitutes a "series tank circuit." It is resonant at a certain frequency depending upon the values of L and C.

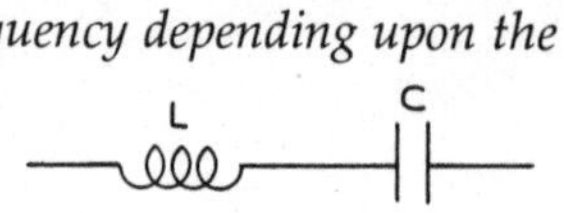

L C

low, the reactance of C is high and the reactance of L is low. Raise the frequency to a high level and now the reactance of C is low while the reactance of L has become high. Obviously, at some midpoint in the frequency travel the reactances will be equal in number but opposite in phase. At that instant the reactances cancel out and the only barrier to the flow of current in the circuit is the resistance of the connecting path. Maximum current flows because the circuit is resonant at the critical frequency. The circuit is tuned to that frequency.

A correct deduction from the foregoing is that an electronic circuit may be tuned by varying either L or C, either inductance or capacitance, or both. Which parameter is varied is a matter of convenience and design. In the early days of amateur radio, tuning was done almost exclusively by varying the inductance. The common homemade inductor then was a large, round cardboard container (originally filled with oatmeal and now wound with wire). A slider traveled along a bare path on the wire coil and connected more or less inductance into the circuit. It is even possible that the then universal habit of eating oatmeal for breakfast was sparked by myriads of young radio amateurs constructing tuning coils. (See Figure 6–2.)

Modern practice in tuning methods is to vary the capacitance. This is much easier to do, both mechanically and space-wise. The variable capacitor shown in Figure 6-3 is typical of those employed in radio circuits.

Figure 6-2. Wire wound closely on a round cardboard oatmeal box served as a tuning coil for many young experimenters. The sliding contact varied the amount of wire in circuit and hence the frequency.

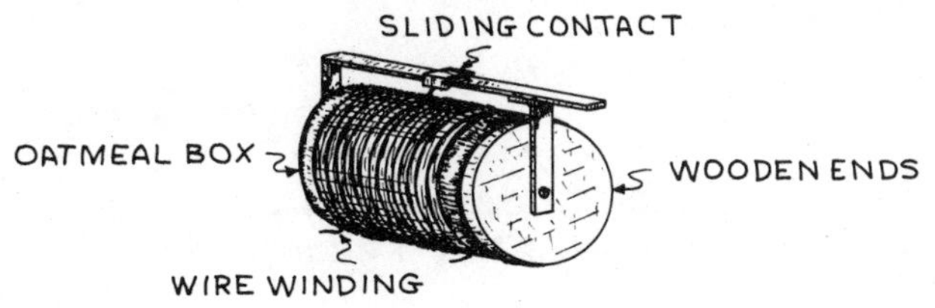

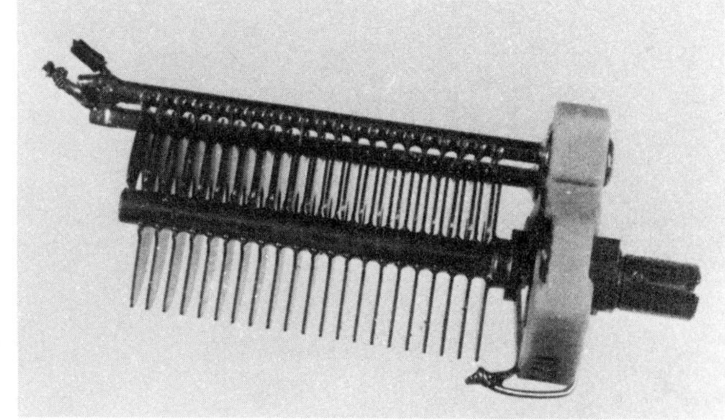

Figure 6-3. The amount of interleaving between the fixed and movable plates determines the capacity of this variable capacitor.

A series of fixed plates is interleaved by movable plates when the knob is turned. The capacity increases as more of the two sets of plates are meshed. The dielectric is air.

There are practical limits to the range of frequencies that a single capacitor and inductor can cover. To overcome this, additional capacitors and inductors (preferably inductors) are switched into and out of the circuit as needed. Generally, the range of frequencies is divided into bands—for instance, the broadcast band goes from 550 kHz to 1600 kHz inclusive. Some receivers eschew switches in favor of compact coils that may be plugged in as needed.

In modern radio receivers it is necessary to tune a number of circuits simultaneously in order to achieve the desired selectivity. These circuits must all change at identical rates in order to "track." Tracking is accomplished by careful design of the capacitors and inductors involved and by limiting the range of frequencies to be covered. Several tuning units that are connected together for tracking are said to be "ganged." (Figure 6–4 pictures ganged capacitors and ganged inductors.) Often small, trimmer capacitors are added to the gang to refine the tracking of variable capacitors and obtain a longer range of frequency coverage.

Figure 6-4. Several variable capacitors are ganged together when it is required to adjust a number of tuned circuits simultaneously.

The inductance of a coil of wire is affected by the core about which it is wound, be this air or metal. If iron be added to the core, the inductance is increased. Add brass or copper and the inductance is decreased. These additions to the core are called "slugs"; movable slugs tune the ganged inductors shown in Figure 6–4. (Iron is not used as a solid metal, but rather as a powder formed into a slug with a suitable adhesive binder.)

The series resonant circuit drawn in Figure 6–1 has a counterpart, the parallel resonant circuit of Figure 6–5. The same reasoning that earlier explained the change in reactance of the capacitor and the inductor with change in frequency applies here. But the end result is different because the interconnection of the two elements is different. Whereas the current flowing in the series circuit was heaviest at resonance, in the parallel circuit the current at resonance is least. Both types of circuit are common and are used often in electronic devices.

Figure 6-5. The parallel combination of an inductor L and a capacitor C constitutes another commonly used tank circuit often found in transmitters. The frequency depends upon the value of L and of C.

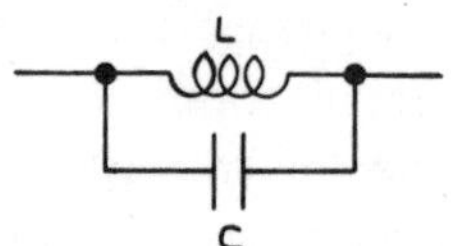

Since the current flowing at resonance is independent of capacitor and inductor and dependent only upon the resistance of the path, it behooves designers to make this path adequate. They do this by using wire of large cross-section, often plated with gold, and by keeping connections short and solid. Circuits that are efficient in this manner are said to "have a high Q," and manufacturers mention this in their literature. Incidentally, in the lingo of electronics, the combination of capacitor and inductor is known as a "tank circuit"; the name derives from the ability of the circuit to store energy.

The ideal, theoretically perfect tank circuit resonates at one sharp frequency. Unfortunately, this circuit does not exist and is only a goal toward which to strive. An actual circuit at resonance tunes to a small band of frequencies instead of just one. The width of this band is determined by the resistance in the tank, hence the striving for low resistance and high Q. How added resistance destroys the sharpness of response is shown by the diagram in Figure 6–6.

Many tuned circuits in electronic devices are fixedly tuned and internal, beyond the adjustment of the user.

Figure 6-6. These curves show the deleterious effect of resistance in a tuned circuit. Low resistance permits the resonant response to be sharp as shown, while high resistance blunts the response curve and makes it broad.

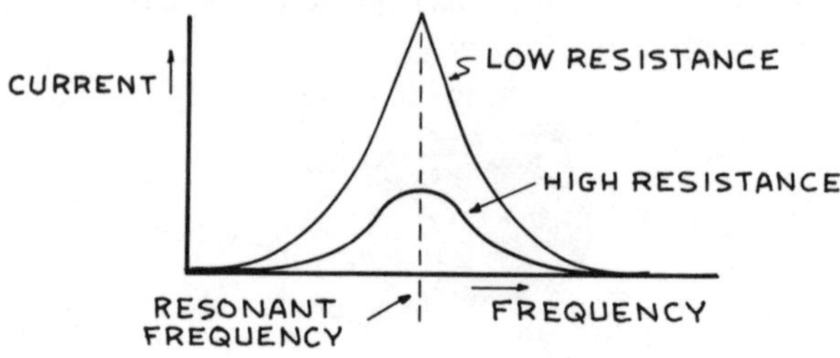

These insure that only signals of the desired frequency are able to pass from one stage to another. Receivers and transmitters have many such circuits in order to refine the signal response down to a very sharp peak. Such selectivity is necessitated by the crowded condition of the radio spectrum. Internally tuned modules are found in depth sounders and abound in Lorans and radars.

The user is provided various methods for tuning electronic equipment. The simplest and most common is a knob with a graduated dial. This is also the most inconvenient and the least accurate. The newest and easiest to use system has a bank of ten numbered buttons. Touching the buttons in sequence for the numerical value of the desired frequency automatically tunes the instrument and flashes the digital setting. An incorrect setting is even announced by a flashing of the digital image.

One disadvantage of knob and dial tuning appears when the desired frequencies are numerically close together—they are crowded too closely on the dial. Some equipment overcomes this by what is known as "band spreading," accomplished either mechanically or electrically in the basic design. Radio amateurs, who are confined to very crowded channels, have refined band spreading to a high degree.

As electronic devices move up into ever higher frequencies, sharp tuning becomes increasingly more difficult. The reason is that tiny changes in capacity or inductance at these elevations make great jumps in frequency. A change in the length of a connecting wire may alter inductance sufficiently to subvert tuning. The approach of the user's hand can vary capacitance away from the critical. Vibration, as on a boat, could affect both inductance and capacitance and, consequently, tuning.

Tuning dials that display frequency are graduated either linearly or logarithmically. The linear system spreads frequencies evenly across the dial for easy reading. The frequencies at the upper end of a logarithmic dial are crowded together, much as are the numbers on a slide rule, and exact reading is difficult. Of course, the linear marking is preferable. (See Figure 6-7.)

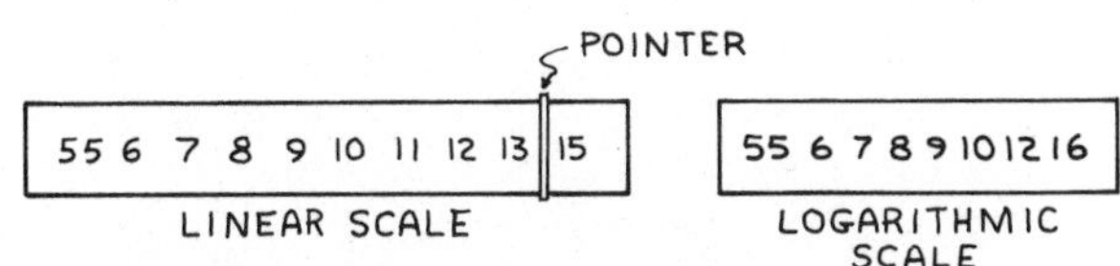

Figure 6-7. Linear tuning permits the frequencies of the various received stations to be distributed evenly across the dial. Logarithmic tuning crowds the stations at one end.

Figure 6-8. Radio frequency transformers are a part of many tuned circuits. Shown here are the relative sizes of the old and new. The unit used in modern transistor radio receivers is so small that it barely covers a penny.

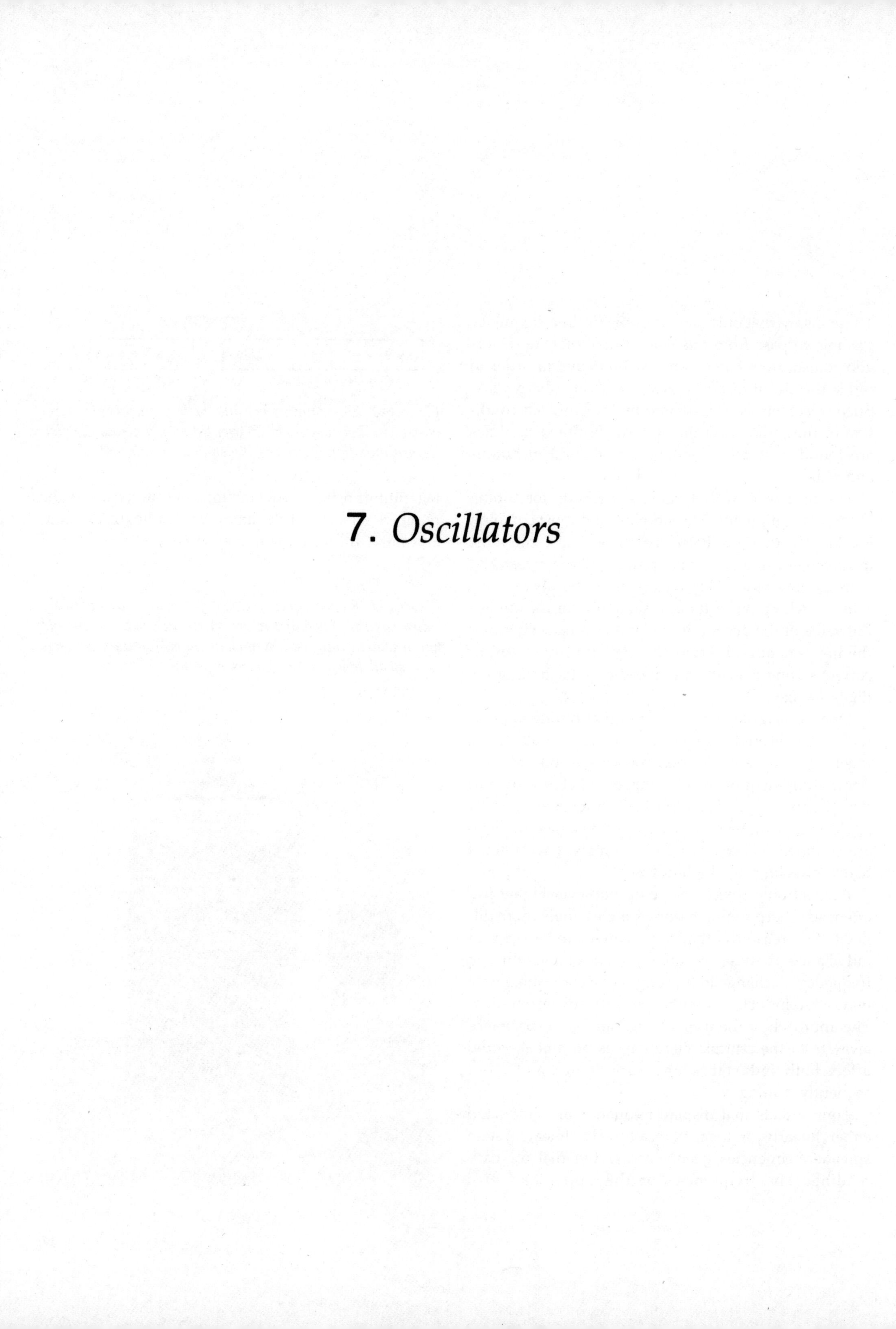

7. *Oscillators*

The basic oscillator is a key module in most navigational electronic equipment. It is at the heart of receivers, transmitters, depth sounders, Lorans, and radars. The oscillator circuit itself generally is either of two types: the Hartley or the Colpitts, both named after their inventors. Compared on the basis of energy output, the oscillator in a receiver generates miniscule power while its counterpart in a big transmitter or radar will develop tremendous wattage, usually with the aid of amplifiers. The active element in any oscillator circuit may be a vacuum tube or a transistor, with modern design opting almost universally for the latter.

The idea of an oscillator may be explained by examining a simple pendulum. Push a pendulum into motion and its swings become gradually shorter until it stops. But give the pendulum a light tap at the end of each swing, and the same frequency of motion continues as long as the taps are supplied. The mechanism of a clock keeps its pendulum going by administering these correctly timed taps. The important points to glean from this example are that the length of the pendulum determines the frequency of its swings and that pushes at the right time keep it swinging.

The inductor and the capacitor in the oscillator circuit of Figure 7–1 are the equivalent of the pendulum length because they determine the frequency of oscillation. They tune the circuit exactly as is discussed in Chapter 6 and set the point of resonance, the point at which oscillation will occur at the permissible frequency. For each combination of inductance and capacity there is only one frequency of oscillation. A new frequency of oscillation is established when either inductance or capacity, or both, are changed. A tuning dial varies frequency because it makes such an internal change when it is rotated.

Figure 7-1. This diagram illustrates the basic concept of an oscillator. Essentially, an oscillator is an amplifier a portion of whose output is returned to the input in correct phase relationship.

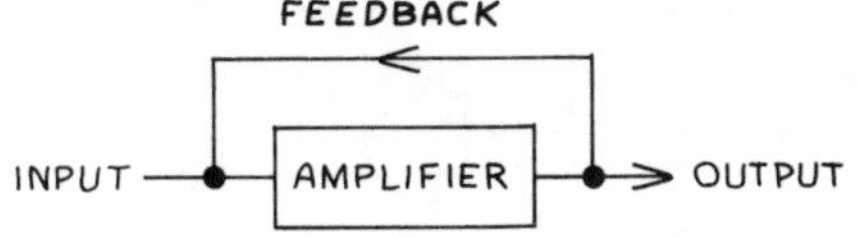

The tuned circuit of the oscillator alone will not sustain oscillation; it needs the electrical pushes equivalent to the mechanical pushes given the pendulum. A property of the circuit called "feedback" supplies these pushes by diverting a portion of the output energy back to the input. Feedback may be "positive," meaning that it induces oscillation, or "negative" to prevent oscillation. Obviously, an oscillator can function only with positive feedback.

The difference between positive and negative feedback may also be explained by reference to the pendulum. A helpful tap at the top of each swing is positive. A retarding or snubbing action at this same point would be negative. (Negative feedback finds use in audio amplifiers, where it reduces distortion. See Chapter 11.)

When the positive pushes to the pendulum occur at exactly the top of each swing, they are said to be "in phase." The electrical push supplied to the oscillator by its feedback circuit must likewise be in phase. This phasing is done by the portion of the circuit that links the plate of the vacuum tube or the collector of the transistor (the output) with the input to the device, and depends upon the characteristics of inductors and capacitors.

An oscillator must be stable and must remain exactly at its set frequency if the electronic instrument that contains it is to be considered reliable and of high quality. Natural causes may provoke instability and must be guarded against. One such cause is a change in temperature that creates expansion and thereby varies the clearance between sensitive elements. Heating and cooling occur slowly in conformance with ambient conditions and the slow changes in oscillator frequency that result are called "drift." Drift in a receiver is annoying to the listener; drift in a transmitter makes its signals difficult to receive and may cause interference to other users of the radio spectrum.

The possibility of drift has become less likely with the advent of transistors and their almost heatless operation. Vacuum tubes discharge so much heat that drift in units employing them is a factor to be considered in design. The layout must keep sensitive elements away from the tubes and provide for ventilation, either natural or fan-induced.

Another source of unwanted frequency variation is fluctuation in the voltage fed to the oscillator. When power is derived from the shore, these fluctuations could result from the ups and downs of pier voltage as boats connect and disconnect their loads. Good design counters this with voltage regulators in the power supply. Here too, transistors have stepped in to lessen or remove the annoyance because most transistor-powered electronic devices run off the boat battery, a very steady source.

An oscillator is essentially an amplifier with a positive feedback arrangement. When the current is switched on, thermal noise or random electron movement creates a tiny input to the vacuum tube or transistor. Amplifying action magnifies this to a larger value at the output. A portion of this output goes back to the input via the feedback and the circulating pulses become ever greater until they reach the steady state oscillation determined by the constants of the inductor and capacitor. The process is cumulative and continues as long as current is supplied.

Figure 7–2 has wiring diagrams of the Hartley oscillator (a) and the Colpitts oscillator (b). Even cursory examination shows that these two circuits are very similar. In both, the desired frequency is obtained by setting the combination of the inductor (L) and the capacitor (C) to the necessary resonance. In both the feedback voltage is obtained by the connection (F) to the output, the plate in the case of the vacuum tube and the collector of the transistor. The mandatory division of voltage between input and output is obtained

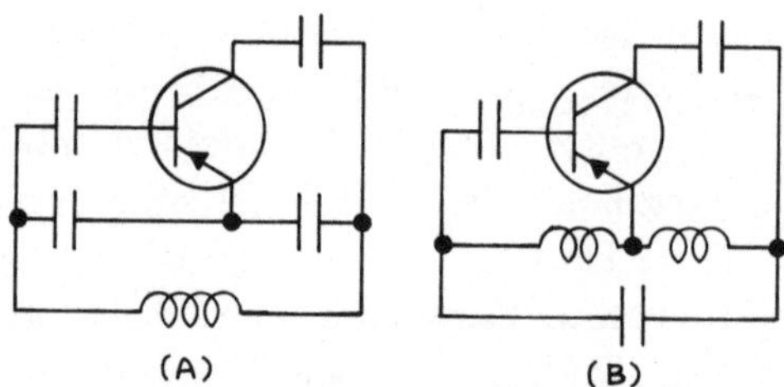

Figure 7-2. The Bartley and the Colpitts, the two basic oscillator circuits, are compared in these schematic diagrams. Resistors and power supplies are omitted for clarity. Note that the Colpitts, at (a), uses a center-tapped capacitor while the Hartley, at (b), makes use of a center-tapped inductor in order to achieve the feedback condition necessary for oscillation.

by tapping the midpoint of the coil in the Hartley and the midpoint of the two capacitors in the Colpitts. (Both vacuum tube and transistor oscillators of each type function identically.)

A slightly more complicated oscillator circuit is shown in Figure 7–3. It is called the "tuned plate-tuned grid" and, as implied, has a separate resonant circuit of inductor and capacitor in both input and output. The path of feedback in this circuit is not obvious. Feedback is achieved through the elements of the vacuum tube itself, aided by the resonance at the grid and at the plate. Since two separate resonances must be adjusted simultaneously and precisely for each frequency, this circuit is not as amenable to single dial control.

Certain capacitance exists between the electrodes in a vacuum tube and the elements in a transistor. This capacitance is always present in the oscillator circuit,

Figure 7-3. The tuned grid-tuned plate oscillator circuit is a very stable form, with resonant tanks in both the input and the output. A schematic adapted to a transistor is shown above. Resistors and power sources are omitted for clarity.

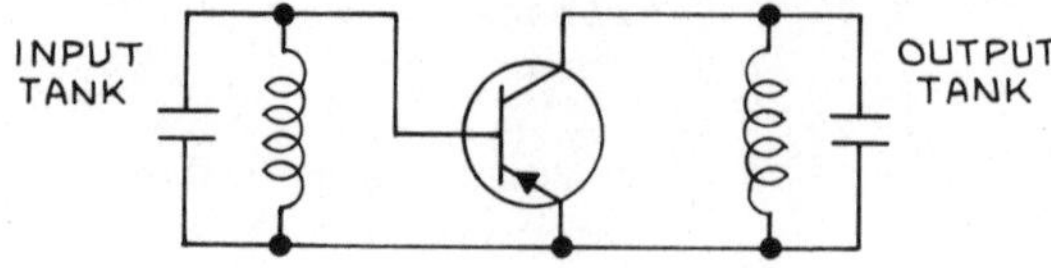

regardless of the tuning of the main inductor and capacitor, and is called the residual capacity. The residual capacity limits the highest frequency at which the oscillator will function. Special designs of vacuum tubes and transistors intended for ultra-high frequency operation pare the residual capacity down to as close to zero as possible.

As already explained, the criterion for oscillation is that the feedback be exactly in phase with the input in order to bolster it in a cumulative manner. It follows that any circuit able to accomplish this phasing should be a basis for an oscillator. This is borne out by the "phase shift oscillator" diagrammed in Figure 7–4. In this circuit, the inductor is absent and the combination of resistors and capacitors keeps the feedback in correct phase. The advantage lies in the elimination of the inductor because such wire coils are more difficult to make and more expensive than capacitors and resistors.

In considering the actions that take place in an oscillator circuit, it is well to bear in mind nature's rule of "nothing for nothing." The circuit itself does not produce any power. The circuit merely controls and switches the energy available from the battery or power supply. And the total energy supplied to the circuit always is greater than the transformed energy that may be drawn from it because of unavoidable losses. Stated technically, the overall efficiency never can reach 100 percent, and often is abysmally low.

Figure 7-4. The phase shift needed to cause oscillation is accomplished by resistor-capacitor modules (a,b,c) in the phase shift oscillator. A minimum of three are needed because each is able to shift phase only by 60 degrees and 180 degrees are needed (see text).

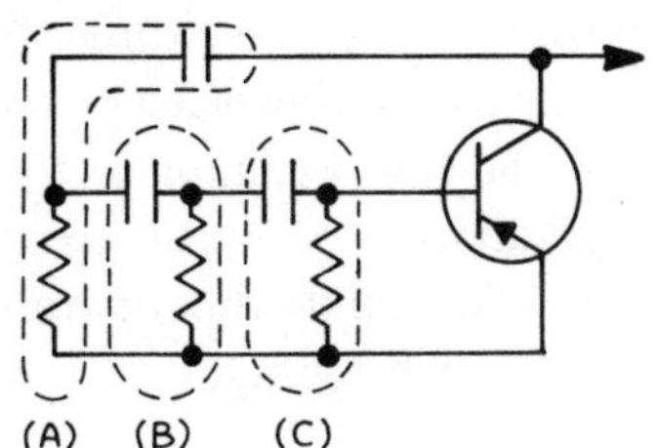

The nature of the output frequency of an oscillator, the shape of the wave, can be important in electronic circuits. The ideal wave shape is sinusoidal as shown in Figure 7–5(a); the smooth undulations prove that no distortion is present. The wave may be square (b); such a wave finds use for timing, as the "clock" in the lingo of electronics. The output of the oscillator also may be a series of evenly spaced spikes (c).

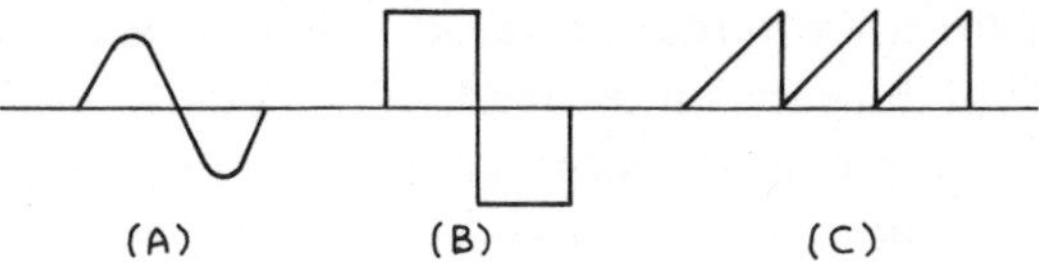

Figure 7-5. The output wave shape of an oscillator depends upon the nature and constants of its associated circuit. A sinousoidal wave is shown at (a), a square wave at (b), and a sawtooth at (c).

Certain conditions in the oscillator circuit may be propitious for the generation of "harmonics." A harmonic is a multiple of the basic frequency. Thus an oscillator set for 1,000 hertz may simultaneously be producing output at 2,000 hertz, 3,000 hertz and so on up to a high multiple. Some electronic equipment that requires ultra-high frequencies achieve them by starting with harmonics of lower frequencies because they are easier to come by.

The feedback in an oscillator is critical, not only as to phase but also as to amplitude. Insufficient feedback will not sustain oscillation. More than the optimum amount entails a waste of energy and generally increases the harmonic content of the output. The amount of power that the oscillator is called upon to deliver to its following circuits, called its load, also is critical. An increase in this load beyond the design point generally will throw the oscillator out of oscillation.

The active element of an oscillator, be it vacuum

tube or transistor, is of course dependent on electron travel for its action. This electron transit requires a finite amount of time. The frequency at which the oscillator is running must allow this amount of time between its cycling. If the frequency desired is so high that this time is not available, then obviously there can be no oscillation. So the transit time needed by the active element is one of the limits to the highest frequency at which any oscillator is able to operate. The special vacuum tube shown in Figure 7–6 is an example of how miniaturization brings the internal electrodes closer together to reduce transit time of electrons.

The high precision demanded by modern boat electronic equipment decrees that reliance upon inductor and capacitor alone for frequency setting of an oscillator is not good enough. A master controller of greater accuracy and stability is needed—and this is found in the quartz crystal. A thin plate of quartz cut from a crystal has a natural frequency of vibration that depends upon its dimensions and may reach into the megahertz. By placing this crystal between two metal plates and suitably connecting it into the oscillator circuit, the crystal frequency governs the oscillator frequency. By placing the crystal assembly into an "oven" to protect it from ambient temperature changes, the output frequency is held accurate to within one cycle in millions. Figure 7–7 gives the diagram for a quartz crystal in its holder. In actuality the plates would be snug against the crystal.

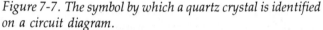

Figure 7-6. The Klystron is an ultra-high frequency tube. The output frequency is varied by small changes in the volume of its cavity resonator that result from turning the tuning screw.

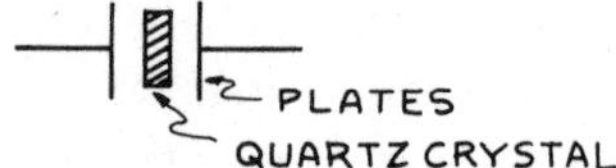

Figure 7-7. The symbol by which a quartz crystal is identified on a circuit diagram.

The theoretically perfect inductor exhibits only inductance and no capacitance or resistance. Likewise, the ideal capacitor has only capacity and no inductance. But these perfect components do not exist. In actuality, an inductor displays some capacity and a capacitor shows some inductance. This lack of perfection becomes a stymie when building extremely high frequency oscillators and other means of providing resonance must be used. One such is the cavity resonator.

The cavity resonator is a small metal enclosure whose internal dimensions permit the existence of an electromagnetic wave of only one exact length or, in other words, of only one frequency. This becomes the resonant circuit that determines the frequency of oscillation. The cavity becomes the equivalent of inductance and capacity in the minute amounts required. Resonant cavities of oscillators produce the microwaves emitted by marine radars.

The gamut of frequencies that oscillators must cover in the many forms of boat electronic equipment is great. The oscillator in a radio receiver may go below 1,000 kilohertz, while the oscillator of a radar will be well up in the gigahertz. With the crowding of the radio spectrum, it is inevitable that operating frequencies will be pushed much higher.

The oscillator circuits of radio receivers, radio transmitters, and the combined transceivers have been improved markedly by a concept called "frequency synthesis." The result is a reduction in the number of basic components, a saving in cost, greater ease of operation, and more compact construction.

In earlier devices the many oscillator frequencies required for multi-channel operation were each controlled by individual quartz crystals. Thus a transceiver intended for transmission and reception on 14 channels contained 28 quartz crystals—a mighty expensive addition. By contrast, a modern counterpart employs only one master quartz crystal and synthesizes the necessary 28 frequencies from it.

The master oscillator is crystal controlled to a super-exact 1 megahertz but its associated circuits encourage the production of harmonics from 2 megahertz through and including 9 megahertz. Each of these frequencies, from 1 to 9, is brought out to a bus and is available to the channel selector. Subsequent dividers and mixers synthesize this into the desired oscillator frequency.

Suppose, for example, that the skipper wants a frequency of 5.87 megahertz and has set his switches or punched his buttons accordingly. The right hand selector takes 7 megahertz from the bus and its circuit divides this by 10 to .7 megahertz. The middle selector abstracts 8 megahertz that is added to the previous .7 and both are divided by 10 to make .87 megahertz. The remaining 5 megahertz is taken directly from the appropriate bus. The final mixer combines all this and the desired 5.87 megahertz is achieved.

The practical advantage of the synthesis system is that if the master oscillator is correct, all derived frequencies cannot be other than correct. The span of frequencies that may be derived is almost unlimited. The synthesizing system is not restricted to digital selection of frequencies. The bus, the selectors, the dividers, and the mixers may just as well be set up for channel selection, and that currently is being done in late model VHF transceivers.

These sets have ten buttons marked from 0 to 9 and the numbers refer to channels and not frequencies per se. The skipper who wants to operate on channel 8 pushes first zero and then 8. Should he desire channel 22 he would push button 2 twice. The two associated frequencies for receiving and transmitting on the channel are selected automatically.

8. *Rectifiers and Power Supplies*

The first central electric generating station in this country was set up in New York City and supplied only direct current to its customers. For many years thereafter, a large section of the city remained on this form of power, even though the surrounding area had switched to the more modern and more versatile alternating current. There were reasons for the switch and eventually the original, die-hard direct current section also switched.

The main factor in alternating current's superiority is the ease with which it may be transformed from one voltage to another. A simple transformer without moving parts does it. The advantage is that copper may be saved in the long transmission lines from central station to user by maintaining high voltage and low amperage. The voltage then is transformed down to the customer's requirement at his location. By contrast, the direct current user must make do with the voltage leaving the station, and the full current traversing the lines means increased copper.

But direct current does have some points in its favor and, in the world of electronics, the need for it is all-pervading. The power supply in electronic devices that run off shore current bridges this gap between AC and DC. Its transformer changes the incoming voltage to the needed high or low amount, its rectifier rectifies this to a pulsating direct current, and its filter finally smooths this to a noiseless output suitable for electronic use.

Each wire in an alternating current circuit is alternately positive and negative many times per second. The graph in Figure 8–1(a) illustrates this as a sinusoidal curve. The current makes equal excursions above and below the zero line into positive and negative areas. Rectification lops off the loops either above or below the line, leaving a series of one-directional pulses, as shown at (b). The result is a pulsating direct current. For electronic purposes these pulses must be melded together into a perfectly smooth output current that induces no unwanted noise into the circuit.

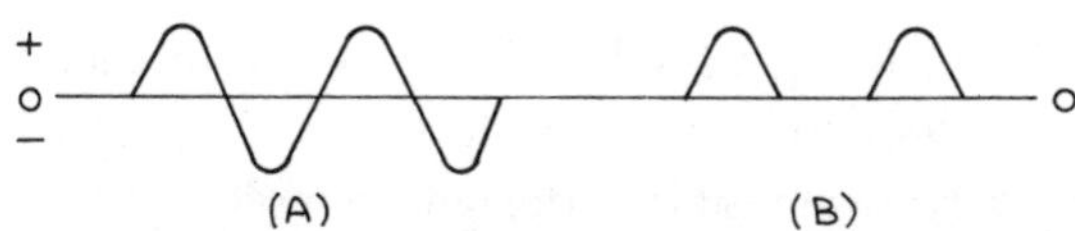

Figure 8-1. How a rectifier lops off half of the alternating current wave is shown in this diagram. At (a) the unrectified alternating current; at (b) after it has passed through a half-wave rectifier.

Luckily, rectifiers are easy to come by. Many materials and combinations possess the property of permitting electric current to pass in only one direction. The galena crystal in the earliest wireless receivers was a rectifier, the Fleming valve and its later counterparts are rectifiers, and the semiconductors make excellent rectifiers. Each of these rectifying substances has a particular niche determined by the current it can carry, the voltage it can withstand, and the reliability of its service.

Early electronic devices intended for shore current attachment employed vacuum tube diode rectifiers exclusively. The overall efficiency of such units is low, and they generate much heat. Their saving grace is that they are able to handle heavy currents and high voltages. Modern design has dropped vacuum tube diodes out of power supplies in favor of semiconductor diodes that exude little heat and are more efficient with almost infinite life expectancy.

All paths for electric current have resistance. The path inside the vacuum tube diode has high resistance. The path inside the semiconductor diode has low resistance. Current traversing a resistance creates heat in proportion to the square of its intensity, hence the reason for the different heating of vacuum tube and semiconductor.

The move from vacuum tubes to transistors in electronic devices has brought revolutionary changes in power supplies. For one thing, most vacuum tubes require high voltages for their plates and screens while transistors are content to work directly off the boat's storage battery. The power that must be handled also

Figure 8-2. Power supplies that change alternating house current to high voltage direct current for electronic devices make use of either vacuum tube diodes or semiconductor diodes—the latter in the more modern units. (The advent of low voltage semiconductor circuits makes the vibrator power supply of only historical interest.)

may differ, being usually high for vacuum tubes and comparatively low for transistors.

The advancements in design sparked by the vast market for automobile radios that had to operate from car storage batteries were a boon to boat electronics before the age of the transistor. The power supplies in these units now took the nominal 12 volt direct current from the boat battery and produced high voltage direct current for the tubes. The seemingly impossible switch of direct current from one voltage to another was accomplished by a vibrator and a transformer in the power supply.

The vibrator is a buzzer with contacts that chop the direct current into a simulated alternating current that a transformer will accept. The output of the transformer then is rectified and smoothed in the usual manner. The vibrators are of two types—synchronous and non-synchronous. The former do away with the diode by performing the rectification mechanically; the latter require the usual diode. A photo of such a power supply is shown in Figure 8-2.

Rectification may be either half wave or full wave, and the name is almost self-explanatory. The half wave passes only one alternation of each cycle of alternating current; the full wave passes both in additive sequence. Figure 8-3 explains this graphically and presents wiring diagrams of simple versions of both systems.

Figure 8-3. Basic half-wave and full-wave rectifier diagrams are given above together with the waveforms involved.

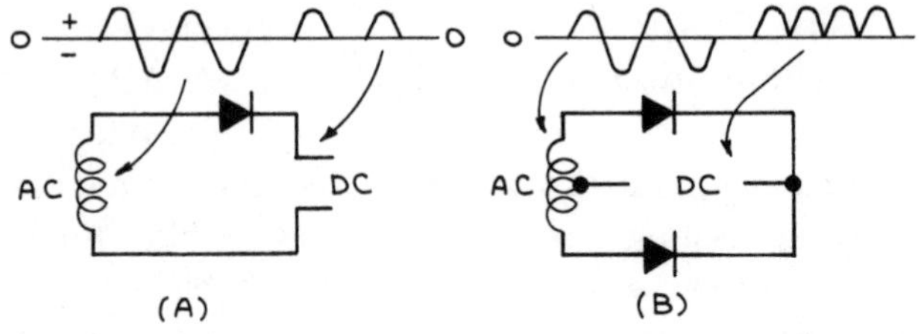

A more advanced and more efficient full wave rectifying circuit, called a "bridge," is shown in Figure 8-4. This utilizes four diodes instead of two, but there is a return for the extra expense. The pulses that must be smoothed regardless of rectifier type are twice as many in the bridge circuit as in the two-diode circuit. The more pulses in a given period of time, the easier the smoothing, and therein lies the advantage of the bridge. In addition, the bridge does not require a center tap on the secondary of the transformer and this permits a reduction in cost plus an increase in voltage.

Figure 8-4. When four diodes are connected as shown, the circuit is a "bridge" rectifier. Note that the secondary of the feeding transformer need not be centertapped, one of the advantages of this circuit.

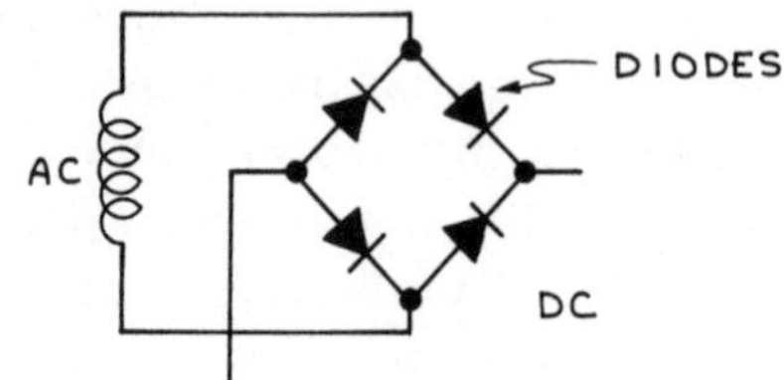

A smooth, ripple-free voltage supply is crucial to all electronic devices, whether vacuum tube or transistor. The ripple is the result of the current void between the rectified pulses and, as the diagrams showed, is greatest with the half wave circuit and least with the bridge. The inductor and the capacitor in the power supply, acting as a "filter," fill these voids more or less completely. How well the filling takes place is dependent upon the complexity of the filter. Good filters can make the output voltage practically continuous, an important factor in high quality audio circuits. (The hows and whys of filters are discussed in Chapter 10.)

Many electronic circuits demand a "regulated" power supply, one whose voltage does not vary regardless of normal fluctuations in the load. The older power supplies achieved this regulation with tubes; today's regulation is with special diodes. The tubes are a gaseous discharge type that maintain a fixed voltage drop. The diodes are known as Zener diodes and they too keep constant the voltage drop across their terminals. Both are available in many nominal values of voltage drop and both may be used with several in series to peg the output voltage at a desired level. (See the wiring diagrams in Figure 8–5.)

As stated earlier, it is routine to change the input voltage to higher or lower values when the supply is alternating current. When the supply is the 12 volt direct current from the boat storage battery, such

Figure 8-5. Many electronic circuits require a regulated voltage that does not vary in amplitude and this may be accomplished with a "zener" diode (see text) as shown.

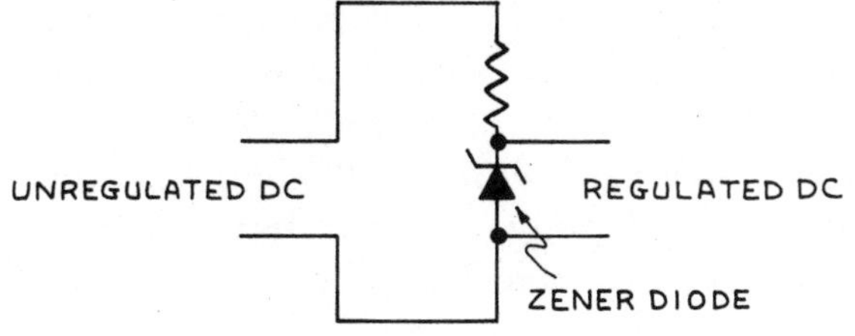

Figure 8-6. Alternating current of 120 volts and 60 Hertz (the equivalent of "house power") may be obtained from the boat storage battery with the aid of an inverter such as shown above.

changing becomes a bit more difficult, but is nevertheless quite common. Inverters and converters do it.

Some equipment, for instance small brushless motors and commercial television sets, operate only on 120 volt alternating current. If these and similar devices are to be used in the absence of a generating plant while the boat is under way, the necessary power must be derived from the battery. A DC to AC inverter whose output is 120 volts at 60 hertz is needed. (A photo of such a unit is shown in Figure 8–6.) Inverters are commercially available for various power outputs. Of course, the battery is the source of the output, and the more power consumed, the sooner the battery will be depleted.

The battery input in these inverters actuates a transistor 60 hertz oscillator that supplies quasi-alternating current to the transformer. The transformer raises the 12 volts to the 120 volts alternating current at the output terminals. The output wave form may be anything from square to sinusoidal, depending on the design. The sinusoidal is preferable but often more expensive. The output could also be any other voltage and any other frequency by choosing other components for the circuit.

When the supply from the boat battery is to be changed to a higher voltage direct current, then a DC to DC converter is needed. As before, the battery input powers an oscillator that feeds a transformer that raises the voltage. But this time the higher voltage is fed to a rectifier and filter to achieve the direct current output. Voltages of 1,000 and more may be generated in this manner.

Power supplies with high voltage direct current output usually terminate in a bleeder resistor and voltage divider such as diagrammed in Figure 8–7. The bleeder stabilizes the system by placing a constant small load across it. The voltage divider and bleeder are combined in a resistor tapped to provide the desired values of voltage. The watts rating of this resistor must be adequate to handle the required current without undue heating.

Figure 8-7. The high voltage direct current output of a power supply may be tapped off at various voltage levels with a voltage divider composed of resistors in series.

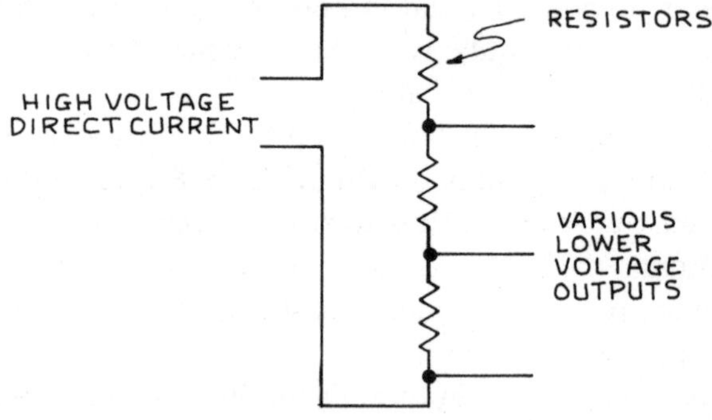

Some power supply circuits accomplish voltage multiplication without transformers by using voltage doublers and triplers. The essence of this idea is to connect two or more high capacity capacitors so that they change in parallel and discharge in series. The diagram in Figure 8–8 traces this scheme. As an example, if two capacitors in a voltage doubler are charged in parallel at 100 volts, they will provide 200 volts if discharged in series. The current such circuits are able

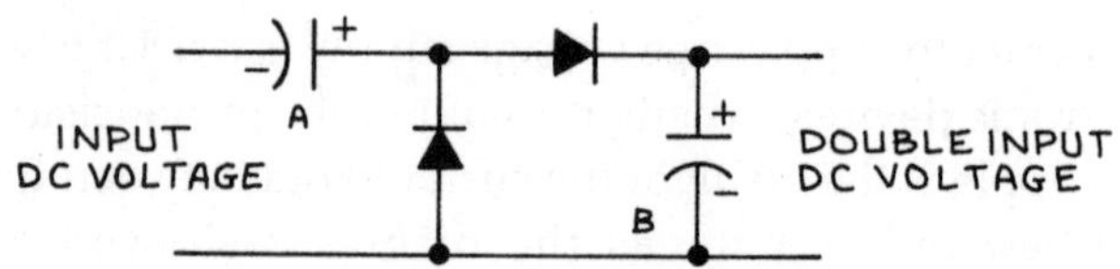

Figure 8-8. A simple circuit for doubling direct current voltage. The high capacity electrolytic capacitor (a) is charged through the first diode. This charge then is added to the charge delivered to capacitor (b) by the second diode and the output is doubled.

to deliver is dependent upon the capacity of the capacitors and generally is nominal.

Short circuit protection may be added to any power supply by placing a resistor in series with the load. This resistor, by its value, fixes the maximum current that may be drawn in the event of an accidental short. Some output voltage is lost across this resistor and this may have to be taken into account.

Commercial power lines, including the shore current installation on the pier, are subject to transients, short spikes of high voltage superimposed upon the regular supply. Transients may be caused by lightning, by customers connecting and disconnecting inductive loads, and perhaps by other factors. Transients can cause the destruction of silicon rectifier diodes. A simple protection is to connect a suitable capacitor across the input of the power supply. This works because the transients are at high frequency and consequently pass more easily through the capacitor than through the power supply. Electricity always takes the easiest path, the one of least resistance.

Modern, transistorized, so-called solid state VHF-FM transceivers are miserly in their demands upon the boat battery. In comparison with the formerly legal AM transceivers, they sip while the latter gulp. Actual figures will corroborate this. A representative VHF-FM unit draws 0.6 amperes while receiving and 5.5 amperes while transmitting. This amounts to 7.2 watts to operate the receiving portion of the transceiver and 66 watts for the transmitter, of which 25 watts will appear at the antenna for actual radiation. An average AM set required 5 amperes for reception and up to 40 or 50 amperes for transmission although, in all fairness, the AM output in watts was greater.

It is often necessary to connect rectifier diodes in series to handle higher voltages and, when this is done, a peculiarity of diodes must be taken into consideration. In a series pattern, that is, one after the other, diodes may not take even shares of the total voltage; excessive voltage to one of them may result. To correct this situation, a high value resistor is connected across each individual diode.

Although semiconductor diode rectifiers are efficient and only a small amount of resistive heat is generated within them, nevertheless this heat must be dissipated through a heat sink to prevent failure. The usual heat sink is the metal chassis. The connection between diode and chassis often must be electrically insulated yet thermally conductive and this may be accomplished with ultra-thin mica washers usually coated with silicone.

Oldtime power supplies for vacuum tubes often contained very high voltages; and safety of the operator was an important facet of design. Transistor power supplies usually are at battery voltage level and electric shock is impossible.

The input cable to a power supply is conveniently made removable. One end of this cable should have a male plug for connection to the power source. The other end should be fitted with a female receptacle that fits on the male plug extending from the power supply. With this scheme, no live terminals ever are exposed.

All rectifier diodes are subjected to an inverse voltage in service. This is the half of the alternating current cycle whose passage the diode opposes in order to create a direct current. Manufacturers issue specifications that state the *p*eak value this *i*nverse *v*oltage may assume for their diodes, the PIV. The PIV is one of the factors that must be checked when replacing diodes in a rectifier. Ignoring this specification may mean arcbacks with vacuum tube diodes and destruction with semiconductors.

When the boat storage battery is the actual direct power supply, as it is with transistorized units, the condition of this battery is all-important for the functioning of the electronic device. A drastically low battery will reduce receiver sensitivity and transmitter output power. With some critically adjusted internal circuits, a low battery may mean total function stoppage. Functioning may also be intermittent—on when the running engine charges the battery and restores voltage, off when the engine stops.

9. *Pulse Techniques*

The giant step forward that eventually resulted in the fantastic achievements of modern electronics was the transition from analog to digital circuitry and the adoption of the pulse techniques that made it all possible. Whereas the analog system utilizes an infinite series of variables to express itself, digital circuits employ only two states, full on and full off. Thus the digital concept largely eliminates ambiguity, because it is much easier to sense the presence or absence of a voltage than to respond to an exact value of it.

This realm of electronics, in which all operations are determined by the presence or absence of a voltage, is based on so-called "logic" circuits that in turn are activated by pulses. The circuits are known as "logic" because results follow logically from causes and these in turn may be manipulated under the rules of a unique system of logical procedure called Boolean algebra. This invention of George Boole is *not* an algebra of numbers and bears no relation to the mathematical algebra taught in elementary schools.

The source of the needed pulses is a specialized class of oscillators called "multivibrators." Despite the inclusion of "vibrator" in the name, these are entirely electronic devices and have no moving or mechanical parts. Multivibrators differ from the oscillators discussed in Chapter 7 whose outputs are designed to be as nearly sinusoidal as possible. Multivibrators generate sharp bursts of voltage with waveforms that may be rectangular, spiked, or otherwise tailored to some mathematical function. Figure 9–1 shows these waveforms in contrast with the sinusoidal.

Figure 9-1. The sinusoidal waveform from a standard oscillator is compared with the square and spiked outputs of multivibrators.

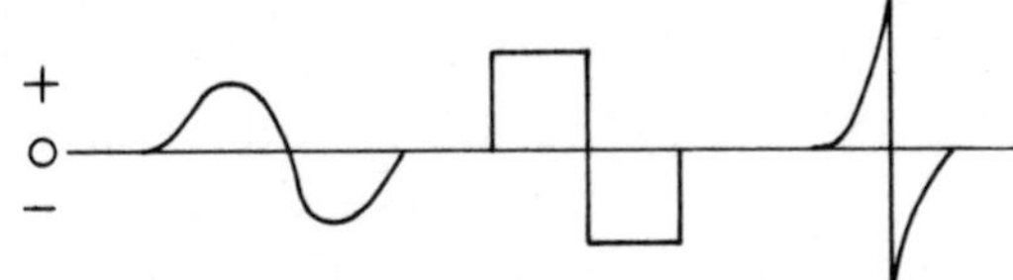

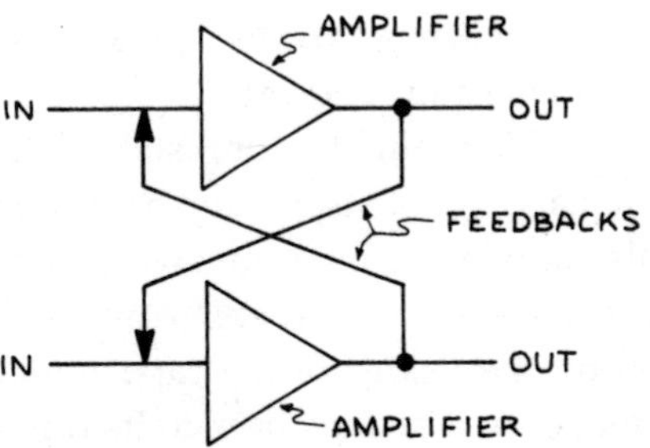

Figure 9-2. This block diagram explains how two amplifiers function together to become a multivibrator. A portion of the output of each amplifier is fed to the input of the other. Circuit constants determine the type of multivibrator that will result.

A glance at the schematic diagram in Figure 9–2 reveals that a multivibrator is actually a two-stage amplifier with a simple addition that makes all the difference. The addition is a "feedback loop" from the output of one transistor to the input of the other. The feedback supplies the kick that swings the circuit between its two possible states of rest. Although transistors are shown in the diagram because modern electronics is powered almost exclusively in that manner, vacuum tubes would function equally if necessary additional voltages were provided.

There are three classes of multivibrators—astable (unstable), monostable, and bistable. The differences entail minor changes in the circuitry. Each class has well-defined jobs to perform in electronic instruments and devices.

The astable multivibrator, as its name implies, is not stable in either of its two states and will not remain at rest in either one. It leaps into action instantly when current is applied and stops when current is cut off. The continuous output of equally spaced pulses constitutes what to technicians is known as a "clock." Such clocks time and synchronize the many operations that must take place in complicated circuits, such as those in computers, for instance. The frequency of the clock depends upon the values of the capacitors and resistors used in its makeup. Ideally, clock pulses are rectangular with sharp on and off.

The monostable multivibrator is stable in only one of its two states and is able to remain at rest only in that one. A pulse applied to its input swings it to its other state, but it does not stay there. After a finite time, determined by the values of the capacitors and resistors in its makeup, it swings back to its original condition without any further triggering.

The bistable multivibrator, as may readily be deduced from its name, can remain indefinitely in either state. A pulse received while it is at rest activates it into its other state. The next pulse simply moves it back. This ability to seesaw has earned this multivibrator the name of "flip-flop," and that is how it is generally found in the literature. The flip-flop is the basis for counting, memory, digital, and other logic circuits. Incidentally, the name "multivibrator" for this genre of oscillators derives from the fact that their output is rich in harmonics, is multi-harmonic.

An example of how flip-flops are able to count is shown schematically in Figure 9–3. The very nature of flip-flops, as has just been described, requires two pulses to restore them to their original state from which they can emit a triggering pulse to the next flip-flop. Thus two pulses to A result in one pulse to B. Two more pulses are required at A to give B that second pulse that enables it to trigger C. It has taken four pulses to activate C and it will take four more before C

Figure 9-3. This block diagram shows four flip-flops (a,b,c,d) that are able to count to sixteen. It takes two input pulses before any flip-flop can trigger the next one and this adds up to the numbers shown below the flip-flop. Thus sixteen pulses must enter A before an output pulse occurs at D.

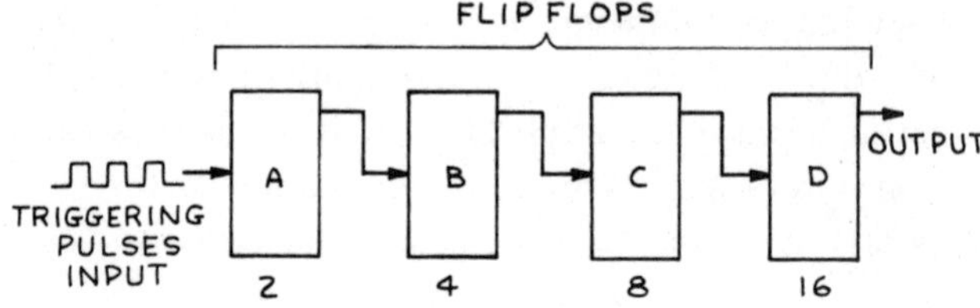

can resume its original state and send a pulse to D. The count at the input of D is now eight—and obviously it will take eight more before D can have an output, at which time the circuit will have counted to sixteen.

The previous counting was accomplished by the binary system, that is, in powers of two. Binary is supremely adapted to the pulse technique because only two circuit conditions, on and off or high voltage and low, can be made to account for any number. By contrast, the decimal system would require ten discrete levels of voltage to identify the digits from 0 to 9. Such discretion is beyond the ability of present day practical electronics.

Two digits, 0 and 1, are used in writing the binary system and any number is represented by a series of 0s and 1s. The 0 and 1 have absolutely no ordinary mathematical significance and are merely labels for the presence (1) or absence (0) of a pulse. Each position of such a series of 1s and 0s is assigned a value to a power of 2, beginning with 2 to the zero power which equals one ($2^0 = 1$) on the right. This is exactly how we read the decimal system, except that there each position of a digit is assigned a power of ten.

Figure 9–4 explains how the binary number 11011 is translated into the decimal number 27. The illustration breaks it down to one times two to the zero power equals one ($1 \times 2^0 = 1$), one times two to the first power equals two ($1 \times 2^1 = 2$), zero times two to the second power equals zero ($0 \times 2^2 = 0$), one times two to the third power equals eight ($1 \times 2^3 = 8$), and one times two to the fourth power equals sixteen ($1 \times 2^4 = 16$). Adding 1 plus 2 plus 0 plus 8 plus 16 equals 27. To add the finishing touch, as remembered from school, Q.E.D.

Converting a decimal number to binary is no more difficult. It is done by successively dividing the number and its remainders by 2. How this is done with the number 27 as an example is shown in Figure 9–5. Each remainder, be it 1 or 0, forms the binary number. Thus the first division of 27 by 2 yields 13 with a re-

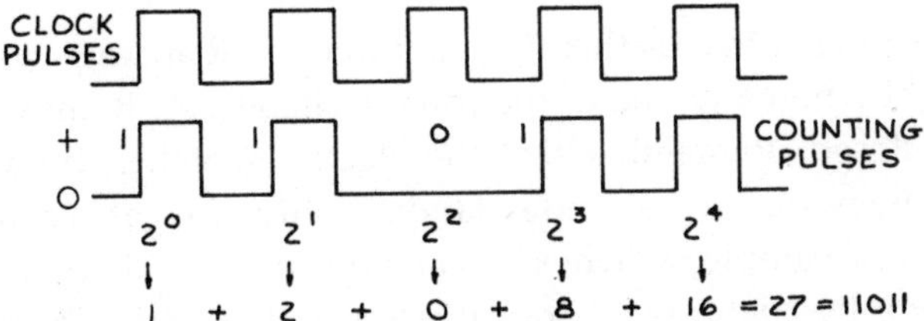

Figure 9-4. Four 1 pulses and one 0 pulse were received during five ticks of the clock. How these are translated into successive powers of 2 to become 27 is explained above.

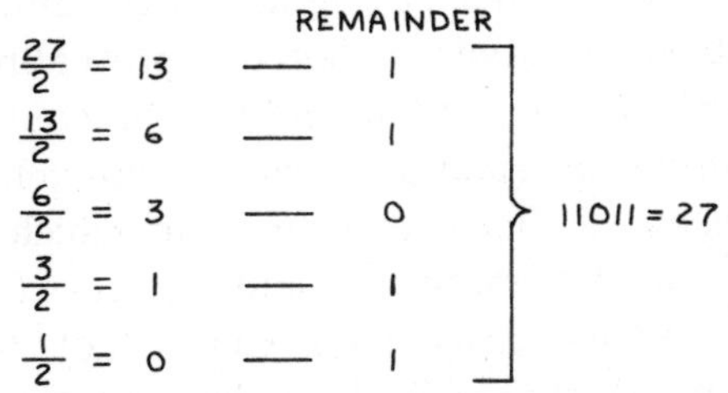

	REMAINDER	
$\frac{27}{2} = 13$	1	
$\frac{13}{2} = 6$	1	
$\frac{6}{2} = 3$	0	11011 = 27
$\frac{3}{2} = 1$	1	
$\frac{1}{2} = 0$	1	

Figure 9-5. Converting the decimal number 27 into the binary number 11011 is done by successively dividing by two as shown above. The remainders determine the binary equivalent.

mainder of 1. Dividing 13 by 2 equals 6 with a remainder of 1, while 6 divided by 2 equals 3 with a remainder of 0. The two divisions left each have a remainder of 1. Assembling the remainders give 11011, the binary of 27. It is far less involved than this written explanation would make it appear.

The binary system is extremely simple, but its simplicity is achieved at a penalty. It takes two places to write 27 in decimal, five places to do the same in binary. As the decimal numbers get greater, the binary places expand exponentially. However, while this may be a problem for manual manipulation, it is nothing to the high speed circuits of a computer that can digest an operation in as little as a billionth of a second. When there is possibility of misunderstanding, the "most significant digit," meaning the one at the highest power of two, is indicated to be at the left or right end of the binary number.

Pulse techniques have done much to improve the ease of readability of the outputs of electronic instruments. The readout in the analog method usually is through the medium of a pointer moving over a calibrated meter scale or else of a numbered dial against an index. What with parallax and the vagaries of human sight, the possibility of error is great. By contrast, the digital system flashes its answers onto a screen in arabic numerals and the chance of misreading is almost non-existent. The internal accuracy of the instrument itself also is generally enhanced because unavoidable slight changes in voltage do not affect the results that depend only on whether or not the voltage is there.

Boolean algebra, mentioned earlier, is used to predict and interpret the operations of an important group of logic circuits known as "gates." Familiarity with these gates is basic to an understanding of modern electronics because they are found so frequently in various circuits. This group comprises AND gates, OR gates, NAND gates, and NOR gates plus inverters.

Figure 9-6. Logic gates are depicted on schematic diagrams by the symbols illustrated: (a) inverter (b) AND gate (c) OR gate (d) flip-flop. The small circle at the output of (e) changes the AND gate to a NAND gate and at (f) the small circle changes the OR gate to a NOR gate. (g) is an exclusive OR gate. (ARRL Radio Amateur's Handbook)

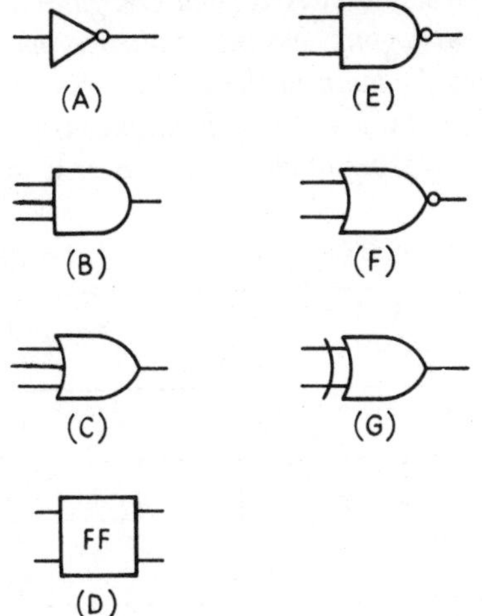

These are all pulse-activated devices whose outputs are the already explained 1 and 0 or the equivalent presence or absence of a voltage. (NAND is a contraction of NOT AND and NOR is derived from NOT OR.) The illustrations in Figure 9–6 show the standard symbols by which gates are represented on wiring diagrams. The small circle at the output end of the symbol changes an AND to a NAND and an OR to a NOR.

The AND gate shown has three inputs—A, B, and C. by definition, an input pulse must be present at A *and* at B *and* at C before an output will be generated.

The OR gate also has three inputs in this example. Again by definition, an input pulse at A *or* at B *or* at C will trigger an output.

A NAND gate will invert the output of an equivalent AND gate and a NOR gate likewise will invert the output of an equivalent OR gate. The inverter has only one input and its output inverts whatever is applied to that input.

The on/off nature of pulses makes it possible to explain the functioning of these gates with an analogy of simple electric switches in open and closed positions. The drawing in Figure 9–7 does this. It can be seen that no current will flow to light the lamp in the AND analogy unless all three switches, A, B, and C are closed. It is equally obvious that the lamp in the OR analogy will light when any one of the three switches, A, B, and C is closed. Remember that the electric switches are an analogy for ease of understanding only; the gates contain no simple switches. Switching in these circuits is done by high speed transistors and the gates themselves usually are in the form of integrated circuits known as IC's. (See Chapter 14.)

Figure 9-7. The action of AND and OR gates may be illustrated with analogous electric switches as shown. It can be seen that the lamp in the AND circuit will not light unless A and B and C are closed simultaneously. It is equally evident that the closing of either A or B or C will light the OR lamp.

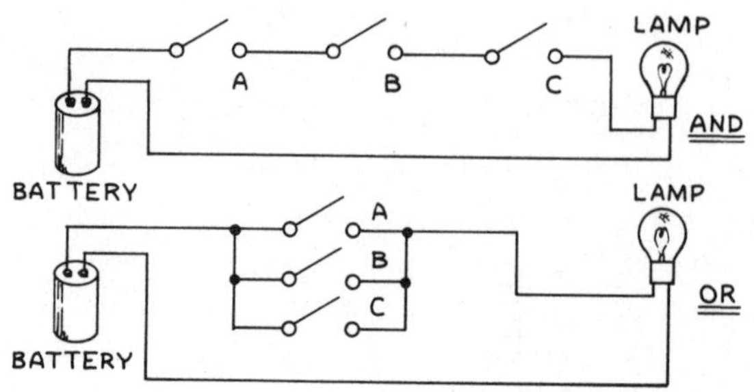

The results of a Boolean manipulation are tabulated in a "truth table" like the one shown in Figure 9–8. The truth table is a graphic method of representing all possible combinations of input and resulting output of the circuit under observation. It is also possible to write these results in Boolean notation in which a dot (·) between two letters means "and," a cross (+) means "or," and a bar over a letter is read as "not" that letter. The further depths of Boolean algebra are arcane and not necessary for the understanding of marine electronics.

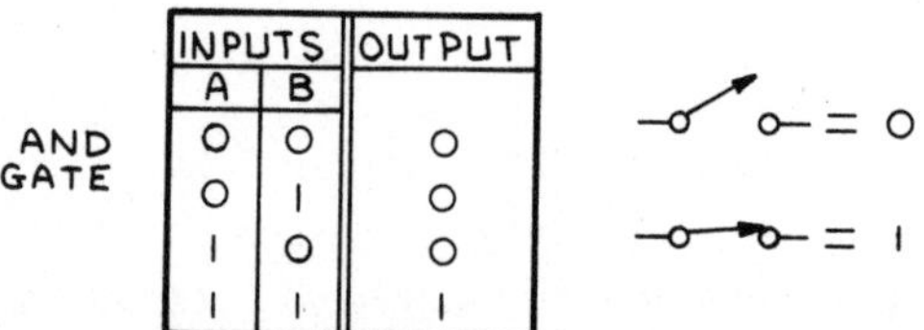

INPUTS A	INPUTS B	OUTPUT
0	0	0
0	1	0
1	0	0
1	1	1

Figure 9-8. The actions of gates may be summarized in so-called "truth tables" like the one above that refers to two switches, A and B, in an AND circuit. An open switch is 0; a closed switch is 1. Note that there is no output until both A and B are closed.

Transistors in these logic and pulse circuits function differently from those in standard oscillators and amplifiers. They are always in either of two states—conducting or non-conducting, meaning "on" or "off." This is accomplished by biasing them to saturation for the conducting state and to cutoff for the non-conducting condition. The response curve of transistors in this service is unimportant because they function only at their two extremes and their intermediate behavior is not utilized.

The length of time a transistor remains in its saturated or in its cutoff state is determined by the "time constant" of its circuit. This time constant results from the combined effects of a resistance and a capacitor because the capacitor takes a finite amount of time to discharge through the resistor. Obviously, the greater the charge, the longer its dissipation will take through a given resistor. (A large tank takes longer to empty than a small one through the same pipe.) Likewise, the higher the value of the resistor, the longer the discharge time for a given capacitor. (A tank takes longer to empty if the pipe is made smaller.) Summed up, the time constant is the product of the capacitance and the resistance, expressed as RC. (See Figure 9–9.)

Because the time constant governs the duration of saturation and of cutoff, it automatically sets the frequency of pulse repetition in the multivibrators. It also affects the output waveform. If the time constants immediate to each of the two transistors are identical, then the output waveform will be rectangular. Almost any other waveforms, from slopes to spikes, may be achieved by making the two time constants unequal.

Time constants alone often govern the frequency of the clocks described earlier when the requirements for accuracy are not too critical. But temperature and other variables may affect the resistors and capacitors sufficiently to change the time constant and in turn allow the frequency to stray beyond allowable limits. A piezoelectric crystal is added to the multivibrator circuit in such instances, as shown in Figure 9–10. The crystal most in use is quartz.

Figure 9-9. The "time constant" RC for the tank to empty is dependent upon the capacity of the tank C and the diameter and length of the pipe R.

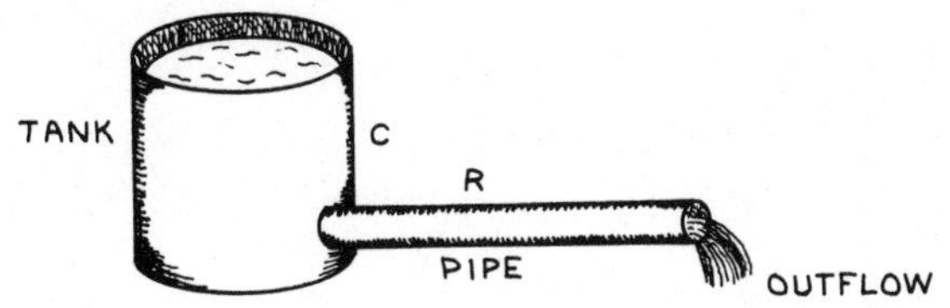

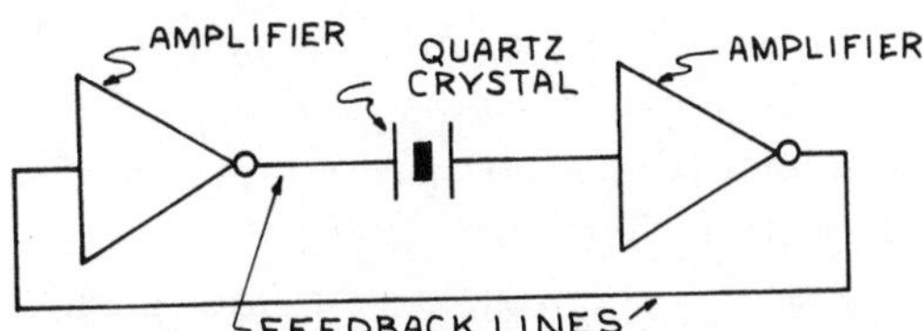

Figure 9-10. By placing a quartz crystal in one of the feedback lines of a multivibrator, the frequency may be held accurate enough to serve as a "clock."

A thin slice of quartz crystal, closely mounted between two flat plate electrodes, exhibits an electrical and a mechanical resonance at a frequency set by its dimensions. Its affect on the circuit resembles that of a flywheel on a machine—it resists change in speed or frequency. When extreme accuracy is desired, the quartz crystal is housed in a thermostatically controlled "oven," beyond the reach of ambient changes. Crystal slicing is performed with reference to optical axes, similar to what is done with diamonds.

The circuits discussed to this point function at both input and output with pulses of voltage rather than with sinusoidally varying voltages. The Schmitt trigger is a circuit, similar to the multivibrator in configuration, in which a sinusoidal voltage at the input triggers a square wave at the output. It senses the level of the input voltage and at predetermined points instigates the rise and fall of output voltage that constitutes a square wave. The Schmitt trigger is the interface that makes it possible for varying voltage circuits to operate digital pulse equipment.

The drawings of square waves with ideally sharp, rectangular corners is an ideal concept not attainable in commercial practice, although some carefully designed circuits may come close. The actual square wave is like the sketch in Figure 9–11. The "instantaneous rise" is not quite instantaneous and therefore the leading edge is slightly inclined instead of absolutely vertical. The fast rise causes an overshoot before the levelling off, hence the bump at the top frontal corner. These aberrations, expressed as time, may account for only millionths or perhaps billionths of a second, but even these unimaginably short periods are important to modern electronic circuits.

Technicians divide logic circuits into positive logic and negative logic. The designation refers to the polarity of voltage pulse that instigates action. Positive logic is the more common and it is more adaptable to the requirements of npn transistors. (See Chapter 3.) The word "positive" in this connection refers to the more positive of two voltages. Thus zero volts is more positive than any negative voltage.

The Boolean concept permits only two opposite conditions—true and false. Translated into electronics, this becomes respectively 1 and 0, a high voltage and a low voltage, a voltage and the absence of voltage, or any other two easily discernible opposite states.

The pulses studied so far are usually at low level and confined to the internals of electronic equipment. However, there are certain categories of pulses purposely made strong for propagation into outer space. Radar pulses are one example. A radio pulse containing up to many kilowatts of energy is sent out to strike whatever targets may be in its path. The reflections bounced off as echoes return to the sender to pinpoint the targets on the radar screen.

Another example is a depth sounder. Here the pulses are high level sound directed at the sea bottom. The returning echoes are calibrated against time and appear on the indicator as water depth.

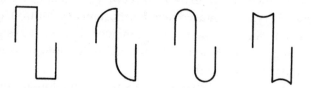

Figure 9-11. At left is the perfect square wave. The remaining drawings show the aberrations that may be caused in the square wave by less than perfect circuit and load conditions.

10. *Electric Filters*

The filter on the kitchen faucet and the filter incorporated into an electronic circuit have similar reasons for their existence. Both are there to prevent the passage of undesired elements. The kitchen filter accomplishes its purpose by blocking the stream with partially impervious screening material. The components of the electric filter have a similar screening effect that selectively shuts out voltages that are not wanted. The selectivity of both filters may be increased—of one by using more impervious material, of the other by adding additional components.

Electric filters may be as simple and inexpensive as a single resistor with an associated capacitor. The cheapie AC/DC compact radio receivers sold a decade ago got by with such an embryonic filter; the resistor at the output of the power supply got rid of most of the hum. Then again, the filters in a high quality electronic instrument may be quite complex; repetitive modules of capacitors and inductors may follow each other in ladder fashion. Such complex electric filters generally are engineered to affect a precise band of frequencies.

All true electric filters are classifiable into either of two groups with rather arcane names—one is "constant-k" and the other is "m-derived." Both names stem from the mathematical formulae that govern the filter design. One formula manipulates its components to maintain a parameter called "k" constant. The other formula computes the value of a filter parameter designated "m." Each group is thus specified by the formula that designs it, a great convenience for technicians.

The classification of filters is broken down further by filter function. There are "high pass filters," "low pass filters," "band pass filters," and "band rejection filters." The names are self-descriptive. The high pass filter offers little obstruction to frequencies above a selected point, but blocks those below. The low pass filter does the opposite and permits passage only to frequencies below a given value. The band pass filter is open only to a narrow band of frequencies, while the band rejection filter blocks this same band. Each type of action has a specific purpose in electronic circuits.

The basic components of electric filters are resistors, capacitors, and inductors. How these are interconnected determines the group into which the filter will fall. The values of these three components then set the frequencies that will be affected by the filter. Three common arrangements are in vogue for connecting together the resistors, capacitors, and inductors that make up a filter, and these are shown schematically in Figure 10–1. The resemblance to an inverted capital L, a capital T, and a Greek Pi gave the filters the names used by technicians to describe them.

The manner in which resistors, capacitors, and inductors react to the passage of electric current is the basis for the effects these components have in filter circuits. A resistor, at least the theoretically ideal resistor, does not discriminate between direct and alternating current; whatever its resistance, it is exerted equally against both. A capacitor will not pass direct current at all and its impedance to alternating current decreases with increase in frequency. An inductor acts as a resistance to direct current and as an impedance to alternating current that increases with an increase in frequency. An electric filter balances these differing reactions to achieve a desired total effect.

It is helpful to understand the difference between the terms "resistance" and "impedance," both of which imply an obstruction to the passage of electric current. Resistance is a property of conductive materials that hinders the passage of electricity; its value is expressed in "ohms" and it does not vary with frequency. Impedance also is a hindrance, but it is exerted only against alternating current and it consists of the combination of resistance and "reactance." Reactance is a hindrance to the passage of electric current and is a property unique to capacitors and inductors. The reactance of a capacitor is called capacitive reactance and it decreases with frequency. An inductor has inductive reactance that increases with frequency.

Because capacitive reactance and inductive reactance exert their effects in opposite directions, they bear opposite signs and may be added algebraically. When both are in a circuit they tend to cancel each other out. Actually, when they are equal in magnitude, their sum is zero, and the circuit of which they are a part is said to be in resonance. (See Chapter 6.)

Electric filters may be found in many sections of electronic circuits and they are especially numerous in radio receivers and radio transmitters. A low pass filter is common in their power supply; its purpose is to weed out ripples and transients that could interfere with clean operation. A filter circuit functions as a tone control in a radio receiver and permits the operator to adjust the quality of the received sound. Filters in both receivers and transmitters maintain the desired response curves. The spurious responses in the detector output of the receiver that could degrade clarity are eliminated by filters.

The frequency at which the output of a filter begins to decline markedly in the desired direction is called the "cutoff frequency." This cutoff frequency is not sharply defined with any simple filter such as those shown in Figure 10-1. However, it may be sharpened by adding additional similar filter sections. High quality components with low inherent electrical resistance also sharpen the cutoff. Incidentally, a component that has low resistance is said to have a high "Q." Many listings of components by suppliers will state this Q.

Much may be learned about a filter's action by examining its circuit diagram and noting the position of the capacitor or the inductor that is its main element. Remembering how capacitors and inductors act with relation to frequency then tells the story. When a capacitor is in series between input and output (as in the sketches of Figure 10–1), it becomes evident immediately that high frequencies will pass through easily while low frequencies will not. Ergo, it is a high pass filter. As noted earlier, a capacitor increases its impedance as the frequency lowers.

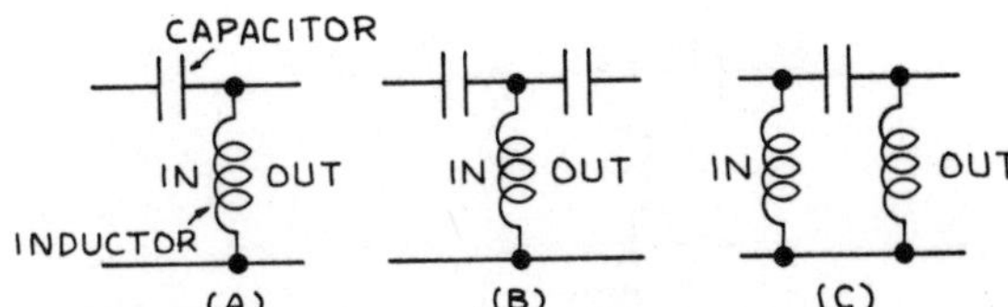

Figure 10-1. These are all high pass filters consisting of capacitors and inductors. (a) L type filter, (b) T type filter, (c) Pi type filter.

The same reasoning is apropos when an inductor is in series between input and output as in the diagrams of Figure 10–2 that picture L, T, and Pi circuits of this type. Since an inductor exhibits decreased impedance as the frequency lowers, the circuit will favor low frequencies; it is a low pass filter.

A handy rule of thumb to employ when inspecting electronic instruments and especially in communication-quality radio receivers. In such service either the capacitor or the inductor is made variable and this permits the active band to be moved over a wide range. Now certain bands of a received signal may be amplified or eliminated to improve clarity. Often noise is removed from the signal in this manner.

In these discussions of high and low pass filters, the terms "high" and "low" are purely relative and, as such, could refer to widely separated points in the frequency spectrum. The specific frequencies at which the filters become active depend entirely on the values of the components.

The T filter may be considered as two L filters placed back to back. The two resistors that would thus be side

Figure 10-2. These are all low pass filters with capacitors and inductors; note the transpositions from Figure 10-1. (a) D type filter, (b) T type filter, (c) Pi type filter.

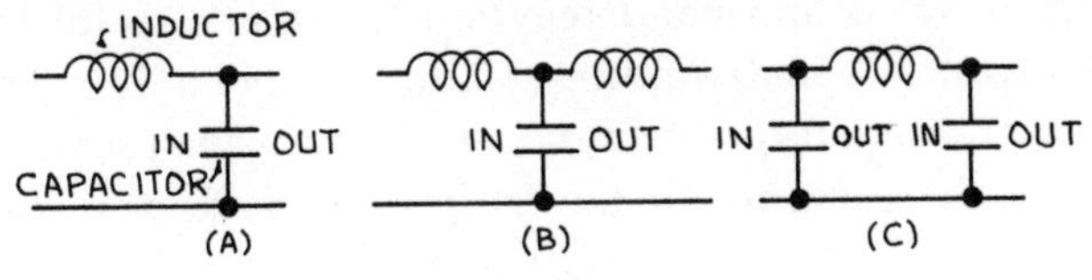

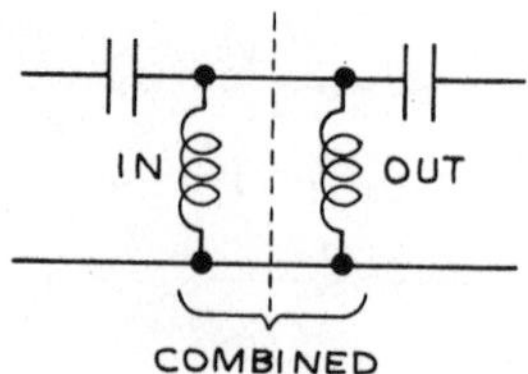

Figure 10-3. A T type filter is produced from two L types by combining the two inductors into one.

by side are simply combined into one of twice the value. Figure 10-3 illustrates how this evolves.

Similarly, the Pi filter circuit may be thought of as two L filters placed face to face as in Figure 10–4. Now the two capacitors are together and are replaced by one of twice the capacitance.

Figure 10-4. A Pi type filter is essentially two L's with the common capacitors combined into one.

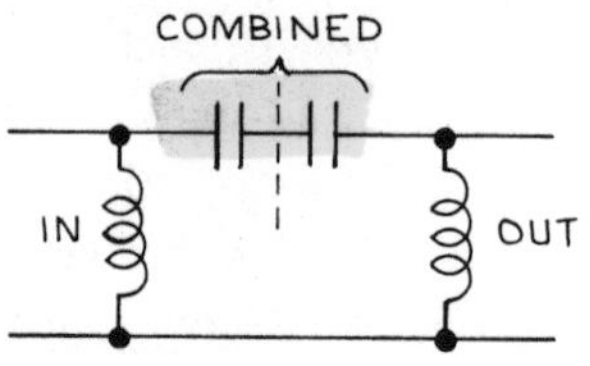

Band pass filters are more complicated but yet not necessarily more difficult to understand if a glance be taken back to the subject of resonance in Chapter 6. The basic idea is to establish a resonant circuit tuned to the center of the band of frequencies that are to be passed. The components then are chosen to widen this center point to the required span.

When the position of the resonant unit in the filter circuit is changed, as in Figure 10–5, the action of this band filter is reversed. The band of frequencies that previously was passed is now rejected. In effect, the band rejection filter cuts a swath out of the frequency spectrum.

Band pass and band rejection filters are useful in electronic instruments and especially in communication-quality radio receivers. In such service either the capacitor or the inductor is made variable

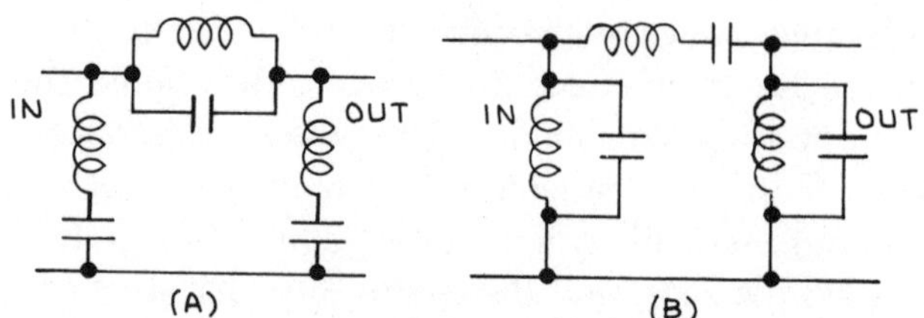

Figure 10-5. The band pass and band rejection filters employ modules of inductors and capacitors that are series and parallel resonant. The positions of these modules in the circuit determine whether a given band of frequencies will be passed or be rejected. At (a) a band rejection filter and at (b) a band pass filter.

and this permits the active band to be moved over a wide range. Now certain bands of a received signal may be amplified or eliminated to improve clarity. Often noise is removed from the signal in this manner.

For the technically inclined, an explanation of the "k" in the constant-k formula mentioned earlier may help general understanding. The impedances of the two arms of the filter are multiplied together and the product is the "k." If one arm be changed, then the other arm must be changed in the opposite direction so that the product remains constant. Filters designed with the constant-k formula do not have as sharp a cutoff as those designated "m-derived."

The "m" in the m-derived formula is an arbitrarily chosen value that may be 1 or fractionally less than 1. Commercial practice generally chooses 0.6. Filters designed with an m of 1 behave similarly to equivalent constant-k filters. With an m of 0.6, the cutoff is much sharper than that of the constant-k.

The RC filters, those with resistance and capacitance, are the least expensive to construct. Next in order of expense are the RL, that comprise resistance and inductance. Most costly are the CL filters, whose components are capacitors and inductors.

Resistors in any circuit, be it of a filter or whatever, are consumers of power. They turn power into useless heat. Neither capacitors nor inductors consume power;

they store energy in either an electrostatic field or in an electromagnetic field. (That statement assumes the theoretically perfect component—actual units do waste a bit.)

The L type high pass filter often is used as the coupling element between two stages of an audio amplifier. (See Chapter 11.) This is inexpensive and it serves its purpose in an elementary sort of way but, as one would expect, it discriminates against low frequencies and thereby robs the amplifier of fidelity to the impressed sound. Why it does this may be seen from the sketch in Figure 10–6 that redraws the conventional L filter but without changing the electrical connections. At some frequency the impedance of the capacitor equals the resistance of the resistor. At this point the impressed voltage is divided equally between the two components, but only the half across the resistor is passed on. The amount changes with frequency and so there is distortion.

The attenuating effectiveness of a filter often is rated with a unit called a "decibel," one-tenth of a bel. (The name derives from Alexander Graham Bell.) The concept of measuring with decibels is based on a peculiarity of the human ear. The ear hears sounds logarithmically and not linearly—and the decibel accordingly is logarithmic. A sound must be quadrupled in power before the ear will consider it twice as loud, and this logarithmic progression continues through almost the entire range of audibility.

Figure 10-6. An L filter (in dotted enclosure) may serve as the coupling element between two amplifier stages. Circuit wiring has been omitted for clarity.

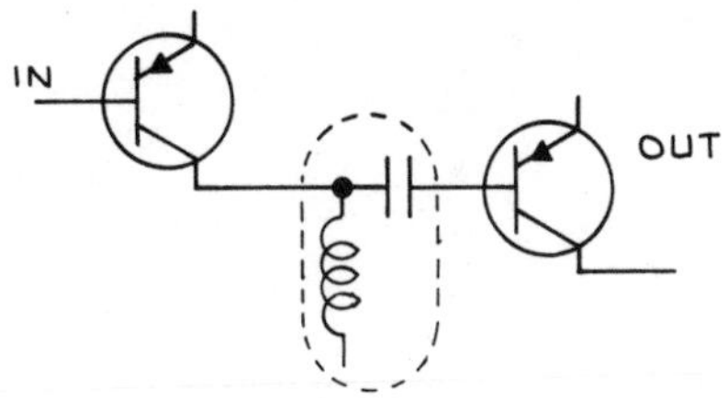

Figure 10-7. The decibel is a logarithmic unit and is calculated with a simple formula. The logarithms are to the base 10.

$$\text{decibel} = 10 \text{ times } \log_{10} \frac{\text{Power out}}{\text{Power in}}$$

dB	*Power Ratio* (approximate values)
0	1.0
1	1.26
2	1.59
3	2.00
4	2.51
5	3.16
6	4.00
7	5.00
8	6.31
9	8.03
10	10.00
11	13.00
12	15.90
13	20.00
14	25.00
15	30.40
16	40.00
17	50.00
18	67.80
19	80.00
20	100.00
30	1,000.00
40	10,000.00
50	100,000.00
60	1,000,000.00

The decibel actually is a measure of comparison between two powers, two voltages, or two currents. But it also can be a measure of actual power if a threshold value for zero decibels is agreed upon. In electronics this agreement has fixed one milliwatt as the power represented by zero decibels. To mark this, the normal abbreviation, dB, has an added m and the unit is written dBm.

Stated mathematically, a decibel is equal to ten times the logarithm, to the base 10, of the ratio of power in to power out. This is explained by the equation in

Figure 10–7, which also features a decibel chart. For example, a filter that attenuates a frequency by 3dB has cut its intensity in half. Incidentally, the logarithm is multiplied by 20 instead of by 10 when the ratio is of voltage or current instead of power.

One use of filters is to narrow the spectrum space required for voice transmission by communication equipment. The telephone people found out long ago that a band width of approximately 3,000 cycles per second (Hertz) was sufficient to insure clarity even though the human voice covers a much wider spread. Filters chop off the excess at top and bottom to maintain this width.

Selective tone ringing, whereby a radio transmitter alerts only desired receivers, is another service made possible by sharply tuned filters. The receiver contains a filter that passes only the tone frequency emitted by the distant transmitter. The audio portion of the receiver is normally squelched so that no sound is heard from the loudspeaker. The tone passes through the filter and unlocks the squelch, putting the receiver into a receptive state.

Strange as it may seem, because the two items appear to be unrelated, filters and transmission lines have much in common from the standpoint of circuit response. This is discussed in Chapter 23.

11. *Amplifiers: Radio and Audio*

The ability to amplify a weak current into a stronger one that may be controlled more easily and can do more work is basic to the functioning of all electronic equipment. Amplifying capability came to electronics with the invention of the three element vacuum tube and has since been furthered by more sophisticated tubes and now by transistors.

Radio reception provides a good example of what amplification can do. The only power available to the early crystal radio receivers was the miniscule amount the antenna was able to intercept from a passing radio wave. This power was in the range of a few millionths of one watt, perhaps the energy a fly uses to take off. It was barely able to actuate the diaphragms of sensitive headphones to produce an intelligible sound. Today that same miniscule amount of received power merely triggers vacuum tubes or transistors that direct locally amplified currents into loudspeakers.

Amplifiers are divided into two general modes—those that are designed to handle radio currents and those that amplify audio currents. The wide disparity in frequency between the two modes and the consequent difference in components are the reasons for the division. Radio amplifiers may be designed to function with frequencies that go upward into the millions or even billions of cycles per second. Audio amplifiers restrict their activity to the much more moderate span below about 20,000 cycles per second.

To be of value in communication, amplifiers must be able to increase the strength of a signal without distorting its original form. Theoretically, the output of an amplifier should be an exact image of the input, but of greater amplitude. This perfection is never attained in actual practice, although the best amplifiers come extremely close. The constant effort to eliminate distortion is at the heart of good amplifier design. In the commercial market, an amplifier's extent of freedom from distortion fixes its price in almost direct ratio.

Amplifiers may be designed for wide band or narrow band functioning, and this refers to the width of the spectrum of frequencies they are able to handle efficiently. Most radio frequency amplifiers are tuned to respond to a comparatively narrow band of frequencies determined by the nature of the equipment of which they are a part. Good audio frequency amplifiers, by contrast, are expected to be able to amplify the entire audible range with imperceptible or minimal distortion.

But how does an amplifier amplify? What takes place within a vacuum tube or a transistor that causes an applied signal to emerge with added strength? The drawing in Figure 11–1 helps to explain this phenomenon.

Shown in the drawing is the dynamic curve of a given transistor. This dynamic curve is simply a graphic method of describing what occurs at the collector of this transistor when a change in input takes place at its base. Since the graduations for the input are in microamperes while those for the output are in milliamperes, the ratio of power becomes quite obvious. (One milliampere is equivalent to 1,000 microamperes.) Both ends of this curve are bent and therefore non-linear, but the midportion is straight and linear and this is the region in which the present amplifying action is made to take place.

The application of 20 microamperes of bias current

Figure 11-1. This hypothetical curve of an amplifier operating in Class A mode shows how a small input signal becomes a strong output through amplification. Note that the scale of the output is one thousand times as great as the scale of the input. Efficiency is low because current flows all the time.

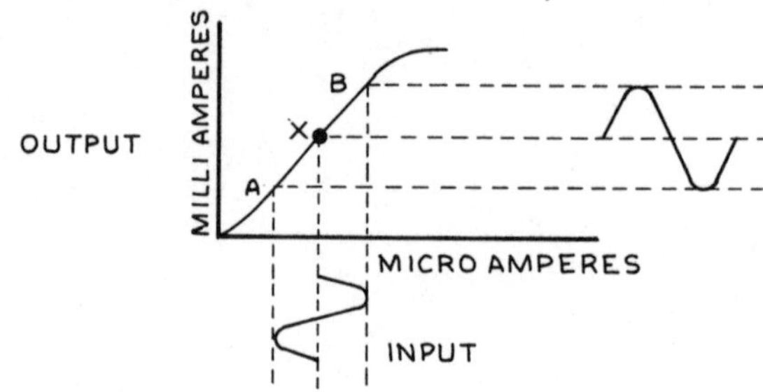

to the base fixes the operating point of this amplifier at the spot marked X on the curve. The applied signal is 20 microamperes wide and thus varies the bias equally on both sides of the operating point from 10 microamperes to 30 microamperes or from A to B along the curve. Extending these two points horizontally to the graduated scale reveals that 4 milliamperes of current flow at A and 16 milliamperes at B. In short, a few *micro*amperes have triggered a greater number of *milli*amperes—surely very potent amplification. Had a vacuum tube been used as the example, instead of a transistor, the explanation would have been identical with a similar dynamic curve.

What has been described is a single stage of amplification. Additional stages may be added so that the output of one becomes the input of the next and thereby increases the amplification exponentially to fantastic final figures. However, there is a practical limit to the number of stages that may be added. The limitation occurs because all signals contain some noise and this noise, unfortunately, also is amplified and sets the limit of acceptability.

Amplifiers also are subdivided into Class A, Class B, Class C, and an intermediate Class AB. The determination is made on the basis of input bias and output current flow. A Class A amplifier is biased so that its operating point falls on the midpoint of the dynamic curve, as in the example just cited, and output current flows continuously whether or not a signal is present.

To achieve Class B, the bias must be set to bring the operating point to cutoff at the bottom of the curve. Now only half the input signal has any effect on the output and output current flows only when a signal is present. A later example will show this graphically.

Class C accentuates the conditions of Class B by moving the operating point even further beyond cutoff. Now less than half of the input signal affects the output and output current flows only during that activated portion. Each class of amplification has a particular niche in electronic circuitry.

Reexamining Figure 11–1 shows that the depicted amplifier is operating in Class A. With no signal present, the operating point is at position X and 10 milliamperes of current is flowing. An input signal varies this output current between 4 and 16 milliamperes. The condition necessary for Class A operation is fulfilled.

An amplifier running Class A introduces the least distortion into the applied signal and thus is especially valuable for audio frequency work. But its output power capability is less than that of equivalent units operating as Class B or Class C. The efficiency of Class A operation is low, usually less than about 30 percent. The poor efficiency is the result of the continuous current flow, most of which eventually is turned into useless heat. However, regardless of these drawbacks, the majority of amplifiers found in marine electronic equipment are operating as Class A.

The operation of a Class B amplifier is depicted in Figure 11–2, and this time a vacuum tube is used as the example instead of a transistor. As required, the operating point X is set at cutoff by the application of 10 volts of negative bias. The signal has a 20 volt swing, equal on each side of the operating point. But only the portion from −10 to 0 has any effect on the

Figure 11-2. This is the dynamic curve of an amplifier working in Class B mode. Note that now only half the input has an effect upon the output and only half the input is reproduced and amplified. Efficiency is higher because current flows only part of the time.

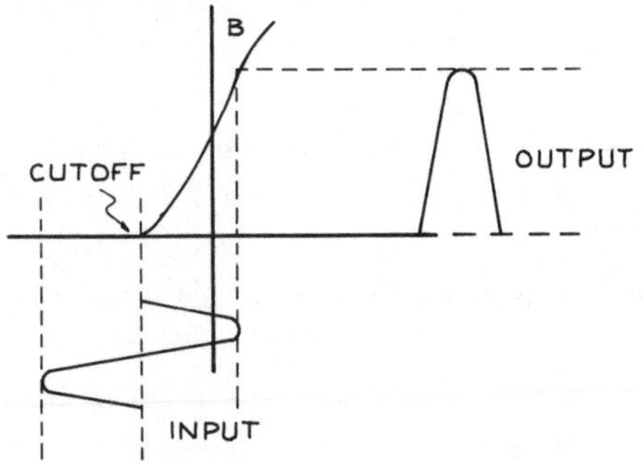

output because the other half takes place during cutoff. The activity extends along the curve from X to B and the degree of amplification is evident.

The striking feature of the illustration is that only half of the input waveform appears in the output. This would cause serious and unacceptable distortion in audio frequency work. Luckily, there is an easy way around this limitation—the use of two vacuum tubes or two transistors per stage, in a form called "push-pull." The second amplifier fills in the missing half and a complete output signal results. Most modern audio amplifiers employ push-pull Class B in order to achieve economical large power output.

In addition to power handling ability, Class B amplifiers are favored because they work at almost twice the effiency of comparable Class A units. The disadvantages of Class B are its requirements for additional bias voltage or current and at the same time the need for much stronger driving signals at the input.

Class C amplification is attained by increasing the bias until it is several times as great as that needed for cutoff. Now output current flows for much less than half the duration of the input and as a consequence the operating efficiency soars to almost three times what it would be with comparable Class A. Of course, this efficiency is obtained at a price; the distortion is horrendous. The distortion is so great that an audio Class C amplifier is impossible. But Class C is greatly in demand for radio frequency amplification where the distortion may be balanced out with proper tank circuits that reinject the portions of the waveform that were lost.

Figure 11–3 portrays Class C operation of an amplifier graphically. Note how far beyond cutoff the operating point X has been placed, and how this allows only less than half the input signal to affect the output. The diagram also emphasizes the great increase in input signal that this form of amplification demands. The factor usually relegates Class C to the final radio frequency amplifier stage preceded by a pre-amplifier able to supply the necessary wide swing.

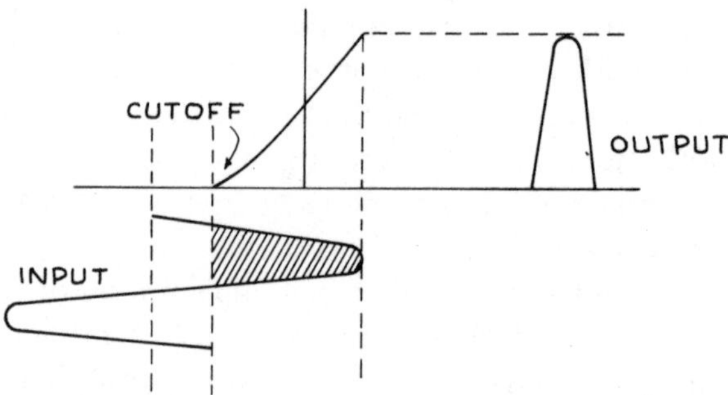

Figure 11-3. The highest amplifier efficiency occurs in Class C operation but at the cost of the greatest distortion. This makes Class C unuseable for audio amplification. Note that less than half of the input waveform affects the output.

When several stages of amplification are required in a circuit, obviously some form of coupling must be employed to connect each stage to the following one. This coupling may be direct, resistive, through transformers, or by means of impedances. The method most in use is resistive, actually a combination of resistors and capacitors. The reason for its popularity is simplicity and low cost plus a reasonable level of quality as far as frequency fidelity is concerned. The wiring diagram of this form of coupling is shown in Figure 11–4.

Studying this diagram with the text of Chapter 10 in mind leads to the correct conclusion that this circuit

Figure 11-4. "Resistance-capacitance" is a common form of coupling between amplifier stages. The resistor and the capacitor that form the coupling are shown in the dotted enclosure. Wiring has been omitted for clarity.

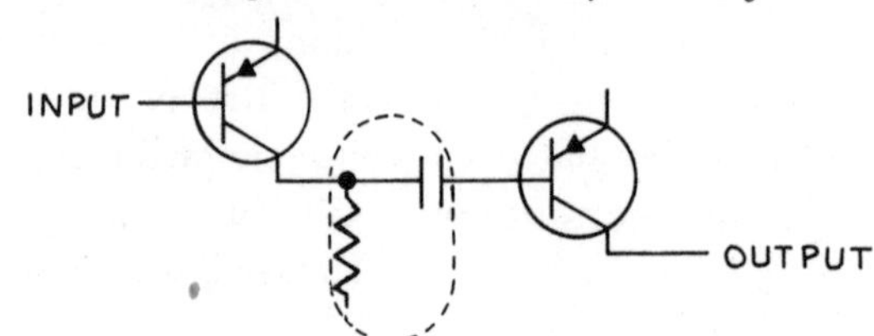

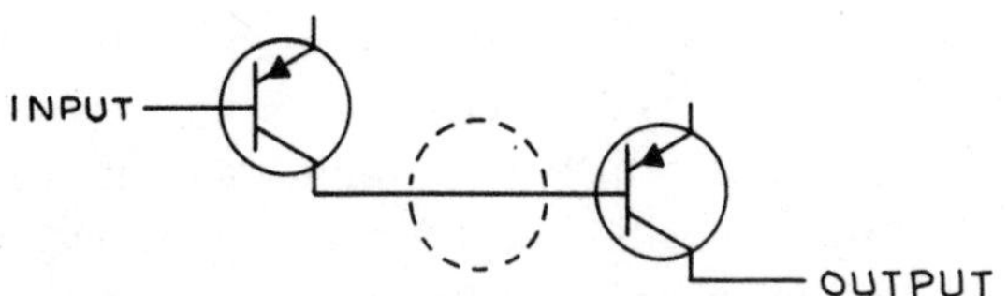

Figure 11-5. Direct coupling between amplifier stages, without any intermediate components as shown, enables the amplifier to handle a wide spectrum of frequencies from direct current to high frequency alternating current. Wiring is omitted for clarity.

favors the passage of the higher frequencies and is obstructive to the very low ones. However, this characteristic is not pronounced enough to prevent satisfactory use in amplifiers of medium quality and price.

Direct coupling is theoretically the best from the standpoint of frequency fidelity, but it loses much of its allure when put into practice. As shown in Figure 11–5, the output of one vacuum tube or transistor is connected directly to the input of the next without any intermediate components. First of all, this requires double the supply voltage needed for one tube or transistor. Second, the circuit is tricky, hard to keep in stable operation, and usually limited to two stages.

Transformer coupling has the advantage of complete isolation between stages because the transfer of energy takes place only through magnetic flux (see Figure 11–6). The transformers may be built with a step-up ratio that increases the output voltage of one stage before it is applied to the next. Transformers also are able to perform impedance matching as, for instance, from the high impedance of a vacuum tube output to the low impedance input of a loudspeaker.

Figure 11-6. Transformer coupling between stages of an amplifier employs the magnetic field between primary P and secondary S of a transformer (dotted enclosure). The transformer may have an air core or a ferric core, depending upon the frequencies involved. Wiring has been omitted for clarity.

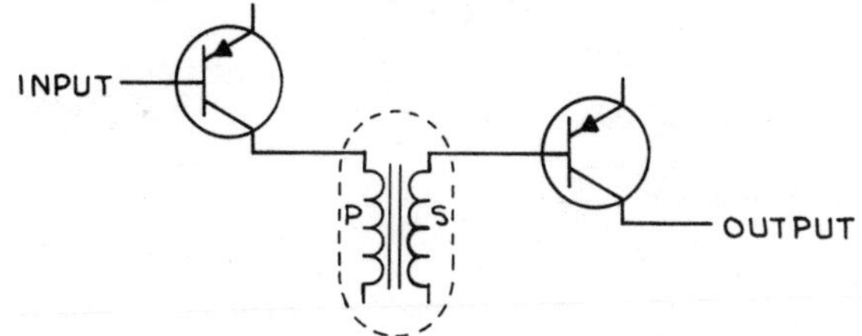

But the transformer story is not all cakes and ale. Transformers are expensive. They are bulky and heavy. They are difficult to design with good frequency fidelity and most discriminate against the higher and the lower frequencies. Transformers also are sensitive to ambient magnetic fields and require shielding for good performance.

Impedance coupling is similar to resistor-capacitor coupling; an impedance is substituted for the resistor. The impedance takes the form of an inductive winding as shown in Figure 11–7. Impedance coupling is highly frequency discriminatory. It is also more expensive than resistor-capacitor coupling and it is not a popular form of construction. The impedance has one advantage over the resistor—it causes a very much smaller voltage drop and therefore permits the use of lower potential power supplies.

As mentioned earlier, distortion is inherent in all amplifiers to a greater or a lesser degree. The distortion may affect the amplitude of the output signal by not applying an equal amount of amplification to all portions of the input. This may occur because the vacuum tube or the transistor is functioning at an operating point that is not on the correct linear section of the

Figure 11-7. The impedance of an inductor plus a capacitor (dotted enclosure) provide the coupling between two stages of an amplifier. Wiring has been omitted for clarity.

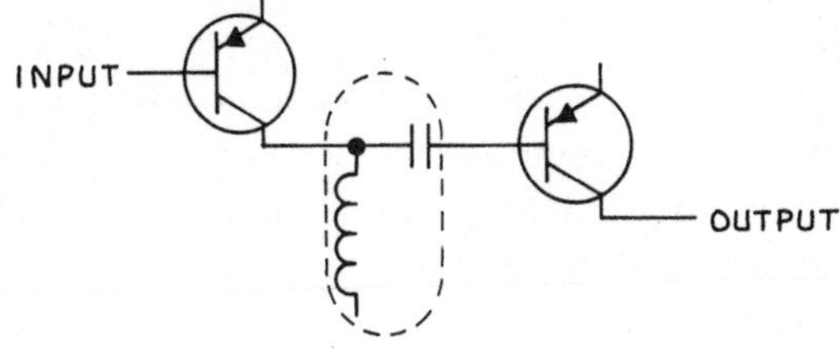

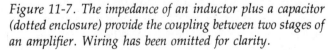

dynamic curve. The reason for this may be improper bias. Distortion also may result because the amplifier is overloaded, is being fed an overly strong input.

The distortion may be relative to frequency. For any of several reasons the amplifier circuit may favor some frequencies in the signal and discriminate against others. The output then becomes discordant and is no longer a replica of the input.

The final form of distortion that may be found in a less than perfect amplifier is phase distortion. This condition is more difficult to explain nontechnically, but the end result is still discordance, an output that alters the frequency relations that exist within the input. True phase relationships have a great deal to do with the unique sounds of various sources by which we recognize them.

Distortion in audio amplifiers is expressed as a percentage of the total output. This distortion may be expressed as occuring individually at various harmonics of the original, or the several values may be added together as total harmonic distortion, THD. Naturally, the lower the percentile, the better the amplifier.

A portion of the output of an amplifier may be fed back into its input; this condition is called "feedback." Feedback may be employed for either of two reasons in either of two manners—positively to cause oscillation or negatively to improve the fidelity of output.

When the feedback signal is in phase with the original input signal (is positive), the result is a continuous building up or regeneration that causes instability and oscillation. Sometimes this condition occurs spontaneously because of incorrect placement of components and this, of course, makes the amplifier unstable and unusable; it becomes an oscillator.

Negative feedback (also called "inverse feedback") is used almost universally in good audio amplifiers as a method of improving frequency response and fidelity. In this case, the signal fed back is opposite in phase to the original input signal and the result is a balancing out or neutralizing of inherent imperfections. Negative feedback reduces the amplifying capability of an amplifier, but this is compensated for in design.

The manner in which an amplifier amplifies is determined by its position and function in an electronic circuit. For instance, in the front end of modern radio receivers the amplifiers are handling radio frequencies. Their purpose here is to increase the strength of the received voltage until it is able to actuate the detector satisfactorily; these are voltage amplifiers.

At the final end of the radio receiver the requirement becomes power to operate the loudspeaker; this is fulfilled by power amplifiers. The difference between the voltage amplifiers and the power amplifiers is, first of all, the type of vacuum tube or transistor that is employed. There are further differences in the circuits, the components, the voltages, and the currents.

Earlier mention was made of push-pull amplifiers. This type of circuit is worth examination in greater detail because it is met with so frequently. (A typical circuit is shown diagrammatically in Figure 11–8.) The underlying idea is that two vacuum tubes or two transistors are excited in opposite phase at their inputs so that their outputs will add together. The desired result is increased output and reduction of distortion.

In the diagram, the opposition of phase between the two inputs is achieved with a center tapped input transformer. The outputs likewise are added by means of a center tapped transformer. Several schemes permit the use of transformers without center taps (which

Figure 11-8. The push-pull amplifier makes it possible to use Class B (see Figure 11-2) for audio amplification because it balances out the distortion.

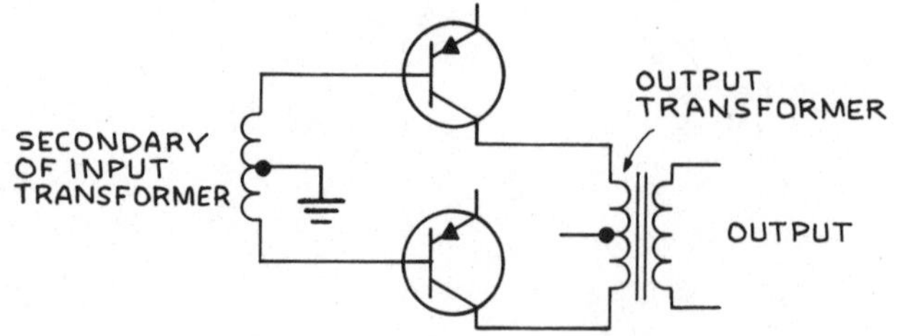

are less expensive) and some even allow the elimination of transformers in favor of less costly components. When the power supply for the amplifier is derived from alternating current, the push-pull circuit brings a bonus in addition to its normal advantages—any residual hum is cancelled out and eliminated.

The concept of eliminating transformers is also illustrated by a circuit known as a "cathode follower" when applied to vacuum tubes and as an "emitter follower" when transistors are used. A typical use is changing a high impedance output to match a low impedance device, a job normally left to transformers. The trick is accomplished as shown in Figure 11–9. Whereas the load normally is fed by the plate circuit, where the impedance is high, this amplifier incorporates the load into its cathode circuit where the impedance is low. Were a transistor substituted for the vacuum tube in this circuit, the load would be placed in the emitter portion.

Figure 11-9. The cathode follower circuit with a vacuum tube (and emitter follower with a transistor) is employed when impedance matching must be done. The output load is taken from the cathode (or emitter) instead of from the plate (or collector).

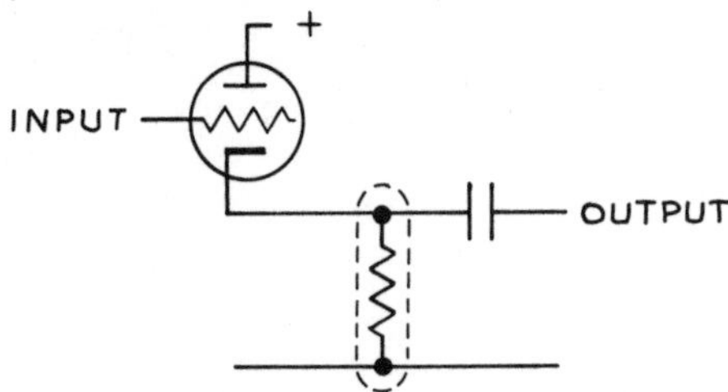

Since the transitions from Class A to Class B to Class C operation are all brought about by changes in the bias applied to transistor or vacuum tube, a knowledge of how bias voltage is obtained will prove handy. The earliest vacuum tube radios had separate batteries for the plate circuit and for the grid or bias circuit. Soon the extra battery was replaced by a resistor whose voltage drop became the bias voltage for the tube. This is self-bias. The resistor is in the cathode circuit and the full tube current flowing through it produces the voltage drop. A self-bias arrangement with a vacuum tube is shown in Figure 11–10. A similar circuit may be used to provide bias for a transistor, but additional precautions are necessary for stable operation because of the transistor's vulnerability to temperature increase.

Figure 11-10. The current flowing in the resistor (dotted enclosure) in the cathode circuit causes a voltage drop that forms the negative bias for the grid of the vacuum tube.

The target of transmission characteristic for a high quality audio amplifier should be a flat response between 20 Hertz (cycles per second) and 20,000 Hz. Manufacturers supply response curves like the one shown in Figure 11–11 to show how closely this target is achieved. The curve allows a reading in decibels at any frequency of how much the amplifier deviates from the ideal.

Amplifiers that handle radio frequency currents are tuned to get greater efficiency of operation. The ideal tuning curve has a small flat top at the desired frequency and sharp fall-offs at both sides. This tuning may be variable, as at the front end of a radio receiver, or fixed, as for an intermediate frequency amplifier in the receiver or a final amplifier in a transmitter.

Figure 11-11. This hypothetical response curve for an audio amplifier shows that there is a drop in amplification at the low frequency end (a) and also at the high frequency end (c). The amplification over the wide central frequency spectrum (b) is flat.

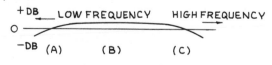

+DB LOW FREQUENCY HIGH FREQUENCY
0
−DB (A) (B) (C)

PART THREE

ELECTRONIC CONSTRUCTION

12. *Hard Wiring*

The components in earlier units of electronic equipment were connected together with individual pieces of wire that were soldered in place by hand. The trade calls this "hard wired." To anyone who actually has done this form of wiring, the description "hard" will seem appropriate. (The opposite of hard wiring, and the method presently in use, is the printed circuit described in Chapter 13.)

Hard wiring is accomplished by first mounting all the components on a metal chassis. The chassis is drilled with holes that permit the wires to be passed from top to bottom at the various connection points. In more approved fashion, the holes allow the binding posts of components to protrude to the bottom. The connecting wires are all run under the chassis. As can easily be imagined, and as the photo in Figure 12-1 proves, the result is a rat's nest.

Of course, the rat's nest is not visible to the eventual purchaser unless he disassembles the equipment, but this form of hard wiring brought more important problems that could affect efficient operation. The proximity of wires from different parts of the circuit often caused undesirable feedback that led to instability. Further, two insulated wires running close together for any distance became an unwanted capacitor.

Figure 12-1. The underside of the chassis of a pre-transistor radio resembled a rat's nest with its many wires running helter-skelter.

Historically, hard wiring was an improvement over the connection method employed by the early experimenters in their home-built radio receivers. The components in these receivers were mounted on wooden boards (actual breadboards were used originally) and long, hard metal rods about as thick as pencil lead, called "bus bars," made the connections. It was a matter of individual pride of workmanship to have all bus bars run absolutely parallel, all bends at right angles, and all curves smooth. The bars were tinned to facilitate soldering.

Actually, hard wiring at the factory was not as difficult as the finished product made it seem. Wire insulation was in many colors. A production line worker would repetitively run a certain color from here to there without the necessity of knowing what he was connecting. Often the wires were precut to length and bundled into a "harness" to bring the job down to utter simplicity.

Hard wiring had one advantage that has been almost completely lost with the printed circuit technique—it made life easier for the service man who had to replace a component. The simplicity of following wires from point to point also facilitated circuit tracing in the event of trouble.

The basic causes of the demise of hard wiring were economic and practical. Wages rose to the point where having a worker place and solder individual wires was no longer affordable, especially when the new scheme of printed circuits could do it all in one shot. But the real death knell was the miniaturization of electronic equipment. The parts became too small for discrete wiring.

The true impact of miniaturization may be seen from the photos in Figure 12–2. An intermediate transformer of the type commonly seen on old metal chassis

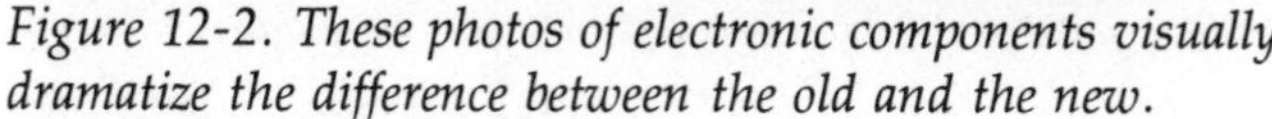

Figure 12-2. These photos of electronic components visually dramatize the difference between the old and the new.

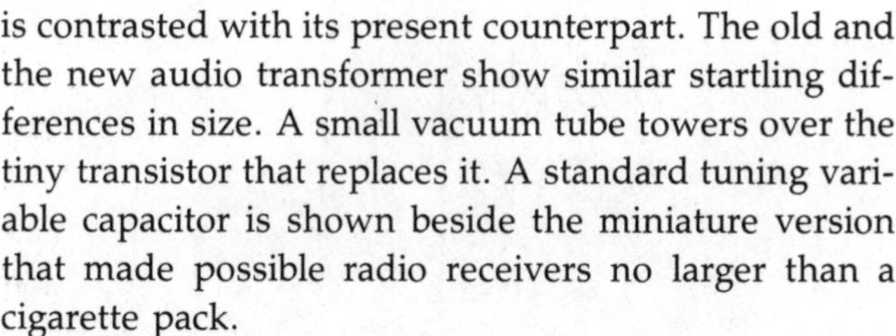

is contrasted with its present counterpart. The old and the new audio transformer show similar startling differences in size. A small vacuum tube towers over the tiny transistor that replaces it. A standard tuning variable capacitor is shown beside the miniature version that made possible radio receivers no larger than a cigarette pack.

Hard wiring remains a justified form of construction for prototypes, for instance when trying out a new circuit. In these early stages, components and connections may not yet have test approval and the more complicated steps in preparing a printed circuit would not be warranted.

Component placement is important in all forms of wiring, whether hard or printed circuit, but perhaps more critical in the former because the individual parts are larger and have a greater field of influence. The heftier magnetic field about a bigger audio or power transformer may affect nearby components. The heat from a vacuum tube or a power transistor needs a means of escape. The capacitative effect of the operator's hand could cause detuning unless a suitable shield were interposed.

The metal chassis forms a convenient "ground" and is used as such in all electronic devices that are constructed in this fashion. This ground becomes a common return for all circuits. It has low electrical resistance and also acts as a shield.

Selection of the correct wire for hard wiring should take into account the voltage and the current to be carried. Voltage itself has become less important with the advent of transistors because the usually high plate voltage required for vacuum tubes is absent. But transistor currents generally are higher and each path must be able to carry them. Insulation is the determining factor on voltage, cross-section on current.

One form of modern construction is a hybrid combination of hard wiring and printed circuitry. A metal chassis is the base upon which a number of printed circuit panels are mounted. Each panel is a complete functioning unit of the total circuit such as the intermediate amplifier, the audio amplifier, the power supply, or whatever. The panels then are hard wired together to form the complete electronic device. The chassis also becomes the base against which the metal front panel is mounted. All this metal around the "works" provides an effective shield, something that the more modern plastic enclosures cannot do. An example of hybrid construction is shown in Figure 12–3.

The connecting wires in chassis-type electronic devices usually are longer because the components are further apart. This increases the possibility of unwanted stray pickup of interfering energy and mandates the use of shielded leads. The shields are grounded to the chassis and thereby protect the internal wire from surrounding fields.

Figure 12-3. Some equipment requires that printed circuit boards be connected together with "hard" wiring. The result is a hybrid.

One small nuisance of hard wiring on metal chassis is that insulated standoffs must be provided for free-standing small components such as resistors and capacitors. (See Figure 12–4.) The standoffs supply locations where soldered connections may be made in addition to acting as insulated supports.

A metal plate closes the bottom of good quality electronic equipment mounted on metal chassis. This plate acts not only as a protection against dust and dirt, but also as a shield against stray magnetic and electric fields.

Figure 12-4. Small insulated "standoffs" like the one shown provide locations where connections may be made.

13. *Printed Circuits*

An average piece of moderately complex electronic equipment may have a hundred or more connections to be soldered. First, imagine these being completed tediously, one by one, with a hand soldering iron. Then switch your vision to a similar printed circuit board, with as many or more connections, floating on a soldering pool. All connections are soldered automatically at one shot. It is not difficult to understand why printed circuits have taken over so completely.

The foregoing is the dramatic side of printed circuit techniques. But a great deal of hard-nosed expertise in several professions is required behind the scene to make the miracle possible. Metal/plastic lamination, photography, printing, silk screening, chemical etching—all combine into the printed circuit reality.

The basic stock for printed circuit boards is a thin sheet of insulating material, one or both sides of which have been covered with copper foil. The insulation usually is one of the common plastics of either the thermosetting or the thermoplastic types, manufactured to a thickness of less than one-eighth of an inch. (A thermosetting plastic cannot subsequently be softened by heat; a thermoplastic can be.) Where rigidity is necessary, the plastics are stiffened by the inclusion of reinforcing agents in the mix.

All plastics are normal insulators, but the degree to which they approach perfect insulation is important in their selection for printed circuit boards. For low voltage, low frequency work, almost any plastic will do. As the voltage goes up, and especially as the frequency goes up, plastic selection becomes critical. Plastic manufacturers supply data on these parameters upon which the technicians rely in making their choices.

The copper foil is cemented to the plastic substrate under high heat and pressure. The bond must be strong enough to resist the delaminating forces that occur in use, principally the deteriorating effects of soldering heat. The copper selected for the foil is almost chemically pure in order to have high electrical conductivity. Various thicknesses of copper foil coating are available to accommodate the strength of current in any particular circuit.

The drilling of holes for component connections is one of the operations to which the printed circuit boards are subjected. Accordingly, the laminate must be easily drillable or punchable. Some plastics have a tendency to shatter during drilling or punching and these must be worked while they are hot. Others, the preferred kind, are able to withstand these operations cold. Heating may also be required for some boards when they are sheared to size. Some plastics are abrasive and these are ruinous to drills and punches.

The flexible cable or ribbon wire is an offshoot of the printed circuit idea. Instead of the standard round configuration of common electric cables, flexible cables are flat and no thicker than a few sheets of paper. Copper foil strips are embedded between two sheets of very thin plastic. The thickness of the copper foil is determined by the current that must be carried, the thickness and width giving the required cross-section. These cables are a favorite form of connection between stacked printed circuit boards. The photo in Figure 13-1 shows a flexible cable for a circuit board.

Figure 13-1. The flexible cable or ribbon wire consists of conductors embedded in plastic strip.

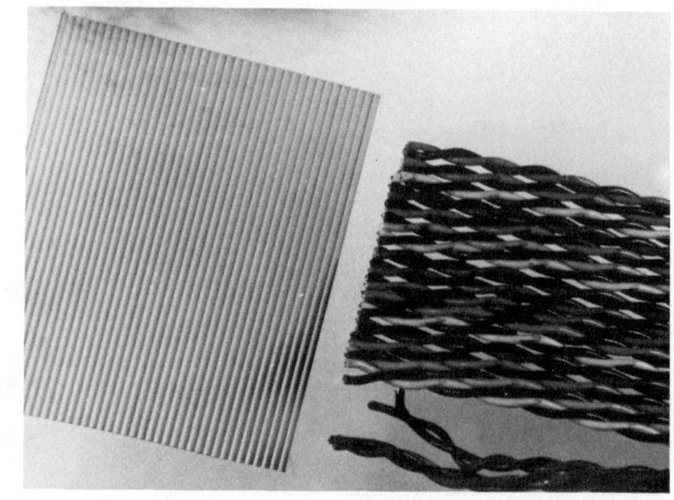

Copper and plastic have different moduli of expansion. When the two materials are laminated together the result is a quasi bi-metallic element that will warp with temperature changes. The action is identical with what takes place in the home thermostat. Consequently, temperature induced warping may be experienced with some circuit boards.

The circuit designer is the first stop on the road to the creation of a printed circuit. He must transform the mass of connecting wires that normally would appear under a standard chassis and reroute them so that all may co-exist on a flat surface without crossing. The job is a combination of ingenuity and trial and error. Usually, he will take scaled paper cutouts of the various components and move them around in a pseudo chess game until everything clicks. Occasionally a cross-over is impossible to eliminate and it will appear on the finished board as a soldered jumper wire.

The miniature components that appear on the upper surface of a completed circuit board have connecting pins protruding from their bottoms. These pins extend down through the foil and are soldered to the copper during the automatic soldering operation. The pigtails of resistors and capacitors also extend through for the same purpose.

Electronic miniaturization has resulted in such small finished circuit boards that it is impractical for the designer to work to actual scale. His steps in laying out will be taken with drawings and scale cutouts that are four or five times the size the finished unit will be. Photography will accomplish precise reduction when all details of the enlarged layout have been completed. There is a further advantage to working on an enlarged scale—any errors in measurement will be lessened by the subsequent photographic reduction.

Printed circuit layout is now such a common occupation that the template makers have stepped in with gadgets that ease the job. Templates to several enlarged scales are available for all the common components of electronic circuits. Transformers, resistors, capacitors, transistors, whatever—printed outlines save the designer the work of drawing his own. Transferrable alphabets make hand lettering unnecessary and tapes take the place of drawn lines. (See Figure 13-2.)

Figure 13-2. Stencils of the major electronic components simplify the draughtsman's work in drawing complicated circuit diagrams.

The layout for the conductor paths on the underside of the circuit board must take certain practical considerations into account. Spacing between the conductors must be sufficient to prevent flashovers at the voltages that will be present. The shape and size of connection points must facilitate solder flow and avoid solder buildup. Solder must not be able to bridge between two conductors.

Efficient mass production soldering demands that the solder be deposited only on the connections and not be spread indiscriminately. To do this requires an additional piece of artwork for a so-called solder mask. This mask, when used in its finished form, will cause everything on the underside of the board except the solder points to be coated with a material that is impervious to solder.

Three finished pieces of artwork now have been

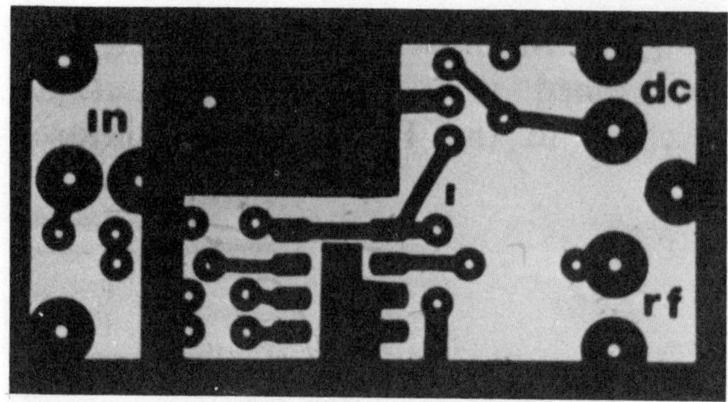

Figure 13-3. After the designers have juggled the components to make printed circuit "wiring" possible, silk screening or other methods of reproduction produce the board shown above.

made from the designer's layout. One, for the copper side of the circuit board, is a series of lines representing the connections. These lines terminate in small disks that will become the solder points. Another, for the top, insulated side of the circuit board, consists of outline drawings of the components that will be assembled there. The production line workers who do the assembling will find this marking a time saver and an error preventer. The third is the solder mask. (See Figure 13–3.)

Since the artwork is many times the size of the proposed circuit board, it must be reduced to very accurate dimensions. This is done photographically by cameras that can be set to exact reduction ratios. The positive or negative actual size film produced by the camera then becomes the basis for various reproduction techniques. The photo in Figure 13–4 shows a typical laboratory with the complex equipment that teams with technical expertise to make printed circuit advances possible. Development of the film from the camera takes place in the standard manner.

The basic idea of a printed circuit is the division of a sheet of copper foil into separate conductive paths to form the "wiring" of a circuit. The division is done chemically by dissolving unwanted areas of the copper

Figure 13–4. Technical expertise and complex equipment such as shown below is the source of the many advances in printed circuitry. (Texas Instruments)

with an etchant such as ferric chloride. The action of the etchant is restricted by means of chemically resistant patterns printed on the copper. The printing is accomplished in either of two methods, both based on the films from the camera.

In one method the copper is sensitized to light by covering it with a coating that reacts photographically, but is etchant resistant. This coating is exposed to light through the film. The sections of coating struck by light become insoluble to the developer while the rest is washed away. What remains is a pattern of lines that protect the copper underneath from being etched away.

In the other method, the film is used to make a silk screen that becomes a primitive printing press. Resistive paint, forced through the prepared silk screen onto the copper surface, forms the protective pattern that keeps the desired connecting paths from being etched away.

Whatever the method, at the completion of etching the copper side of the circuit board looks like the photo in Figure 13–5. The soldering mask is now imprinted on the remaining copper and the board is ready for drilling and for the assembly of components on the insulated side and the subsequent automatic soldering. Of course, in actual manufacturing, certain intermediate cleaning and protecting steps are taken, but these do not affect the technical analysis of circuit boards.

The printed circuit system has spawned a number of hardware accessories to ease manufacture and use. Chief among these are various terminal posts that are fastened to the board mechanically and serve as connection centers. Some of these reflect the newest technique that wraps the connecting wire around the post tightly instead of soldering it. The telephone company was the prime mover in this development. The wrapping is done with a power tool that exerts enough force to cut the sharp edges of the post into the wire. The

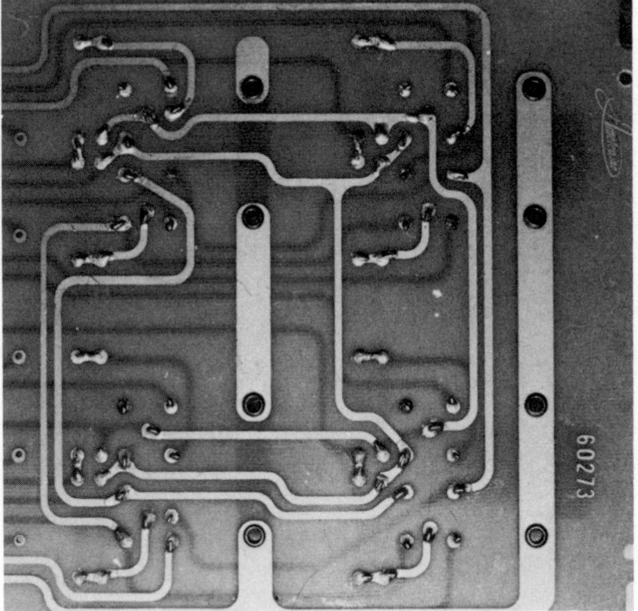

Figure 13-5. The various current paths required for the electronic circuit are produced by etching away the intermediate copper foil.

multi-circuit connector shown in Figure 13–6 is another accessory to the printed circuit board; it is often used with the flexible cable mentioned earlier.

Some complex circuits require intricate connections beyond the scope of the boards with copper foil on only one side. For these, boards are available with copper on both sides. Where connections are needed from one side to the other, eyelets between the two foils are soldered into place.

The circuit boards have also sparked many changes in components. Tiny resistors and capacitors are supplied on long tapes with pigtails bent to fit exactly into the designated holes. Transformers and other upstanding units are held in place by the soldering. Potentiometers, once large on instrument panels, now are the size of a dime and are soldered directly into the circuit.

Figure 13-6. Various modules of electronic equipment are tied together by connectors such as the one above that complete many circuits at one insertion.

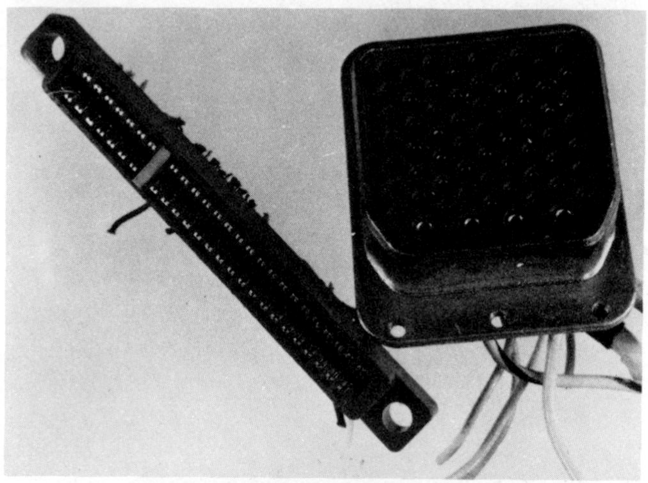

Transistors are temperature sensitive (see Chapter 3) and the provision of heat sinks to protect them is one of the points in printed circuit board design. One scheme mounts a metal plate on the board and then the transistors on the metal with suitable insulation when required. Another scheme simply places a tight metal collar with fins on the transistor.

Soldering is completely automated in mass production systems. The boards pass on a conveyor slightly above the surface of a lake of molten solder. Flux is applied to remove oxides and to permit the solder to flow. Mechanically generated waves of solder wipe the underside of the boards and solder all connections. It is considered good practice to remove the flux remaining after soldering.

Printed circuits are not all beer and pretzels for the service man who must cure the ailments they occasionally develop. Unsoldering a component that must be replaced is difficult. The amount of heat that may be applied must be carefully controlled in order not to damage the copper foil and its adhesion to the board. When a component with several solder leads (such as a transistor, for example), is to be removed, unsoldering all leads simultaneously can be a problem. Some new gadgets have been designed to help.

One of these gadgets is a solder sucker; a bulb sucks molten solder out of the connection until it is dry and loose while a soldering iron applies the necessary heat.

(See Figure 13–7.) Special tips for soldering irons heat the several solder joints of a component at the same time. Extractors put an even pull on items that are multi-soldered.

As with everything else in modern electronics, greater sophistication in equipment requires equal upgrading in the skill of the men who keep it ticking.

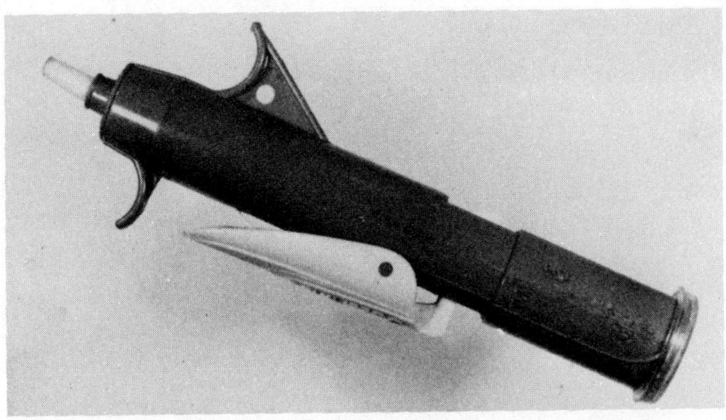

Figure 13-7. Replacing a component on a printed circuit requires that the solder holding it in place be removed first. The solder sucker picks up the solder after it has been melted by the soldering iron.

14. *Integrated Circuits*

The integrated circuit, the so-called IC, is the true wonder child of electronics. Try to visualize scores of transistors and other circuit components on a chip of semi-conductor material one-fourth the size of a dime! It brings to mind the Lord's Prayer engraved on the head of a pin (a feature display at oldtime country fairs). fairs).

By definition, an integrated circuit is one in which the normally discrete components of an electronic combination have been produced simultaneously on a single chip. All internal connections needed to create the desired circuit are made coincidentally and the result is a tiny monolithic chip ready for sealing into a container. The container provides the contact pins that enable the IC to be incorporated into electronic equipment.

The container may be a metal cup, a half inch or so in diameter, or else a half-inch wide plastic coffin an inch or two in length. The chip occupies only a small part of the internal volume; the space is needed for the contact pins and the connections to them from the chip. Sealing compound closes everything off from ambient moisture. Once the IC has been completed, there is no way to determine its circuit other than through its catalog number. The photo in Figure 14–1 shows a variety of finished integrated circuit units.

Figure 14-1. Integrated circuits are available on the market in various types and sizes to accomplish amplification, gating, counting, and whatever.

Figure 14-2. The interior of a pocket calculator contains no moving parts. A number of complicated integrated circuit modules are fastened to the front of the printed circuit board and their connections extend through to the back of the board as shown. The batteries are rechargeable by means of a small plug-in charger not shown.

The combination of fantastic ability and miniature size makes the IC uncanny to the layman and creates wonder even in the mind of the technician. The curious person who digs into an electronic pocket calculator to see the miracle-performing internals and expects to find a forest of transistors finds only a few small IC's on a printed circuit board, as proved by Figure 14–2. Yet the integrated circuit is not difficult to understand if the knowledge gained from Chapter 3 be used as a base.

Integrated circuit manufacture is a sophisticated extension of the techniques developed in the production of semiconductor diodes and transistors. Again the photographic camera plays an important part, with major credit going to the lenses whose astounding resolution maintains precise detail at previously unheard-of reductions. Circuit paths only one or two thousandths of one inch apart remain properly isolated. Chemical etching also is carried out to the same microscopic degree and forms diverse components.

In any system of manufacturing geared to quantity production, the accuracy level required of the finished product determines the rate of rejection of imperfect units. Consequently, an ample rejection rate in the manufacture of highly accurate integrated circuits is not surprising—and naturally is figured into the final cost. Despite this, the price of ICs remains surprisingly moderate and always is less than the total cost would be of the equivalent discrete components.

Reliability is one of the talking points in favor of integrated circuits. If an IC passes its final quality test at the factory, the chances are excellent that it will remain in that optimum operable condition indefinitely. Only abuse, such as overload, will destroy it. This is compared to a hard wired unit, or even a printed circuit, where the points of possible individual failure are many.

Semiconductor devices are temperature sensitive. When they are spread over an area like a large circuit board, each device may react differently from the others because its own ambient temperature may be different. The combined response then becomes unpredictable. The integrated circuit, being monolithic and small, is affected by only one ambient, and its response is predictable.

Integrated circuits are manufactured in two forms—either monolithic or hybrid. The monolithic type is a composite formed at one time upon a semiconductor base. The hybrid type is essentially in two parts; one comprises the passive elements such as resistors, capacitors, and inductors formed upon an insulating base to which the second part, consisting of active semiconductor elements, is connected. At the present time the hybrid form is more versatile because the active components may be selected individually to suit a given set of specifications. Both monolithic and hybrid ICs reach the user in the same completely encapsulated form.

The making of a monolithic integrated circuit begins with a thin slice of semiconductor material upon which the various operations for creating active and passive components are performed. This slice eventually will be cut up into perhaps 100 chips, all identical ICs.

First, an insulating film of silicon dioxide is laid down on the entire slice by photographic or other means similar to those discussed earlier. This film is then etched away at those spots where access is required to the semiconductor beneath it. Microscopically accurate photographic masks locate these spots that will become the diodes, transistors, resistors, capacitors, and even inductors required by the proposed circuit.

Injecting either n or p doping chemicals transforms a spot into a diode or a transistor. The spots are insulated from each other by making adjacent areas opposite in doping characteristics. Resistors are produced by utilizing the inherent resistivity of doped semiconductor material. Separating two conductive layers with the silicon dioxide produces a capacitor and laying down a conductive spiral results in an inductor.

The necessary connectors to all these component spots are made by sputtering or evaporating a thin film of metal over them and then etching the desired paths. Metal evaporation takes place in a vacuum. The metal is heated until its atoms have energy enough to fly off from the surface and deposit themselves as a thin film on adjacent cooler areas. (The metallic "plating" on common plastic objects often is accomplished by sputtering or evaporation.) The cross-sectional view in Figure 14-3 shows a portion of a finished IC and explains what has been described.

Mounting the finished IC chip in its container and making the connections to the contact pins is done under a microscope. Manufacturers maintain strict quality control over the finished product. Each integrated circuit module is tested by automatic circuitry that checks each parameter to verify that it meets specifications.

Figure 14-3. The makeup of one form of integrated circuit is shown in this highly magnified cross section. The carefully located injections of doping materials into the semiconductor substrate produce the transistors, diodes, and other components needed for the complete circuit.

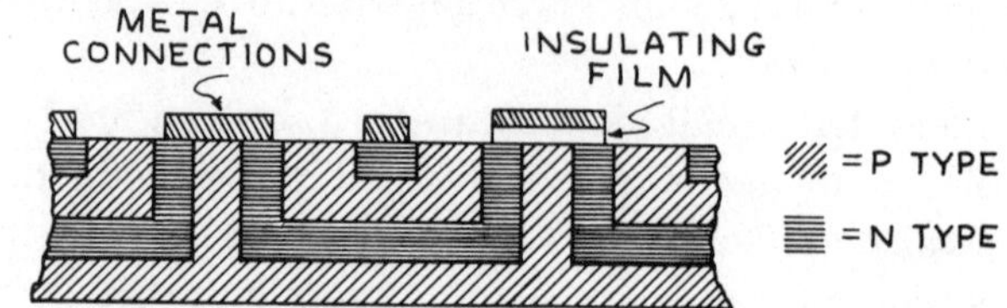

The wide variety of integrated circuits available on the commercial market has greatly reduced the problems of the electronic designer. Where formerly he had to calculate carefully unit by unit to achieve a certain result, now he scans a catalog and selects a readymade IC to accomplish the job in one leap. Many ICs have terminals brought out from individual internal components in addition to the overall input and output. Special connections to these terminals and especially feedback connections, can lean the IC toward many tasks.

Integrated circuit amplifiers are generally high gain. This together with the microscopic spacing of internal parts tends to bring on instability. But this is no problem in practice because the addition of an internal feedback loop tames the IC into a steady, docile worker. (Feedback in relation to oscillation is discussed in Chapter 7.)

The recurring description of the integrated circuit as a miniature encapsulated unit doubtless has brought to mind its one shortcoming—the IC cannot be repaired. If any internal failure occurs, there is only one remedy and that is complete replacement. Luckily, the comparatively moderate prices of integrated circuits make replacement not quite as traumatic as it sounds.

The multiplicity of pins emanating from an IC may make identification of input, output, and other terminals confusing without a manufacturer's data sheet. However, the pins are numbered according to a standard system even though these numbers may not actually be printed on the case. An identifying dot is placed on the face that carries the catalog number. This dot marks the number one pin and the following pins are numbered consecutively going counterclockwise. The system is similar to the one that numbers the pins of vacuum tubes. (See Figure 14-4.)

An integrated circuit amplifier is shown on a block diagram as an equilateral triangle with one side vertical. The inputs are at this vertical side, the outputs at the apex to the right of it. If the output of the amplifier is inverted, this fact is indicated by a small circle at the

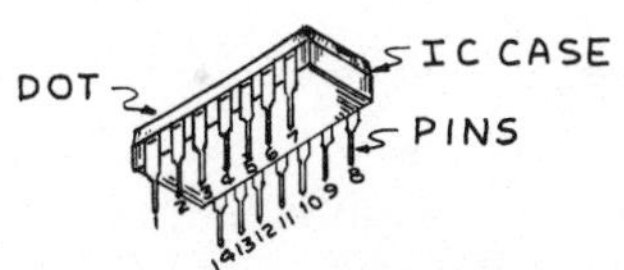

Figure 14-4. Although the multiplicity of pins of an IC may appear confusing, the numbering is to a standard system. A dot identifies pin #1 and the numbers seen from the top proceed counterclockwise from there.

apex. The triangle itself gives no indication of the internal circuit or of how many components make it up. (See Figure 14–5.)

The differential amplifier is a popular circuit for ICs. It has two inputs and one output. In essence, it amplifies the difference between the two inputs, hence its name. The practical value of the differential connection is that unwanted noise is balanced out and only the desired signal appears at the output. Many makers bring leads out to separate pins, allowing the differential IC to be used as two individual amplifiers.

Integrated circuit differential amplifiers have extremely wide frequency capability that ranges from di-

Figure 14-5. This circuit and its components would occupy considerable space in discrete form but it is all contained in the tiny semiconductor chip of an IC. (ARRL Radio Amateur's Handbook)

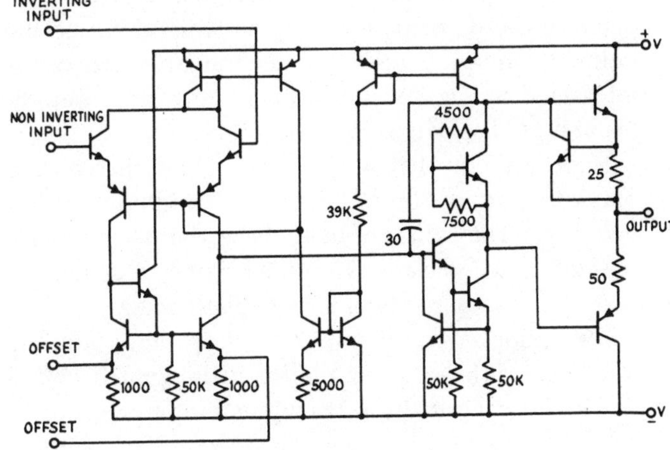

rect current to very high frequency alternating current (VHF). The fact that the internal IC components and leads are microscopic is an advantage at the high frequency end of the spectrum because of the consequent low inductance and capacitance.

Manufacturer's data sheets list the characteristics of their integrated circuits in a technical jargon that may be confusing until one has been initiated. One listing gives the "common mode rejection ratio." Earlier it was stated that one advantage of the differential amplifier is its ability to balance out and reject unwanted noise; the common mode rejection ratio is a numerical measure of how well the circuit does this. The result is expressed in decibels (dB) and is the ratio of the strength of the desired output to what is left of the undesired one. The greater the number of dBs, the better the performance of the IC.

Another term is "input offset voltage." Theoretically, equal voltages applied to the inputs of a differential amplifier should produce equal outputs or no output differentially. In practical circuits this does not always hold. Often a small additional voltage must be added to one of the inputs to bring the outputs to equality. This added voltage is the offset voltage; it is usually in the order of a few *milli*volts.

"Device dissipation" is another parameter. It states the amount of energy, in *milli*watts, that may be dissipated in the IC without raising its temperature to the point of damage. Considering the miniature size of the integrated circuit, it is not surprising that the permissible energy dissipation is small.

Specifications that are understandable in view of previous text are "input impedance" and "output impedance." The values of these two parameters are important to technicians who must match a certain IC with the requirements of a particular circuit.

Operational amplifiers, called "op amps" in the trade, are differential amplifiers in integrated circuit form whose normally tremendous amplifying power has been toned down with feedback to make them stable. The feedback circuit, whether internal or external, channels a portion of the output back into the input to achieve better fidelity in addition to stability. Op amps are widely found in electronic instruments, computers, oscillators, and in circuits that perform mathematical functions. Op amps are true electronic building blocks.

Two inputs generally are provided on op amps. One leads to an amplifier that multiplies the signal in the normal way. The other also leads to multiplication, but inverts the signal at the output. The ideal op amp has a very high input impedance. This reduces to a negligible amount the power the op amp draws from a previous circuit.

Arrays of diodes also are available in integrated circuit form. The internal connections enable these to be functional as bridges, balanced modulators, and other active elements. These arrays assure the similarity of all the contained diodes because they are made simultaneously on the same substrate of semiconductor material. Arrays of IC transistors are likewise available and they too have this advantage.

A perusal of manufacturer's literature often encounters the term "epitaxial." The word, of Greek derivation, denotes a process in which material is laid down in a layer and retains a given crystalline structure. Applied to semiconductor production, an epitaxial layer is produced in a warm vapor of the desired impurity when the atoms of the vapor deposit themselves on the inserted surface.

One apparent technical contradiction in the operation of the differential amplifier IC is that a circuit resistor high enough for best performance reduces emitter current to the point of instability. This is neatly overcome by adding a constant current source while yet retaining the high value resistor. An additional transistor is the current regulator that assures emitters the optimum current.

"Direct connection" is a favorite internal method of having each amplifying unit of an integrated circuit

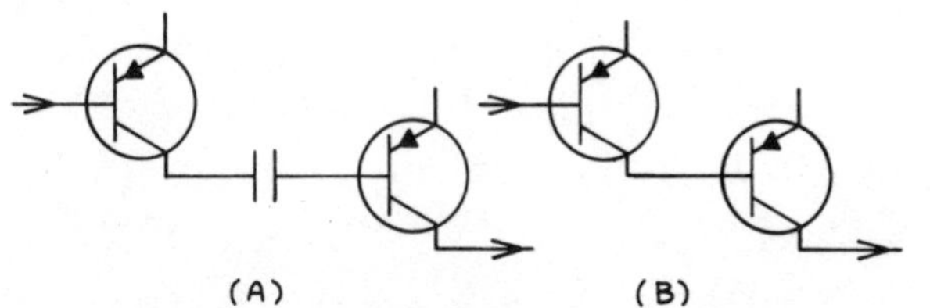

Figure 14-6. By removing the coupling capacitor (a) and substituting a direct connection (b) the amplifier becomes insensitive to frequency.

feed its output to the next in line. Direct connection, as explained earlier, does away with the customary capacitor between stages; as a result of this, the amplifier is able to handle direct current in addition to the total frequency spectrum its design permits. But direct connection has its woes as well as its advantages. The fact that each unit directly affects the others may lead to instability. The difference between direct connection and capacitive connection is shown in Figure 14–6.

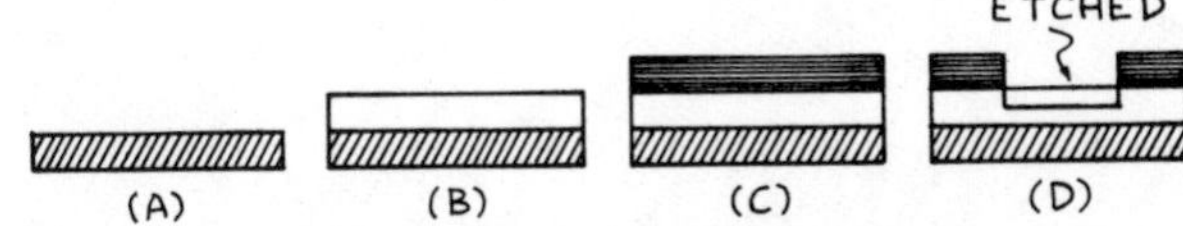

Figure 14-7. The epitaxial process begins with a substrate of p semiconductor material (a). A layer of n material is laid down epitaxially upon this (b). There follows a layer of silicon dioxide (c). This is etched away as shown at (d) for further required doping in restricted areas.

In summary, the standard integrated circuit structure is the culmination of the art of building transistors and diodes by the planar epitaxial process. Everything is accomplished by depositing upon a semiconductor substrate specific epitaxial layers of n material, of p material, of insulating oxide, or of metallic connections, as required. Successive layers are built up by first etching the necessary hollow and then filling it with the new composition. The step by step drawings are shown in Figure 14–7.

15. *Calculators for Navigation*

The electronic pocket calculator, that wondrous offspring of the integrated circuit technique, can save the navigator a great deal of pencil pushing and brain wracking when determining the vessel's position. This electronic help is available not only for the comparatively simple problems of pilotage, but also for the more complex manipulations of celestial navigation. A few pushes of the calculator keys provide answers in seconds that would take many minutes with normal mathematical procedures.

Best of all, most of these position finding operations may be performed on simple calculators that are moderate in price. The basic requirement is that the calculator be capable of working sines and cosines and their inverses. The solutions become even easier and quicker with the more sophisticated (and more expensive) calculators that are programmable to make many of the intermediate calculations automatically, without step by step key pushing. Incidentally, much of this navigational mathematics is also within the capability of the old-time slide rule that has been pushed into oblivion by the pocket calculator.

Figure 15-1. This simple, so-called "slide-rule" calculator may be used to solve many navigational problems.

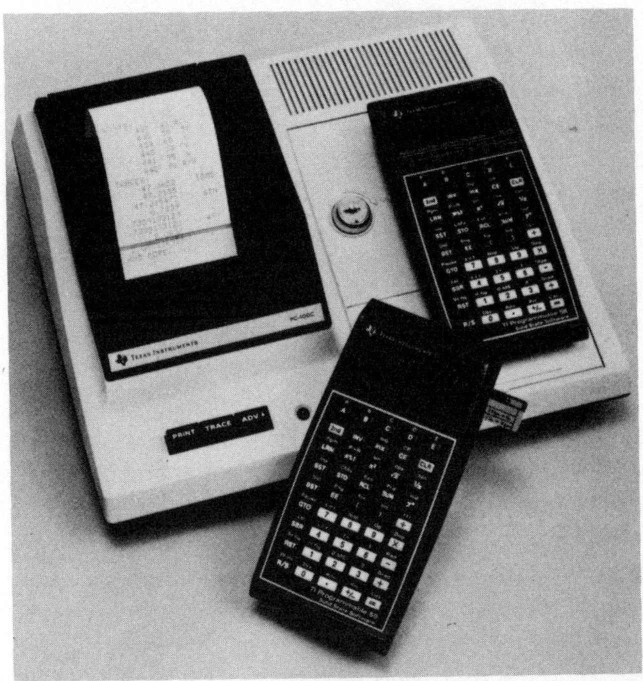

Figure 15-2. These latest generation calculators have memories and are programmable to solve celestial navigation problems almost automatically. They contain their own crystal-controlled "chronometer" that delivers near-perfect Greenwich Mean Time. A printer is also shown. (Texas Instruments)

From the skipper's standpoint, a very rough division of available calculators would be those that can work trigonometric functions but are not programmable, and those that are. (A representative example of the former is shown in Figure 15–1 and of the latter in Figure 15–2.) The most sophisticated of the programmable calculators, developed especially for navigational use, contain complete tables in their memories and enable the sight reduction of celestial observations without reference to books.

All solutions to the problems of pilotage and celestial navigation depend upon the science of trigonometry, a discipline that concerns itself with the fixed relationships of the angles and sides of triangles.

Trigonometry was born two thousand years ago in the mind of a Greek genius, Euclid, yet today's electronic calculator is particularly suited to the rapid solution of its equations.

A quick review of the highpoints of schoolday "trig" will be helpful. It will enable the skipper mentally to transform a pilotage problem into the triangle form that permits its solution. It will also remove the mystery from the terms "sine," "cosine," "tangent," and "cotangent" (the "functions") by showing how they are derived.

The triangle shown in Figure 15–3 is labelled in the standard custom that fixes the relation between angles and sides. Thus a capitalized letter denotes an angle and the same letter in lower case pertains to the opposite side. The functions may now be written as simple equations of one letter over another—and as easily solved. Note that the functions are based on so-called right triangles that include one angle of ninety degrees and that the side opposite this right angle is called the "hypothenuse." Remember, too, that the three angles of a triangle always total 180 degrees, enabling one unknown angle to be found by subtraction.

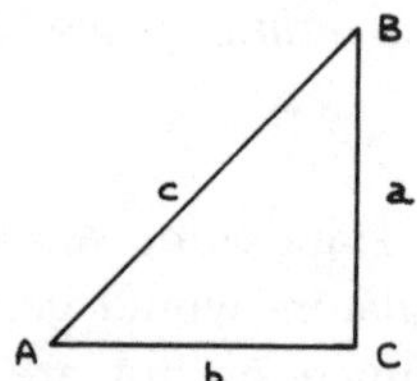

Figure 15-3. Mathematical custom requires the above-shown standard nomenclature for triangles. The angles are designated by capital letters and the opposite sides by lowercase letters.

The "sine" of an angle is defined as the quotient of the side opposite the angle divided by the hypothenuse. This may be "shorthanded" as sinA = a/c and reference to the drawing of Figure 15–3 will make this clear. Normally, the actual division would not be necessary because recourse to a table such as shown in Figure 15–4 would produce the numerical value of the sine. As an example, the sine of 29 degrees 10 minutes is given as 0.4874. Conversely, if a solution yielded the number 0.4874, a glance at the table would reveal the corresponding angle of this "inverse sine" to be 29°10′.

Now comes the wonder of the pocket calculator. The tables no longer are necessary. The foregoing

Figure 15-4. Trigonometric functions are listed in many textbooks in tables similar to this.

	29°						30°				
	sin	cos	tan	cot	′	′	sin	cos	tan	cot	′
0	0.4848	0.8746	0.5543	1.8040	60	0	0.500	0.8660	0.5774	1.7321	60
1	51	45	47	028	59	1	03	59	77	309	59
2	53	43	51	016	58	2	05	57	81	297	58
3	56	42	55	1.8003	57	3	08	56	85	286	54
4	58	41	58	1.7991	56	4	10	54	89	274	56
5	0.4861	0.8739	0.5562	1.7979	55	5	0.5013	0.8653	0.5793	1.7262	55
6	63	38	66	966	54	6	15	52	5797	251	54
7	66	36	70	954	53	7	18	50	5801	239	53
8	68	35	74	942	52	8	10	49	05	228	52
9	71	33	77	930	51	9	23	47	08	216	51
10	0.4874	0.8732	0.5581	1.7917	50	10	0.5025	0.8646	0.5812	1.7205	50
11	76	31	85	905	49	11	28	44	16	193	49
12	79	29	89	893	48	12	30	43	20	182	48
13	81	28	93	881	47	13	33	41	24	170	47
14	84	26	5596	868	46	14	35	40	28	159	46
15	0.4886	0.8725	0.5600	1.7856	45	15	0.5038	0.8638	0.5832	1.7147	45
16	89	24	04	844	44	16	40	37	36	136	44
17	91	22	08	832	43	17	43	35	40	124	43
18	94	21	12	820	42	18	45	34	44	113	42
19	96	19	16	808	41	19	48	32	47	102	41
20	0.4899	0.8718	0.5619	1.7796	40	20	0.5050	0.8631	0.5851	1.7090	40
21	4901	16	23	783	39	21	53	30	55	079	39
22	04	15	27	771	38	22	55	28	59	067	38
23	07	14	31	759	37	23	58	27	63	056	37
24	09	12	35	747	36	24	60	25	67	045	36
25	0.4912	0.8711	0.5639	1.7735	35	25	0.5063	0.8624	0.5871	1.7033	35
26	14	09	42	723	34	26	65	22	75	022	34
27	17	08	46	711	33	27	68	21	79	1.7011	33
28	19	06	50	699	32	28	70	19	83	1.6999	32
29	22	05	54	687	31	29	73	18	87	988	31
30	0.4924	0.8704	0.5658	1.7675	30	30	0.5075	0.8616	0.5890	1.6977	30
31	27	02	62	663	29	31	78	15	94	965	29

example would be worked on the calculator shown in Figure 15–1 as follows:

First, the 10 minutes would be changed to decimal degrees. Push keys: 1,0, ÷, 6,0, =. The display: 0.1666666. The 29°10′ now is handled as 29.166666 degrees. To find the sine: Push keys: 2,9,.,1,6,6,6,6, F, sine. The display: 0.487352. If only the inverse sine (0.48732) were given, the procedure would be: Push keys: 0, . ,4,8,7,3,2,F,sin⁻¹. The display: 29.16666. The key F is needed to differentiate between the two purposes of the 7 key in this calculator.

All the other needed trigonometric functions are obtained from the calculator in the same manner as above, merely substituting cosine, tangent, or cotangent keys for the sine key. (This particular calculator does not have secant and cosecant functions, but this presents no problem for pilotage use.)

Just as the sine is the opposite side of the right triangle divided by the hypothenuse, so the cosine is the adjacent side divided by the hypothenuse, expressed as cosA = b/c. The tangent is the opposite side divided by the adjacent side (tanA = a/b) and the cotangent is the adjacent side divided by the opposite side (cotA = b/a). In all of these, refer again to Figure 15–3 for clarification.

The problems of pilotage concern a relatively small area and for this the earth may be considered flat (a plane) without causing appreciable error. Thus plane trigonometry becomes the method of solution. For celestial navigation the section of the earth and sky taken into consideration is so vast that the curvature of the earth must be reckoned with. Now spherical trig becomes the method of solution. Although spherical trigonometry is more involved than plane, working with it is hardly more difficult for the skipper, thanks to the calculator.

One of the more common pilotage problems is the so-called "bow and beam." In the example of Figure

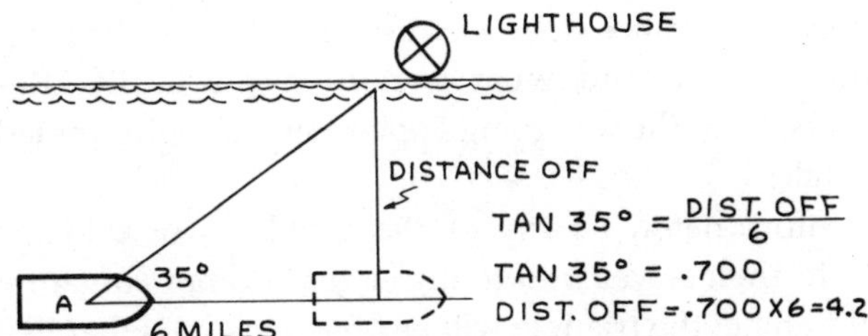

Figure 15-5. The so-called "bow and beam" relationship is illustrated above. The skipper first sights the lighthouse at 35 degrees. He finds he has run 6 miles when he comes abeam. The simple trigonometry shown (plus the text) tells him that he is 4.2 miles off.

15–5 a vessel running parallel to the shore wishes to know how far off it is. The skipper draws the triangle shown in order better to visualize the problem. While still some distance away, he takes a sight of the lighthouse by handbearing compass or other means and finds the angle A to be 35 degrees. From that moment on he keeps careful count of the distance covered until the lighthouse bears exactly 90 degrees, and he finds it to be six miles. He enters the information on the sketch to visualize the problem. Remembering that the tangent of an angle is the opposite side divided by the adjacent side, the situation may be stated as "the tangent of the angle A is equal to the distance off divided by the distance run. Algebraically,

$$\tan A = \frac{\text{distance off}}{6}$$

Transposing, distance off = 6 times tan 35°.

Now to the calculator. Push keys: 3,5,F,tan,×,6. The display now reads 4.20 (miles), the distance off the lighthouse.

An excellent habit in these problems is to form the appropriate triangle, either mentally or graphically, and then to place the known facts in their proper positions. This will suggest which of the trigonometric functions explained earlier to employ for the solution.

A few key pushes does the rest—and to a decimal accuracy far beyond what is needed practically. Remember that the foregoing applies only to right-angled triangles.

Unfortunately, many of the problems of pilotage will fit themselves into triangles that do not contain a right angle; the triangle will be oblique, somewhat like the one shown in Figure 15–6. The solutions to oblique triangles are slightly more complex, but from the standpoint of pushing the keys of a calculator the additional difficulty is negligible.

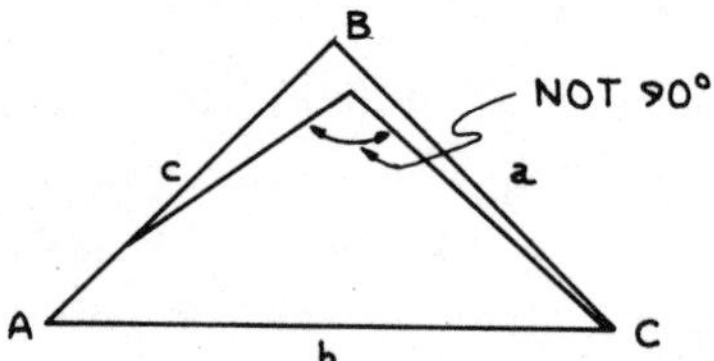

Figure 15-6. An oblique triangle is one that does not contain a right angle of 90 degrees. It is labeled as shown.

As with right triangles, listing the functions that apply to oblique triangles makes it possible to choose the most advantageous path to solution.

Oblique triangles maintain a unique relationship between the sines of their angles and their sides and this is known in the mathematical world as the "law of sines." In general, an oblique triangle may be solved if three independent parts of it are specified such as, for instance, as two angles and an included side or two sides and an included angle. Expressed algebraically on the basis of the standard trigonometric labelling described earlier:

$$a = \frac{\sin A}{b \sin B} \qquad b = \frac{\sin B}{c \sin C} \qquad c = \frac{\sin C}{c \sin A}$$

By algebraic manipulation this leads to:

$$\frac{a}{\sin A} = \frac{b}{\sin B} = \frac{c}{\sin C}$$

Any of these relationships may be chosen to effect a solution, the choice depending on the knowns and unknowns in each problem. The following pilotage situation will illustrate this:

The skipper in the previous example, who wanted to know how far he is from the lighthouse when arriving abeam, would rather have this information before he gets there so he can check his chart for possible dangers. He takes a reading with his handbearing compass or by other means while he is still some distance away and notes that the angle to the lighthouse is 27 degrees (angle A). He stays on the same course for seven miles and takes another reading that turns out to be 52 degrees and subtracting this from 180 degrees (the straight line) shows the angle B to be 128 degrees. The third angle of the oblique triangle therefore is 25 degrees. All this is shown in Figure 15–7.

The skipper's final interest lies in the triangle BDC, and especially in the length of the line CD because that is the distance to the lighthouse when abeam. However, the oblique triangle ABC must be solved first in order to get the length of side a, needed for solving the triangle BDC, a right triangle whose angles now are known.

An inspection of the data now known for the oblique triangle suggests that the relationship

Figure 15-7. This skipper would like to know in advance how far off the lighthouse he will be when abeam. His first sight is at 27 degrees at A and after running 7 miles he takes a second sight at B and finds it to be 52 degrees. The simple trigonometry above (and the text) advises him that he will be 5.92 miles off the lighthouse when abeam.

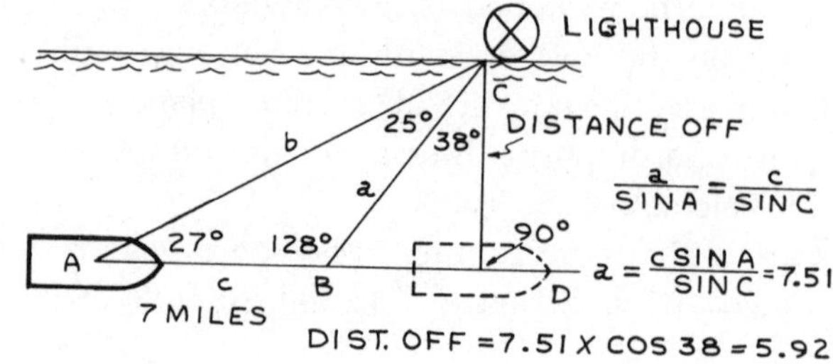

$$\frac{a}{\sin A} = \frac{c}{\sin C}$$

would serve because the length of side a is needed. For convenience this becomes

$a = \frac{c \sin A}{\sin C}$ or a equals c times sinA divided by sinC.

Keys pushed for sinA: 2,7,F,sin Display: 0.45399

Keys pushed for sinC: 2,5,F,sin Display: 0.42261

Keys pushed for a: 7,×,.,4,5,3,9,=,÷,.,4,2,2,6,1,= Display: 7.518

The third angle of the right triangle BDC equals 180 minus 90 minus 52 equals 38.

The cosine of 38 degrees equals the line CD divided by 7.518 or CD equals cos38 times 7.518

Keys pushed: 3,8,F,cos,×,7,.,5,1,8, = Display: 5.924 (miles) the distance off the lighthouse when abeam.

$$CD = \frac{\text{distance run times sin A}}{\text{sin of B minus A}} \text{ times sinB}$$

The formula for achieving the answer to the above example may be simplified in the interest of faster computation while underway. This is possible because of a unique condition that prevails in right triangles: The cosine of one acute angle always is identical to the sine of the other acute angle. Thus, instead of working out the remaining angles of each triangle, only the actual angles of the sights to the target need be taken into account.

The formula now becomes (again referring to the standard marking of Figure 15–3):

$$CD = \frac{\text{distance run times sinA}}{\text{sin of B minus A times sinB}}$$

Punching this new formula into the calculator for the same example:

sinA key pushes: 2,7,F,sin Display: 0.45399

sinB key pushes: 5,2,F,sin Display: 0.0.7880

sinB-A key pushes: 5,2,−,2,7,=,F,sin Display: 0.422618

solution key pushes: 7,×,.,4,5,3,9, =, ÷ ,.,4,2,2,6, =,×,.,7,8, 8,0, = Display: 5.924 (miles) to lighthouse when abeam.

Many of the common problems of pilotage involve simple multiplication and division that often may be solved mentally and are hardly the food for calculators. Nevertheless, the calculator saves time and bestows a feeling of accuracy.

The relationships between time, speed, and distance are often encountered. What is the equivalent speed after a run over a measured course? At a given speed, how long will it take to get from here to there? What speed must be attained if a certain run is to be accomplished in a given time?

Example: What was the speed if it took 7 minutes 42 seconds to run the measured mile? The 42 seconds is first changed into decimal minutes. Key pushes: 4,2,÷,6,0,=Display: 0.7. The time is now 7.7 minutes. Formula: Speed in mi/hr is equal to distance in mi divided by time in minutes multiplied by 60.

Key pushes: 1,÷,7,.,7,×,6,0,= Display is: 7,792 mi/hr

(Of course this is more theoretical than practical because neither wind nor current has been taken into account. The course should be run at least in both directions and the results averaged.)

Example: How long will it take to run 17.4 miles at the just calculated boat speed of 7.792 miles per hour, disregarding wind and current and steering errors?

Formula: time in hours equals distance in miles divided by speed in miles per hour.

Key pushes: 1,7,.,4,÷,7,.,7,9,2,= Display: 2.233 hours

Example: The same boat ran 4 hours 16 minutes at its same steady speed. How far did it go through the water? Formula: distance in miles equals speed in miles per hour times time in hours.

Key pushes: 1,6,÷,6,0,= Display: 0.2666. Time now is 4.2666 hrs.

Key pushes: 4,.,2,6,6,6,×,7,.,7,9,2,= Display: 33.245 miles

The foregoing examples have listed the necessary key pushes in minute detail and have related this to the anatomy of trigonometry upon which all is based. This is excellent training for the skipper whose calculator is newly come abroad. It makes it possible for him to reason out each application without the necessity of memorizing silly doggerel about dead men or whatever. Once skill is acquired through practice, the keys will be pushed almost subconsciously.

For the purposes of pilotage the earth is flat, but when it comes to celestial navigation the earth is undeniably spherical. Computations become complex and lengthy and the simple slide rule type calculator, such as the one used in the previous examples, is outclassed even though it could arrive at the answer after many manipulations. The programmable calculator with adequate memories now comes into its own.

One of the computations employed in celestial sight reduction is illustrative. It consists of a string of sines and cosines, parenthetically related, that must be added and multiplied in various orders. To accomplish this with the simple calculator requires a score of separate steps with intermediate notations of results that must be reinserted. A programmable calculator properly set up finds the answer with minimal key pushes and retains the intermediate values automatically in its memory.

No detailed operating sequence for these super calculators is given here because each manufacturer supplies his own unique instructions. Some makers supply the needed programs in module form. Snapping the module in place automatically programs the calculator. The quintessence of these electronic marvels even contains its own "chronometer" set to Greenwich time and functions without any outside data other than the details of the celestial sight.

The maintenance requirement for electronic pocket calculators is practically nil. Most calculators contain rechargeable batteries, and a polarized receptacle for the charger is provided in the case. The charger generally is supplied with the calculator and connects to house current. Chargers that work from the 12 volt direct current system on boats are also available. (Some mini-calculators contain camera-type batteries that are not rechargeable.)

Note: The F key mentioned above in some of the calculations changes a numeral key temporarily into a function key.

PART FOUR
BOAT ELECTRONIC EQUIPMENT

16. *Receivers, Transmitters, and Transceivers*

The "coherer" that Marconi employed to detect and transform radio waves into audible tones was by itself larger than a complete modern marine radio transceiver. Yet the coherer was only the detector portion of a large assortment of connected gear—and a pea-sized diode does an infinitely better job of detection today. Only by such comparison can one truly appreciate the compact, efficient, effective radio equipment the pleasure boat skipper currently has at his command. Technology has soared and the price has come within reach.

Marine radio traffic is confined to a number of groups of frequencies reserved for that particular usage by international agreement. These groups are spotted along the radio spectrum from 2 to 22 megahertz and they are allocated for voice, code, facsimile, and teletype transmissions. However, the average yacht skipper finds his greatest interest not in these groups, but rather in a band of frequencies (extending from 156 megahertz to 163 megahertz) that have been set aside for short range ship-to-ship and ship-to-shore communications. This is the "VHF-FM" band (*very high frequency-frequency modulated*). The VHF-FM band is divided into some 78 channels allocated for American and international usage, including some reserved for weather information only. The chart in Figure 16–1 lists the groups and shows their relative positions.

Until it was outlawed in 1977 by a Federal Communications Commission decree, pleasure boat radio traffic was carried on at frequencies slightly above 2 megahertz by means of amplitude modulated (AM) transmitters. The calling and distress position was 2.182 megahertz. This marine traffic has been moved, in the main, to the frequency modulated VHF-FM band, with calling and distress restricted to its channel 16.

The radio wave itself, when emitted without superimposed information, is a steady stream of energy at the frequency to which the transmitter is tuned. Several methods exist for harnessing this wave to carry intelligible communication. It may be broken up into the dots and dashes of a telegraphic code. It may be controlled or "modulated" by voice with either amplitude modulation (AM) or frequency modulation (FM). The ubiquitous microphone, near the helmsman's position on almost every pleasure boat, proves that voice is the universal method in use by yachtsmen, with the telegraph code left to the commercial services.

The idea of modulation requires further explanation. How can the human voice, or any sound, be imposed upon an unseen radio wave?

The voice first must be amplified. The actual energy in speech is miniscule, but amplifying this to whatever level is required is a simple process with modern electronic technology. The strengthened sound now becomes a varying electric current, increasing and decreasing to mirror the voice at the microphone. These variations exert control over the radio wave—they modulate it to make it conform. The steps in accomplishing this are shown in Figure 16–2.

Each block in the diagram represents an action that takes place in the several sections of an amplitude modulated (AM) transmitter. The oscillator generates and maintains the desired frequency. The power amplifier steps up the low output of the oscillator to the rated power of the transmitter. Meanwhile the output of the microphone has been amplified in the modulator. Now the modulator acts upon the radio power amplifier and varies the latter's output in step with the voice. The amplitude of the energy fed to the antenna thus varies above and below the amplitude existing when the microphone is silent. The broadcast radio wave, with its maxima and minima, may be processed at the receiver to yield its message.

In frequency modulation (FM) the amplified voice currents affect the radio wave differently. They do not cause changes in amplitude, as in AM; now they cause changes in frequency and the stronger the voice, the

Figure 16-1. The numbered channels of the VHF spectrum are divided into three categories. In List A are the channels authorized for pleasure boat use. List B channels are for the specific purposes named. List C identifies the channels forbidden to pleasure boats.

A

9	public and commercial intership and ship-to-coast
15	environmental (receive only) for weather, sea conditions, notices to mariners, etc.
16	distress, safety, & calling (Coast Guard monitored)
24	marine operator
25	marine operator
26	marine operator
27	marine operator
28	marine operator
68	public intership and ship-to-coast
69	same as 68
70	public intership
71	same as 68
72	public intership
78A	same as 68
84	marine operator
85	marine operator
86	marine operator
87	marine operator

B

6	intership safety
12	port operations, intership & ship-to-coast
13	navigational—commercial ship bridge-to-bridge, locks, bridges
14	same as 12
17	available to all vessels to communicate with ships and coast stations operated by state or local governments.
20	port operations, ship-to-coast (traffic advisory)
22A	Coast Guard liaison with recreational boating (working channel). Authorization required; initial contact made on channel 16
65A	same as 12
66A	same as 12
73	same as 12
74	same as 12

C

7A	commercial intership & ship-to-coast
8	commercial intership
10	same as 7A
11	same as 7A
18A	same as 7A
19A	same as 7A
21A	U.S. government only
23A	U.S. government only
67	commercial intership
77	commercial intership
79A	same as 7A
80A	same as 7A
81A	U.S. government only
82A	U.S. government only
83A	U.S. government only
88A	commercial intership

greater the change. But the amplitude of the radio wave remains constant. The block diagram is shown in Figure 16-3.

As before, the modulator has the amplified varying voice currents, but in this system they act upon the oscillator. The frequency of the oscillator is caused to go above and below the "resting" frequency that is obtained when no sound is applied to the microphone. The varying frequency of the oscillator goes to amplifiers and multipliers that increase its lower level and raise its frequency to the license requirement of the transmitter that broadcasts it. Again, a distant receiver can process the signal and recover its message.

The block labelled "frequency multiplier" may need explanation. A frequency multiplier is an amplifier purposely designed to include higher order harmonics in its output. These harmonics are at several times the input frequency. The desired harmonic is selected by a tuned circuit to achieve frequency multiplication.

The reason for including the frequency multiplier instead of tuning the oscillator itself to the higher figure is a technical one, especially applicable to FM transmitters. The lower the frequency of the oscillator, the easier it is to modulate it, to vary its frequency, in step with the applied voice. The modulated lower frequency then is multiplied to the value required for legal transmission. The multiplication does not interfere with the modulation, in fact accentuates it.

Figure 16-2. The block diagram of an AM vice transmitter. The oscillator sets the frequency for which the transmitter is licensed. The miniscule output of the microphone is strengthened by audio amplifiers. The modulator superimposes the message onto the radio wave by changing its amplitude in conformance with the voice.

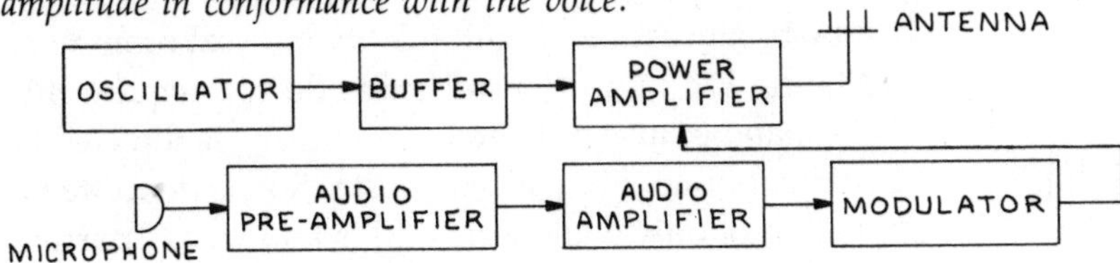

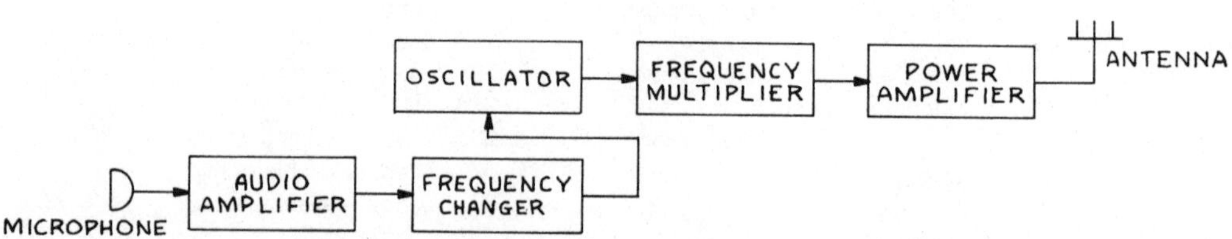

Figure 16-3. The block diagram of an FM transmitter. The voice current from the microphone is amplified and controls the frequency of the oscillator through the medium of a frequency changing component. The varying frequency of the oscillator is multiplied to the high frequency for which the transmitter is licensed. The output of the power amplifier goes to the antenna to be radiated.

A pictorial representation of modulation takes some technical liberties, but nevertheless shows what happens to the radio wave broadcast by AM and FM transmitters. (See Figure 16–4.) The resting output of the AM transmitter, the carrier, is shown to be at constant amplitude. When modulation occurs because of a voice at the microphone, the carrier amplitude increases above and decreases below the resting amplitude. With 100 percent modulation the carrier varies between twice its resting amplitude and zero.

A quick glance at the FM illustration proves that the radio carrier wave remains at constant amplitude whether or not it is modulated. But modulation causes the number of cycles depicted in a given space to increase (higher frequency) and decrease (lower frequency). Contrasting the two diagrams should make a crystal clear mental picture of AM and FM radio transmission.

Figure 16-4. An explanatory sketch of an amplitude modulated (AM) wave is shown at (a) and of a freqnency modulated wave (FM) at (b).

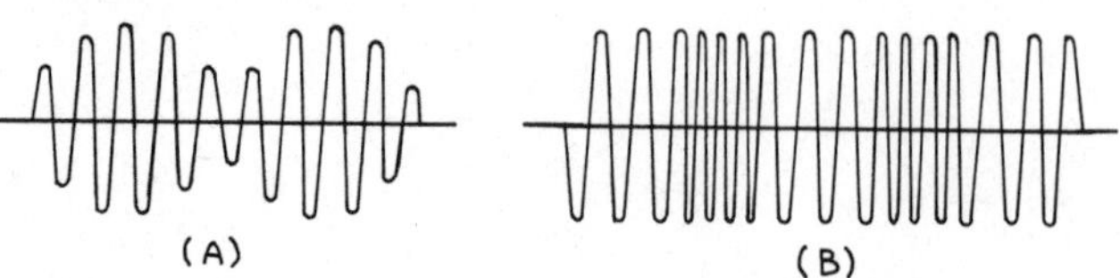

The value of the FM concept lies in the almost complete elimination of noise. The electrical noises that afflict radio reception are entirely amplitude generated and, theoretically and also nearly practically, amplitude variations have no effect on an FM receiver. An AM receiver cannot distinguish between the amplitude variations of noise and of a signal.

The term "sidebands" is heard often. What are sidebands? A simple mathematical example should prove explanatory. Assume a carrier frequency of 100 kilohertz that is modulated by a tone of 5 kilohertz at the microphone. Because of the nature of modulation, the output will now contain the original carrier frequency of 100 kilohertz plus the new frequencies of 105 kilohertz and 95 kilohertz. The 105 kilohertz portion is the "upper sideband" and the 95 kilohertz portion is the "lower sideband." The upper sideband is always the modulation frequency plus the carrier and the lower sideband is the modulation frequency subtracted from the carrier.

The generation of sidebands also takes place in an FM transmitter because of the frequencies created by modulation of the carrier resting frequency. The maximum frequency shift due to modulation, known as the "deviation," is fixed by FCC ruling in the interest of conserving spectrum space. FM broadcasting stations are permitted to deviate their carriers no more than 75 kilohertz above and below the resting figure. This provides a span of 150 kilohertz and makes possible the broadcasting of high fidelity musical programs.

Marine FM radio transmissions do not require this wide span because only voice, no music, is directed at the microphone. FM with this restriction is permitted a deviation of only 5 kilohertz and is known as "narrow band FM" or NFM. A narrow band FM transmitter occupies the same space in the radio spectrum as an AM transmitter with identical microphone input.

The complete radio receiver, or the receiver section of a transceiver, combines the various components that are discussed separately earlier in the text. The method of interconnecting the components and their electrical size determine whether the result is an FM receiver, an AM receiver, an SSB receiver, or whatever, and sets the frequencies to which the unit will be responsive.

The block diagram in Figure 16–5 depicts the operation of an AM radio receiver; as is usual in this form of illustration, each block represents an entire function that may consist of many interconnected parts. The two general types of circuits are the "tuned radio frequency" and the "superheterodyne." The latter is chosen for all high quality radio receivers. The radio signal entering a tuned radio frequency receiver is tuned and amplified without any frequency change, whereas the superheterodyne receiver will change or heterodyne the incoming signal to another frequency at which amplification is accomplished more effectively.

As shown in the diagram, the signal from the antenna enters the radio frequency amplifier (RF) where it is increased in intensity. The desired signal is selected from all others available from the antenna by tuning the RF amplifier and the oscillator with a front panel knob or with a push-button tuning system. The amplified signal goes to the "mixer" where the heterodyning action takes place.

The mixer receives the output of the local oscillator simultaneously with the signal. A numerical example will explain what the mixer does. Assume a signal of 150 megahertz and the local oscillator input of 140 megahertz. The mixer, by reason of its technical nature, will combine these two frequencies additively and subtractively, but the only output of interest will be the difference frequency of 10 megahertz. This difference frequency will enter the "intermediate frequency amplifier" (IF) that is permanently tuned to reject all others.

The heterodyning action not only allows major amplification to take place at a lower frequency, where it can be done more efficiently, it also prevents local adverse interaction between input and output of the various amplifiers. The standard intermediate frequency for UHF-FM receivers ranges from 10 to 17 megahertz; for AM broadcast receivers it is 455 kilohertz.

The by now greatly amplified signal in intermediate frequency form is fed to the detector that removes the radio envelope and recovers the audible variations that went into the original microphone. The simplest form of detector is an ordinary diode. Manufacturers add various sophistications to eliminate unwanted noise.

The best radio receivers make use of double heterodynes and change the frequency of the signal not once but twice, all in the interest of greater sensitivity and stability. The first change is as indicated in the numerical example. The next heterodyne brings the 10 megahertz resultant signal down to a new intermediate frequency of 455 kilohertz (the same as in the

Figure 16-5. The path the received signal takes from antenna to loudspeaker. See text.

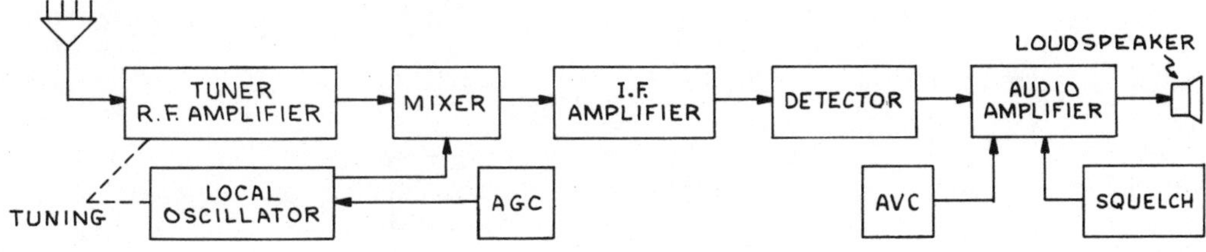

broadcast receivers mentioned above) at which level great amplification at firm stability can take place with little danger of destructive oscillation. This second heterodyning is accomplished with a second mixer and a second local oscillator whose difference in frequency is 455 kilohertz, both located ahead of the detector.

The output of the detector is usually insufficiently strong to activate the loudspeaker that is a built-in part of most marine receivers, and the audio amplifier (AF) becomes necessary. This is a straightforward amplifier (see Chapter 11) with no undue attention paid to high quality because of the limited range of the voice.

The squelch control is found on all marine radio receivers. It raises the minimum level of signal to which the audio amplifier will respond and thereby keeps the loudspeaker quiet for all input below this level. This control is a boon to the man on continuous radio watch; it relieves him of the annoyance of the myriad noises that the antenna picks up in the absence of a desired message.

The action taking place in an FM radio receiver is identical with what has been described—as far as the detector. The detector is different. As noted earlier, the difference between an AM receiver and an FM receiver is that the former is responsive to amplitude changes and the latter responds only to changes in frequency. In each case, the nature of the detector makes the difference. FM detectors are of various types, the result of constant research to make them completely insensitive to AM signals and noise, and thereby bring about the clarity and freedom from background noise that is a feature of FM reception. Incidentally, the detector is also known technically as a demodulator or a discriminator. A representative type is shown in Figure 16-6.

One of the recent developments in FM detection or demodulation is the "phase locked loop" (PLL) shown in block diagram form in Figure 16–7. The circuit constantly compares its frequency to the applied frequency and adjusts its voltage-controlled oscillator to stay in synchronism. The changes in voltage needed to do this are an exact image of the microphone input at the transmitter that caused the frequency variations in the received signal. The PLL voltage therefore reproduces the voice and the phase locked loop becomes an FM detector. It is a more efficient detector than its predecessors and requires no adjustments. The PLL has many other uses in modern electronics.

Early FM detectors, despite theory to the contrary, exhibited considerable unwanted response to amplitude modulation. The solution was to place limiting circuits ahead of the detector. These limiters lopped off the tops of amplitude modulated inputs and brought them all to a constant level before they reached the detector that recovered the FM modulated message.

Figure 16-6. This block diagram emphasizes the similarity between AM and FM receivers. The difference arises at the limiter stage and the following detector that discriminates against AM modulation.

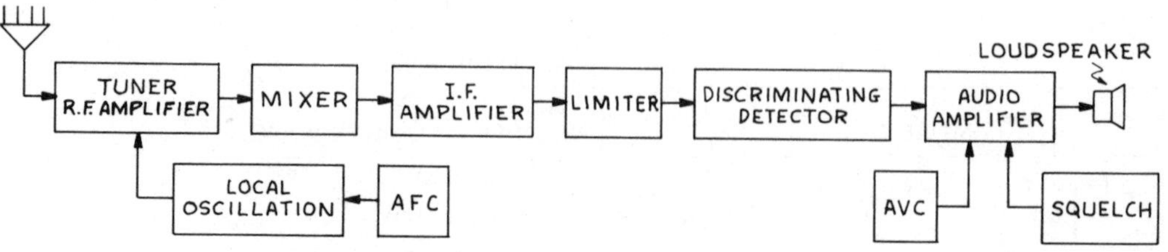

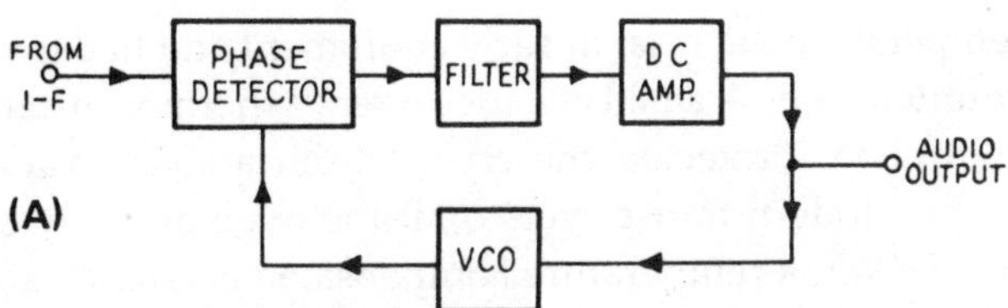

Figure 16-7. The phase locked loop is one of the new results of semiconductor wizardry. By following the deviations in frequency of an FM reception the phase locked loop recreates the original voice modulation. (VCO is a voltage-controlled oscillator.) (ARRL Radio Amateur's Handbook)

The waves reaching the receiving antenna are of constantly varying strength, because of the natural hindrance to propagation they must overcome in their travel. Without compensation this would cause a constantly varying output from the loudspeaker, certainly an annoyance to the listener. Modern radio receivers have automatic gain control (AGC) and automatic volume control (AVC) circuits to counteract this. These circuits control the amplification supplied by the radio frequency and audio frequency amplifiers in the changing amounts needed to maintain a constant output. In a well designed receiver the speaker volume will hardly change even though the signal at the antenna falls to one-thousandth of its original strength.

Another sophistication found in the modern marine radio receiver is automatic noise limiting (ANL). This circuit removes the electrical crackles, bangs, and pops that assail the antenna from natural causes such as distant lightning or static electricity from clouds and turbulent atmosphere. A refinement called noise blanking (NB) is often added—this is more effective on man-made electrical noise such as that from engine ignition. The blanking puts silence where the noise would be, but the interval is too short to be noticeable.

The amplitude modulated (AM) signal arrives with two sidebands and a central carrier, as already explained. This takes a much wider portion of the radio spectrum than is necessary to carry speech because the two sidebands contain identical information and the carrier bears none. In short, the message would be carried equally well if only one sideband were transmitted and the rest rejected at the transmitter. In fact, the signal would arrive many times stronger because now the total energy of the transmitter would be concentrated in the one sideband instead of being distributed over the combination. This is the genesis of single sideband (SSB) operation, now moving strongly into the marine radio communication field.

Figure 16–8 explains this graphically. At A the carrier and its two sidebands are shown; consider that the output energy of the transmitter is spread to cover the entire area. At B, one sideband and the carrier have been eliminated; now that same energy is concentrated into the small remaining area. The transmitter's effectiveness has been increased approximately eight times.

Three modes are legally permitted for SSB transmissions. One, called A3H, allows a full carrier and one sideband. Only a small amount of carrier plus one sideband qualifies as A3A. A single sideband with no carrier is A3J. The choice of modes is dictated by the ability of the distant marine receiver to decipher the message.

The standard AM receiver does not respond to SSB without the carrier, namely A3J. However, A3H transmissions can be received by AM as well as by SSB equipment and this mode of operation is mandatory on the 2182 kHz distress channel. The purpose of the A3A mode is to enable automatic distress receivers,

Figure 16-8. The energy in an amplitude modulated (AM) double sideband transmission covers the entire area shown in (a). With single sideband suppressed carrier transmission, this same energy is concentrated in the area of one sideband, as shown at (b).

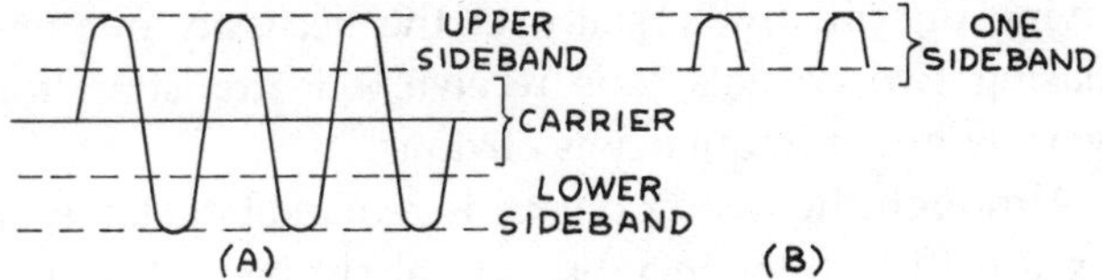

such as those in use by the Coast Guard, to track SSB transmissions.

The normal, double sideband, and carrier output of a transmitter may be changed to single sideband by either of two methods—a filter or a balanced modulator. The filter is sharply tuned to lop off either upper or lower sideband and the carrier; the sharpness is usually attained either with mechanical filters or by the use of quartz crystals. (See Chapter 10.) The balanced modulator accepts the radio frequency carrier and the audio frequency message, but balances out the unwanted portion so that only the upper or lower sideband appears at its output. Block diagrams of the two methods are shown in Figure 16–9.

The radio transceivers available to the boatman in the commercial market offer a choice of frequencies, or channels, at which to communicate with others. The number of channels is a function of price. Low-cost VHF-FM units may have as few as five channels while the deluxe sets will provide all the American and international frequencies. By FCC edict, all VHF-FM radios must include safety channel 6 and channel 16 in their lineup because these are the distress, calling, and intership frequencies. By the same ruling and for the same reason, all transceivers operating in the 2 to 23 mHz spectrum must be operable on 2182 kHz in a mode that is receivable on standard AM receivers.

The manufacturer of the equipment has two options in the manner in which he provides multiple frequency choice. He can include a series of quartz crystals to stabilize the various frequencies to the required precision. Or else he can resort to the synthesis of frequencies described in Chapter 7. In the frequency synthesis system one master crystal is the source of precision, the primary standard, for all outputs. The economy of having one inbuilt crystal over the necessity of purchasing two crystals (one receive, one transmit) for every working frequency is obvious.

Although the quartz crystal is remarkably stable, a slight drift in frequency brought about by changes in temperature did exist in early equipment and had to be counteracted. A small electric "oven," thermostatically controlled, protected the crystal from ambient variations. Modern transceivers omit the oven and rely on the TCXO, a *t*emperature *c*ompensated crystal (X signifies crystal) *o*scillator, to meet FCC standards. Channel selection is made with a dial in sets that have limited frequency choice and with a numbered keyboard in fully synthesized units. Representative units are shown in Figure 16–10. Incidentally, the stability of the well constructed TCXO is unbelievable—less than one part in one million variation!

The wide spread in operating frequencies of SSB and other transceivers that function in the 2 to 23 megahertz range brings some problems with antennas. An antenna is most efficient in radiating energy when it is excited at its resonant frequency. (See Chapter 24.) But an antenna, of itself, is resonant at only one frequency—and becomes greatly inefficient when

Figure 16-9. Single sideband suppressed carrier transmissions may be accomplished by the two systems shown in the block diagrams. At (a) the carrier and one sideband are filtered out before the transmission is presented to the antenna. At (b) the carrier and one sideband have been suppressed by shifting the phase at critical points in the circuit. (ARRL Radio Amateur's Handbook)

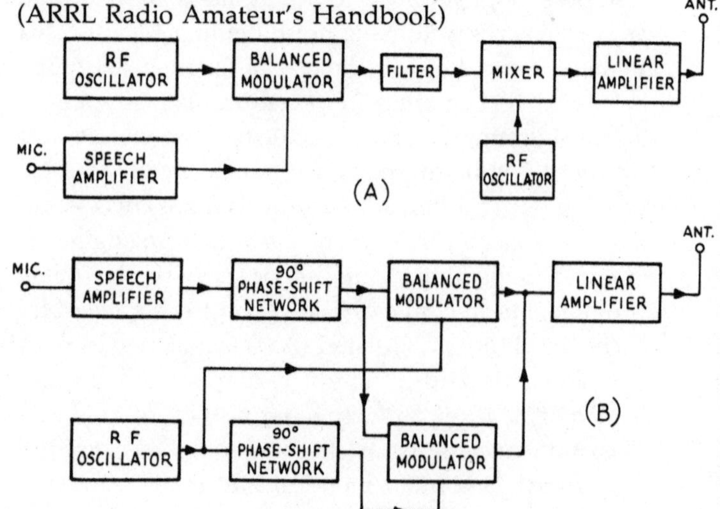

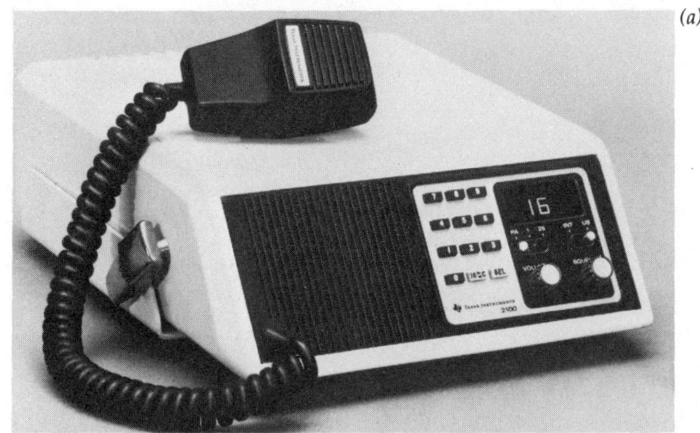

(a)

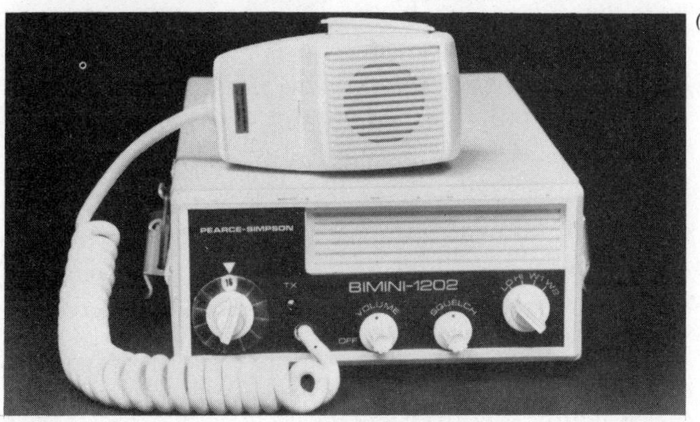

(b)

Figure 16-10. VHF radio transceivers available on the market provide station tuning by either of two ways. One is with a dial (a), and this usually is limited to no more than 25 channels. The other is with push buttons (like modern telephones), and (b) this system does not limit the number of channels that may be tuned in. (A: Pearce Simpson; B: Texas Instruments)

forced to work at higher and lower frequencies. Yet each marine band requires operation over a wide spread that goes well beyond antenna resonance. The answer lies in antenna tuning.

Oldtime equipment required the operator to do the tuning of the antenna to resonance at whatever frequency he needed to use. Present-day deluxe transceivers accomplish antenna tuning automatically. An antenna coupler situated close to the antenna senses the lack of resonance and immediately adds whatever inductance and capacitance is needed to bring things into balance. The addition of these characteristics makes the antenna act as though it were physically longer or shorter and permits it to perform at its most efficient resonant point. A weatherproof cabinet houses the antenna coupler as close to the antenna as possible. The automatic tuning feature of the antenna coupler also continuously corrects the small variations in transmitter to antenna matching caused by changing conditions around the antenna. A coupler is shown in Figure 16–11.

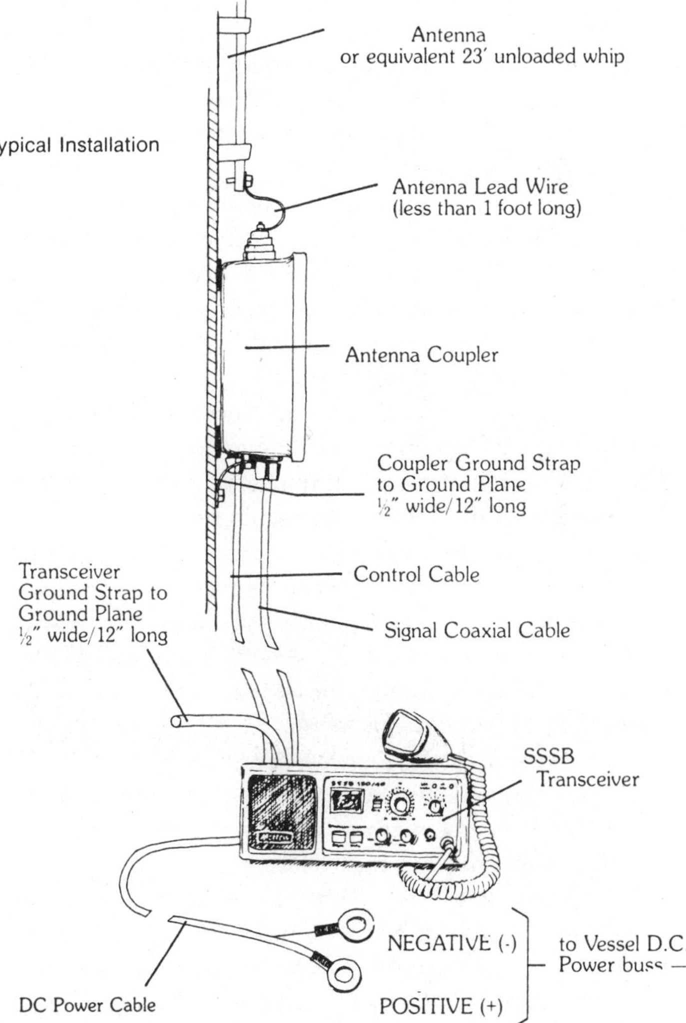

Figure 16-11. It is important that the antenna be tuned to resonance at the frequencies from 2 to 20 megaHertz used for single sideband transmissions. The antenna coupler does this automatically in conjunction with the frequency setting of the transceiver. The coupler must be installed as close to the antenna as possible.

The need for matching the antenna by tuning it arises not only from the desire to transfer the maximum amount of energy for best long distance communication. There also is the necessity of protecting the transmitter from damage than can be caused by a mismatch. When a mismatch exists, a portion of the energy sent to the antenna is reflected back unused to the transmitter and raises havoc in the output circuits. Some transceivers include internal circuits that protect them from this backward surge of power, but the waste and inefficiency remain.

The degree of success in matching transmitter to antenna is measured by the "voltage standing wave ratio," the VSWR, that is described fully in Chapter 24. A ratio of 1 to 1 is an ideally perfect match. Most good marine installations hover around 1 to 1.5 and even 1 to 2 is tolerable, but not desirable.

Luckily for skippers who do their communicating with VHF-FM, antenna matching is not a problem for them. The antennas sold for this service are resonant at the center of the 156 to 158 megahertz band, and the frequency difference from the center to both ends of the band is small enough to be negligible in its effect. As a comparison, the frequency spread on VHF-FM is approximately 1 percent of the carrier, whereas in 2 to 23 megahertz operation the spread on some bands could be 10 percent and more.

The comparatively low cost of Citizen's Band transceivers, coupled with the feeling that their style of boating does not warrant regular expensive marine radio equipment, has motivated many skippers to install CB on their boats. Whether or not this decision provides these boatmen with adequate safety "insurance" is a moot question. The citizen's band is not monitored constantly by the Coast Guard and by other vessels, as are Channel 16 and 2182 kilohertz. But countless CBers are listening at home and in their cars and this may provide a link to shore for the CB user in trouble afloat. The overriding difficulty is that the CB channels are jungles of discourtesy and filled with so much good buddy yak-yak that a call for help becomes a chancy thing. (See Figure 16–12.)

One manufacturer states in his literature that his marine radio is as excellent as the latest technology can make it, but that its successful operation depends upon the merit of its installation. It is a true statement. A transmitter may develop its full power, but it has no communication value until that power has reached the antenna and is radiated. There should be only negligible power loss along the entire path from transceiver to antenna.

The first consideration, of course, is to have the correct antenna for the frequencies at which the transceiver operates. The choice here is either a simple whip a full quarter wave long at the dominant frequency or else shorter whips that are "loaded." The loading consists of small inductors built into the antenna and placed either at its middle or at either end. The connection between transceiver and antenna usually is a coaxial transmission line. (See Chapter 24.)

Antenna location is next in importance. It should be as high as possible and with as much free space about it as possible. Since high radio frequency voltages may develop on antennas, it is wise to place them where normal human contact is not likely.

Figure 16-12. Citizen's Band (CB) transceivers are found on many pleasure boat consoles. However, CB is not a reliable marine communicator. See text.

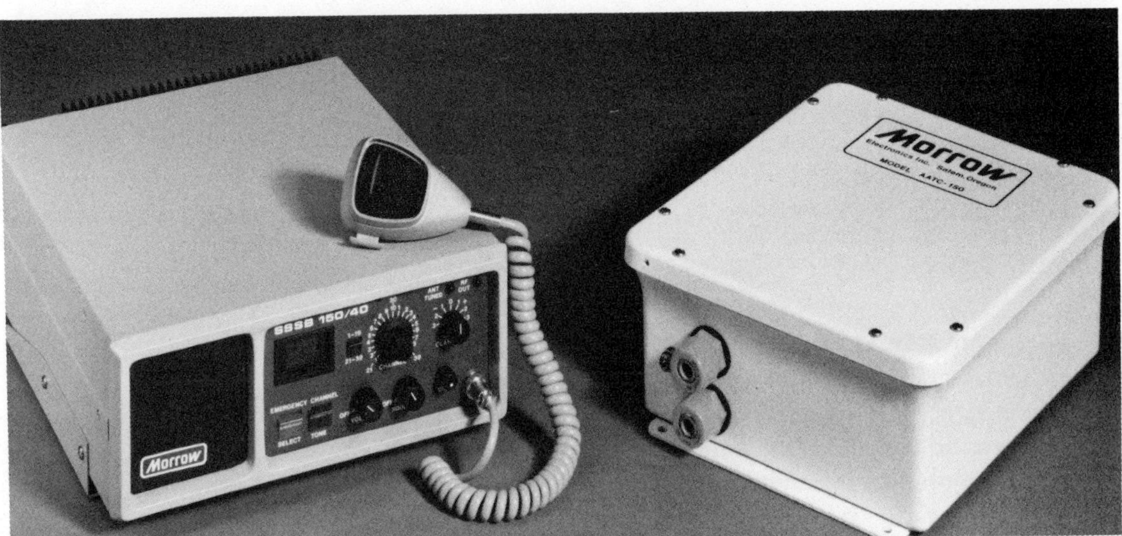

Figure 16-13. A representative SSB (single sideband) transceiver that operates over the medium frequency spectrum of 2 to 20 megaHertz. The antenna coupler also is shown. (Morrow Electronics, Inc.)

Proper grounding is as important as the antenna, and in some installations, even more so. The actual ground of the system often is not at the point where visual inspection would place it. In such cases the antenna would become only a part of an inefficient radiating system composed of everything from the false ground up, including the transceiver.

VHF-FM transceivers are delivered to the user ready to go on the air. Since no internal adjustments are required, it is legal for a skipper to install the equipment himself without the aid of a licensed technician. The instructions in the owner's manual are usually specific enough for a creditable do-it-yourself job. Common sense backed by the basic knowledge available in this book should be a sufficient guide.

The installation of an SSB transceiver does not permit this do-it-yourself freedom. A licensed technician must install and test the equipment and must leave evidence of his having done so by signing the log. Any subsequent service that could affect operation must again be attested to by his signature. Aside from the legal requirement, the complexity of multi-channel SSB units makes the employment of an expert equivalent to low cost equipment insurance. Locating the antenna coupler and grounding it properly are included in the work the technician does to install the SSB.

Installing a CB transceiver also is classed as a do-it-yourself-permissible project because the unit is factory tuned and ready for use. Again, the owner's manual is the source of the needed directions. Most citizen's band mobile antennas are designed for use on automobiles where the metal car body acts as the ground plane; these are not optimum for installation on a boat, although they will function there at lowered efficiency. (A ground plane or counterpoise could be added if space and location are available on the vessel. See Chapter 24.) Antennas correctly designed for marine use are preferable, of course.

Adequate current supply, meaning circuits that are able to supply the full specified amperage, is a requirement that a good installation must meet. The overall efficiency of radio transmitters is very low; this may be used as a guide in determining current supply. For example, a representative VHF-FM unit draws approximately 8 amperes at 12 volts (96 watts) when putting 25 watts into the antenna; similarly, an SSB transmitter delivering 150 watts to the antenna is consuming approximately 50 amperes at 12 volts (600 watts). The copper wire table in Chapter 27 translates amperes into wire sizes.

Single sideband transceivers provide practically world-wide radio communication. The many frequencies available make it possible to overcome most atmospheric deterrents such as static and other electrical noises. The SSB set is more complicated, radiates more power, and requires slightly more expertise on the part of the operator than the much simpler VHF units. (See Figure 16–13.)

Since VHF transmissions function as line-of-sight, the operating range at these very high frequencies is determined largely by the distances to the radio horizons. These distances are a function of the height of the transmitting antenna, the height of the receiving antenna and the curvature of the earth. A nomograph that makes it possible to quickly determine the radio horizon for any set of circumstances is shown in Figure 16–14.

All VHF receivers have provision for listening to the official weather reports broadcast continuously by NOAA. This up-to-the-minute weather information is localized and therefore of great help to skippers on whatever water they may be. The broadcasts are on three different frequencies. The complete list is in Figure 16–15.

Radio transmissions may legally be made only by licensed equipment operated under the supervision of licensed individuals. The equipment or station license is issued by the Federal Communications Commission on the basis of a properly completed FCC Form 506A. Provision is made for a temporary license that permits immediate operation while waiting for the regular license. This is done under the aegis of Form 506A. A state-licensed boat takes the letter K and follows it with the first two letters and the next four numerals of its state license. A documented vessel takes the letters KD and adds the first four numerals of its official number. These temporary self-made call signs are good for up to sixty days and the official license should arrive within that time. The skipper, or anyone, may apply for a Restricted Radiotelephone Operator Permit by completing FCC Form 753; no examination is required and the permit is valid for life. (Most manufacturers supply Form 506A and Form 753 with the transceiver.)

Even a short stint of monitoring any radio channel used by pleasure boats will quickly convince the listener that most skippers follow a sloppy procedure that reduces clarity. Yet the proper manner of handling radio communications is learned easily and pays con-

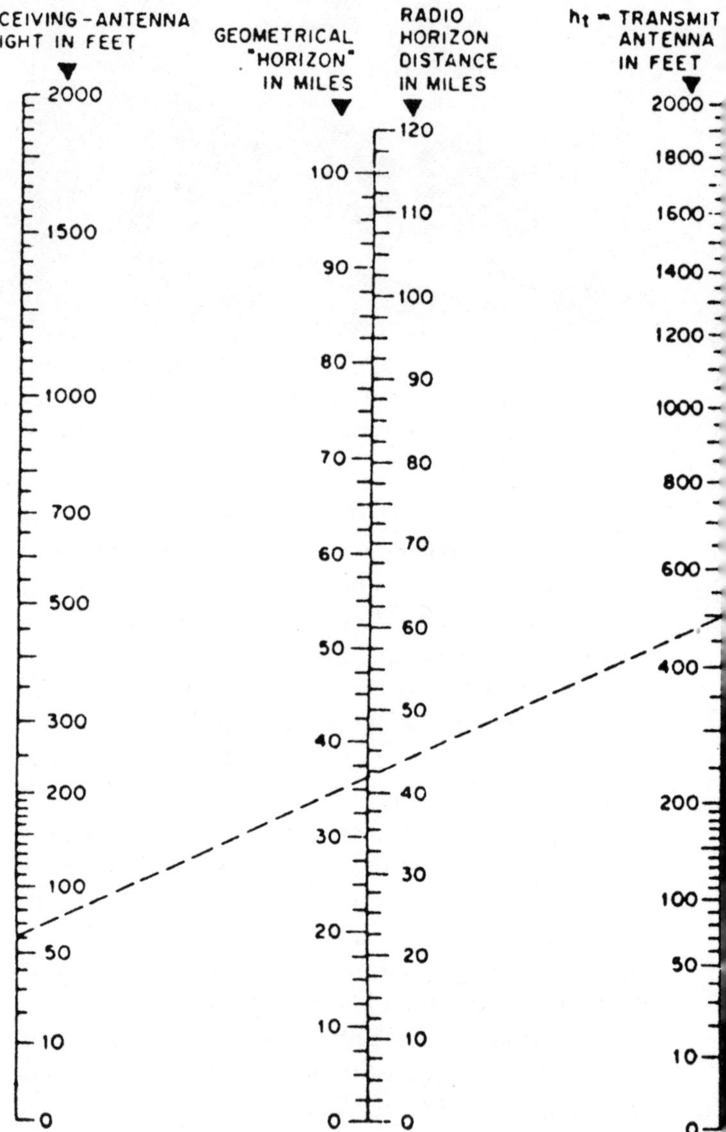

Nomogram giving radio-horizon distance in miles when h and h_t are known.
Example shown: Height of receiving antenna 60 feet; heigh of transmitting antenna 500 feet; maximum radiopath lengt =41.5 miles.

Figure 16-14. The very high frequencies used by VHF transceivers function as line-of-sight and communication range often is governed by the distance to the radio horizon. The nomograph above allows this distance to be found easily. (SBE Inc.)

Location	Frequency
ALABAMA	
Anniston	3
Birmingham	1
Dozier	1
Florence	3
Huntsville	2
•Louisville	3
Mobile	1
Montgomery	2
Tuscaloosa	2
ALASKA	
Anchorage	1
Cordova	1
Fairbanks	1
Homer	2
Juneau	1
Ketchikan	1
Kodiak	1
•Nome	1
•Petersburg	1
Seward	1
Sitka	1
Valdez	1
Wrangell	2
•Yakutat	—
ARIZONA	
Phoenix	1
ARKANSAS	
•Alco	—
Fayetteville	3
Fort Smith	2
Gurdon	3
Jonesboro	1
Little Rock	1
Star City	2
Texarkana	1
CALIFORNIA	
Coachella	2
Crescent City, CA/ Brookings, OR	1
Eureka	2
Fresno	2
Los Angeles	1
Monterey	2
Point Arena	2
Sacramento	2
San Diego	2
San Francisco	1
San Luis Obispo	1
Santa Barbara	2
COLORADO	
Denver	1
CONNECTICUT	
Hartford	3
Meriden	2
New London	1
DELAWARE	
Lewes	1
DISTRICT OF COLUMBIA	
Washington, D.C.	1
FLORIDA	
Daytona Beach	2
•Fort Meyers	—
Gainesville	3
Jacksonville	1
Key West	2
Melbourne	1
Miami	1
Orlando	3
Panama City	1
Pensacola	2
Tallahassee	2
Tampa	1
West Palm Beach	2
GEORGIA	
•Albany	1
•Athens	2
Atlanta	1
Augusta	1
•Chatsworth	2
•Columbus	2
•Macon	—
Savannah	2
•Waycross	—
HAWAII	
Hilo	1
Honolulu	1
Kokee	2
Mt. Haleakala	2
IDAHO	
•Boise	—
ILLINOIS	
•Champaign	1
Chicago	1
Moline	1
•Mt. Vernon	—
Peoria	3
•Quincy	—
Rockford	3
Springfield	2
INDIANA	
Evansville	1
Fort Wayne	1
Indianapolis	1
•Lafayette	—
South Bend	2
•Terre Haute	—
IOWA	
Des Moines	1
•Dubuque	—
•Waterloo	—
KANSAS	
Wichita	1
KENTUCKY	
Ashland	1
Bowling Green	2
Covington	1
Hazard	3
Lexington	2
Louisville	3
Mayfield	3
Somerset	1
Elizabethtown (Translator)	2
LOUISIANA	
•Alexandria	—
Baton Rouge	2
•Burras	—
•Lafayette	—
Lake Charles	1
Morgan City	3
New Orleans	1
Monroe	1
Shreveport	2
MAINE	
Ellsworth	2
Portland	1
MARYLAND	
Baltimore	2
Salisbury	2
MASSACHUSETTS	
Boston	2
Hyannis	1
MICHIGAN	
Alpena	1
Detroit	1
Flint	2
Grand Rapids	1
Marquette	1
Sault Sainte Marie	1
Traverse City	2
MINNESOTA	
Duluth	1
International Falls	1
Mankato	2
Minneapolis	1
Rochester	3
Thief River Falls	1
Willmar	2
MISSISSIPPI	
Ackerman	3
Booneville	1
Bude	1
Gulfport	2
Inverness	1
Jackson	2
McHenry	3
Meridian	1
Oxford	2
MISSOURI	
Camdenton	1
•Columbia	3
•Joplin	1
Kansas City	1
St. Joseph	2
St. Louis	1
Springfield	2
MONTANA	
Great Falls	1
Helena	—
NEBRASKA	
Omaha	2
NEVADA	
Reno	1
NEW HAMPSHIRE	
Concord	3
NEW JERSEY	
Atlantic City	2
NEW MEXICO	
Albuquerque	2
Clovis	3
Farmington	3
Hobbs	2
Ruidoso	1
Santa Fe	1
NEW YORK	
Buffalo	1
New York City	1
Rochester	2
NORTH CAROLINA	
Cape Hatteras	1
New Bern	2
Wilmington	1
NORTH DAKOTA	
Fargo	3
OHIO	
Akron	2
Caldwell	3
Cleveland	1
Columbus	1
Dayton	3
Lima	2
Sandusky	2
•Toledo	—
OKLAHOMA	
•Clinton	—
•Enid	—
•Lawton	—
•McAlester	—
•Oklahoma City	—
Tulsa	1
OREGON	
Astoria	2
Coos Bay	2
Eugene	2
Newport	1
Portland	1
PENNSYLVANIA	
Erie	2
Philadelphia	3
Pittsburgh	1
PUERTO RICO	
San Juan	2
•Mayaguez	—
RHODE ISLAND	
Providence	2
SOUTH CAROLINA	
Beaufort	3
Charleston	1
Columbia	2
Florence	1
Greenville	1
Myrtle Beach	2
TENNESSEE	
Bristol	1
Chattanooga	1
Knoxville	3
Memphis	3
Nashville	1
•Camden	—
•Cookville	—
•Jackson	—
•Shelbyville	—
TEXAS	
Abilene	2
Amarillo	1
Austin	2
•Beaumont	3
Big Spring	3
Brownsville	1
Bryan	1
Corpus Christl	1
Dallas	2
Del Rio	2
•El Paso	1
Fort Worth	1
Galveston	1
Houston	2
Laredo	3
•Lufkin	2
Lubbock	2
Midland	2
•Paris	1
Pharr	2
•San Angelo	1
San Antonio	1
•Sherman	3
Tyler	3
Victoria	2
Waco	3
Wichita Falls	3
UTAH	
Salt Lake City	1
VERMONT	
Burlington	2
VIRGINIA	
Norfolk	1
Richmond	3
WASHINGTON	
Neah Bay	1
Seattle	1
Yakima	1
WEST VIRGINIA	
Charleston	2
Clarksburg	1
WISCONSIN	
Eau Claire	2
Green Bay	1
La Crosse	1
Madison	1
Milwaukee	2
Wausau	3

Figure 16-15. All VHF receivers have provision for listening to the official weather broadcasts. All weather broadcasting stations in the USA are listed above. (Frequencies: #1 is 162.550 MHz, #2 is 162.400 MHz, #3 is 162.475 MHz.)

tinuing dividends in orderly, precise, and effective contacts. The following example is the correct form and should be used without change:

"Calling the Yacht (name), calling the Yacht (name), calling the Yacht (name). This is the Yacht (your boat name), this is the Yacht (your boat name), this is the Yacht (your boat name), Your radio call sign, Over."

The call is initiated on Channel 16 with VHF-FM and on 2182 kHz with SSB after a preliminary listening shows the channel to be clear. Listening after the calling sequence determines whether the called station is answering. If an answer is received, return to the same channel only to state the new channel on which the contact is to be continued. If no answer is received in about 30 seconds, the call is repeated exactly, and listening for a reply is again done for 30 seconds. One more repetition of the call may be made if there be no answer. A mandatory wait of 2 minutes is required before the calling sequence is repeated.

For true distress that endangers life and vessel and requires help, the words "Mayday, Mayday, Mayday" precede the calling sequence and only your own boat name is given. Remember, however, that in distress anything goes; the over-riding requirement of the transmission is to get attention that can bring help. The answering Coast Guard will want information relative to your position, identifying appearance of the boat, number of people aboard, nature of the distress—and these facts should be on the tip of the tongue.

17. *Depth Sounders and Their Use*

When you are cruising in comparatively shallow bodies of water, such as the Intracoastal Waterway for instance, the depth sounder becomes the most valuable instrument on the console. And when fishing is the order of the day, the depth sounder again takes on prime importance. This dual dependence on the depth sounder is based on the ability of an invisible inaudible sonic beam to measure the distance to the bottom and also to spot fish.

Running aground is the great bugaboo of boating in shallow, unfamiliar waters. It is feared because its consequences, although happily often only a nuisance, could build into a tragedy. The need for depth information is all important. The oldtime leadsman at the bow heaved his lead at rythmic intervals and sang out his findings. The depth sounder goes him one better by giving a continuous reading and even discovering underwater mounds that he may have missed. If a graphic depth sounder be chosen, it will draw a diagram of the bottom contour and as a bonus add a few pips to show the location of fish.

As with almost all marine electronics, the modern depth sounder has banished the vacuum tube and relies entirely on solid state devices such as the transistor, the diode, and the integrated circuit. One result is a tremendous reduction in the electric current required for operation. The current drawn by these new depth sounders is now so small that internal batteries can supply it when the usual boat storage battery source is not available.

The skipper is offered a wide choice of depth sounders. He may choose instruments that give their readings digitally, or by flashes on a scale, with a meter pointer, or by lines on calibrated paper. The range may be short, for accurately indicating small changes in shallow waters; it may be an extra-wide range for ocean use that can probe down hundreds of feet. Several ranges, selectable by a switch, may be combined in one unit. Many instruments have alarm circuits that may be set for a critical depth and will alert the skipper when it is reached.

The digital readouts are based on the fact that all digits may be represented in block form by varying the positions of not more than seven short bars arranged in two vertically-connected squares. These bars are lighted when in use. Thus, with all seven bars alight, the digit eight is presented 8. Extinguishing the middle bar makes this a zero, while lighting only two side verticals results in a one. Similar choices build the rest of the ten digits. (See Figure 17–1)

The lighting of the bars is accomplished either with light emitting diodes, liquid crystal diodes, or incandescent filaments. All three forms are employed by one or another of the many instruments on the market. The light emitting diode (LED) is actually a pinpoint source of light in any of several colors. The liquid crystal diode (LCD) functions differently; it becomes opaque to transmitted light when it is energized and the dark areas form the digits. The incandescent unit contains seven individually lightable filaments within its glass bulb, arranged in the already-mentioned vertically-connected square formation.

Opinions differ as to which of these forms of digital presentation is best for marine use. It is claimed that

Figure 17-1. A digital depth sounder proclaims the depth in a manner that cannot be misread. The unit shown is available as a do-it-yourself assembly kit. (Heathkit)

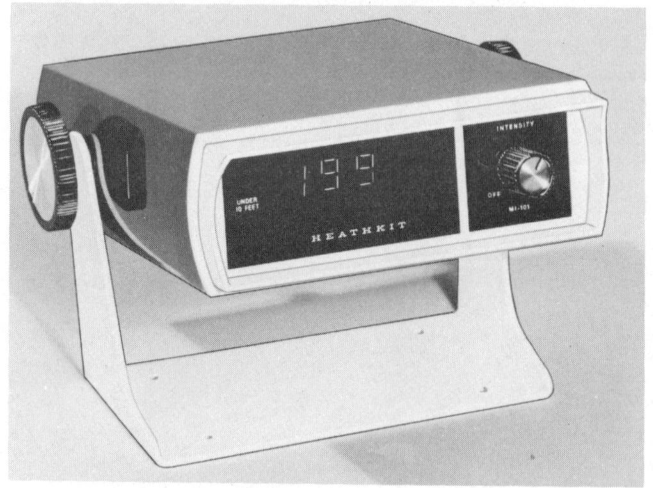

the LEDs are not sufficiently bright for easy reading in sunlight. The LCDs are susceptible to blackout from the high temperatures that may befall them when exposed to the sun on the flying bridge. Even the incandescents have their boosters and their decriers on the basis of readability. Yet reputable manufacturers incorporate all three digital readouts in their equipment, and there is good likelihood of reasonable satisfaction with any one.

The flasher type of depth sounder is perhaps the most common. A miniature neon tube circles rapidly behind a glass panel imprinted with a scale. A flash opposite a scale division indicates the depth and smaller flashes show intervening fish. Theoretically, the nature of the flash reveals the kind of bottom, whether soft, hard, or rocky, but it takes experience and imagination to eke out this detailed information. For most skippers, a flash on the scale at a figure greater than the vessel's draft is reassuring and sufficient. (See Figure 17–2.)

The deluxe depth sounder is the one that draws a graph of the bottom contour on a moving roll of paper and for good measure adds indications of vegetation

Figure 17-2. A rotating neon bulb indicates the depth on a flasher-type depth sounder. (Morrow Electronics, Inc.)

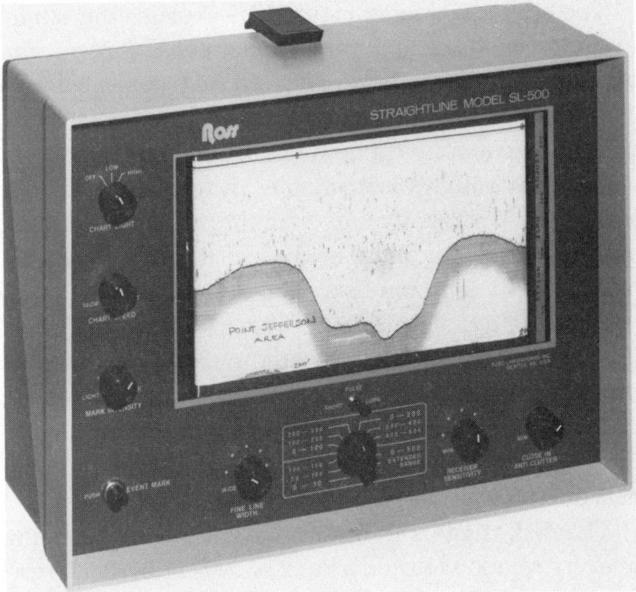

Figure 17-3. The recording depth sounder draws a "picture" of the bottom that the expert fisherman can interpret to his advantage. (Ross Laboratories, Inc.)

and fish. Fishermen, from amateur to professional, favor this and consider it an important tool for effective fishing. Some units combine the flasher and the graphic. A more sophisticated and more expensive model substitutes a cathode ray screen for the paper roll and produces a radar-like representation of the bottom. Figure 17–3 shows a recording depth sounder that uses a paper roll.

The operation of a depth sounder may be understood easily by remembering childhood shouts at a distant palisade and waiting for echoes that came back seconds later. The depth sounder "shouts" down to the bottom through the water, measures the time it takes the echo to return, and translates this into feet or fathoms. The scheme works because the speed of sound through water is known; for purposes of instrumentation it is considered to be 4800 feet per second, although in actuality it is slightly higher and varies with salinity and temperature.

The block diagram in Figure 17–4 explains the action in greater detail. The depth sounder consists of a combined transmitter and receiver and may be considered a transceiver. The transmitter portion includes an oscillator and an amplifier tuned to the oscillator frequency which, in modern instruments, hovers around 200 kilohertz. A similarly tuned amplifier plus a detector and a voltage amplifier make up the receiver. The transducer is common to transmitter and receiver and functions alternately in both circuits. A rotating disk (in the flasher type) triggers the operations and displays the resulting depth reading. An electronic clock (see Chapter 9) does the triggering and measuring in digital depth sounders, while a closely timed rotating belt does the job in graphic recorders.

The disk triggers the transmitter at one fixed point in every rotation to send a burst of 200 kilohertz electric energy to the transducer. The transducer transforms this into ultrasonic sound and directs it at the sea bottom. The returning echo impinges on the transducer, which now turns it back into electric energy that is fed to the receiver. The high voltage output of the receiver causes a miniature neon tube on the rotating disk to flash at a point on the calibration equivalent to the water depth. A previous flash occurred at the time of the original triggering to act as a zero mark. These flashes recur so rapidly, once every revolution, that the eye sees them as continuously illuminated lines on the scale.

The rotating disk is the actual measurer of the time it takes the burst of sound to hit the bottom and return as an echo; it translates this time into feet or fathoms on the scale. Obviously, the accuracy of disk speed control determines the accuracy of the whole instrument. Electronic circuits maintain disk speed within very close limits.

The chosen disk speed sets the limit of depth measurement for one revolution. A numerical example will make this clear. Assume that the dial is to be calibrated at 100 feet for the full circle. To measure a depth of 100 feet, the sound must go down 100 feet and back up 100 feet, or a total travel of 200 feet. It requires 1/24th of 1 second for sound to traverse 200 feet at its velocity through water of 4800 feet per second, and the disk must make its full revolution in that same time. This is equivalent to 1440 revolutions per minute—and that is the exact speed at which the disk in a 100 foot depth sounder rotates.

The transducer might be called a specialized loudspeaker designed to feed its energy into water instead of into air. Early transducers made use of magnetostriction, magnetic induction, and piezoelectric effects. Present units are almost exclusively of the piezoelectric type, the so-called ceramic transducers. The heart of these is a crystal that vibrates at the frequency of the energizing voltage and imparts these vibrations to the water through its case. Commercially available cases are bronze or plastic.

The sound beam from a transducer is similar in shape to the light beam from a searchlight. The market offers a choice between narrow beam and wide beam units and specification sheets usually state the angle of the cone of generated sound waves. There is a difference in bottom coverage between wide and narrow beams. A wide beam spreads the output sound energy over a greater area, thereby lessening the intensity available for echo at any point.

The transducer and its cable form part of the capacitance and inductance needed to tune the output of the transmitter to resonance. For this reason the cable should not be shortened; excess cable length is merely coiled and retained. Add-on lengths of cable are offered by most manufacturers for installations where transducer and console are too far apart for standard cables. The owner's manuals explain the slight internal alterations that must be made in the depth sounder when these add-ons are used.

The triggering by which the rotating disk initiates the sound burst is accomplished by several methods. In one, a small permanent magnet attached to the disk

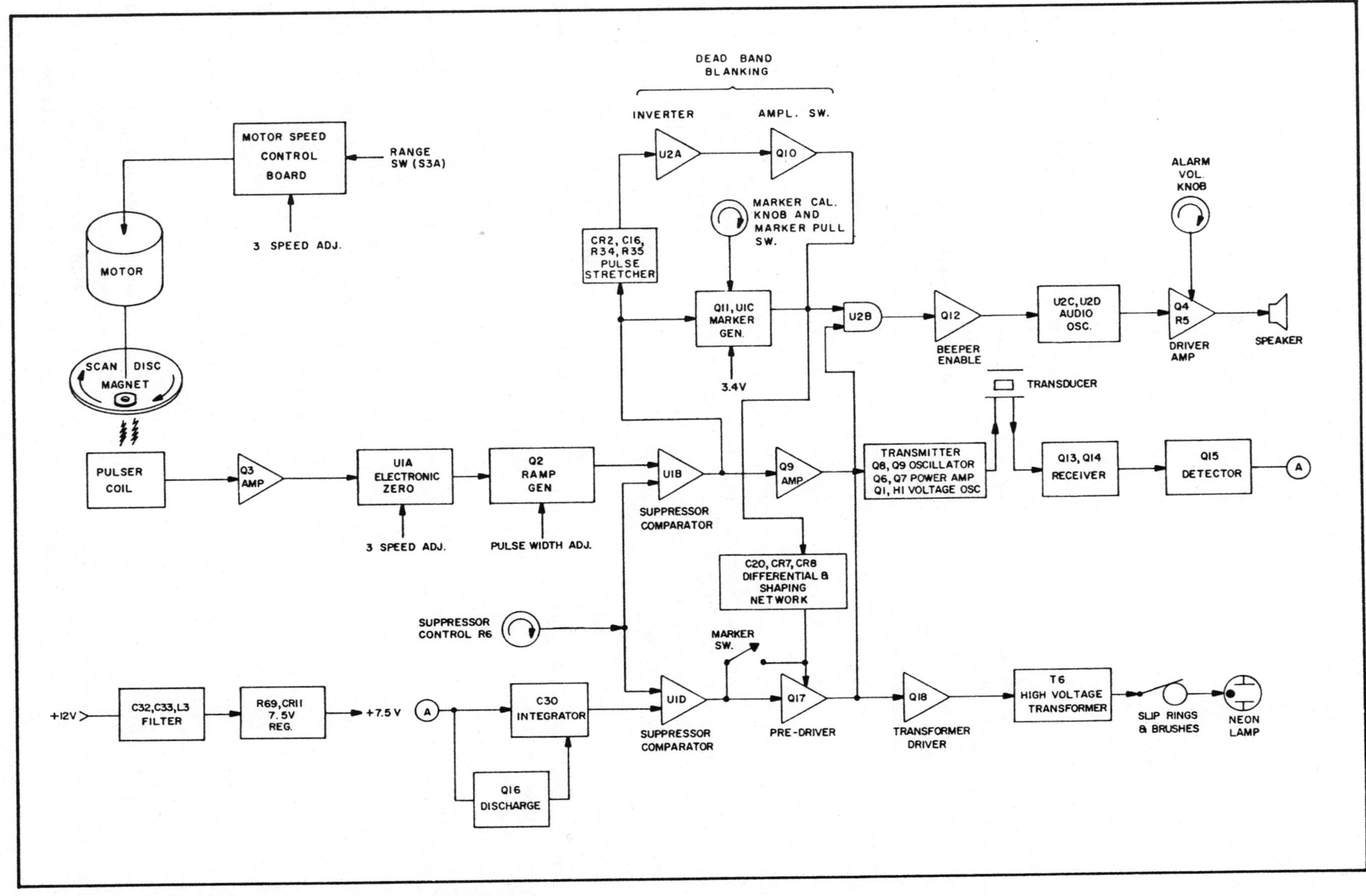

Figure 17-4. The block diagram above shows the sequence of events from the generation of the transmitted pulse to the reception of the echo. This circuit also includes a depth alarm. (Lowrance Electronics, Inc.)

swings closely past a miniature coil of wire. The pulse of voltage this generates is sufficient to set the transmitter in operation. Another scheme interrupts the passage of light from a light emitting diode to a photocell at the instant when triggering is desired. Some older ideas included make and break contacts. The drawing in Figure 17–5 details the trigger in the disk.

The power generated by the transmitter and passed on to the transducer varies in commercial depth sounders almost in direct ratio to price. A low-priced unit may be rated at as little as 20 watts; a top-of-the-line professional instrument may put out as much as 1000 watts. Quite naturally too, the deeper the sounding to be made, the greater the requirement for power because of the losses incurred by the sound waves on the way down and then back up as an echo.

Modern depth sounders are tuned to frequencies close around the 200 kilohertz mark. This is a move up from earlier models and gains some advantages. But there also are disadvantages—although they are largely overcome by special circuits. One disadvantage stems from the fact that many disturbances in the electrical and ignition systems of boats seem to peak at this same 200 kilohertz and consequently work their way into the depth sounder circuitry as confusing "noise." The voltage "spikes" in a tachometer system are prime offenders. The existence of electrical noises leads to the caution that transducer cables should be routed separately, never in conjunction with other wiring.

Figure 17-5. The transmitted pulse for the depth sounder is triggered when the magnet on the rotating disc passes the pulser coil and generates current in it. (Lowrance Electronics, Inc.)

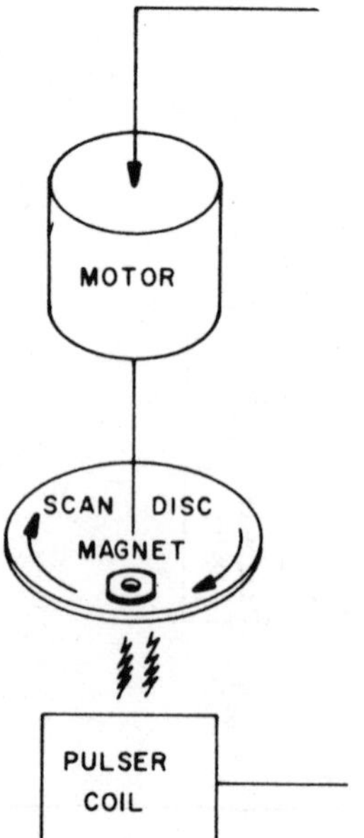

Electrical noise is not the only problem in many installations; acoustic noise may prove a bother too. Under this heading come the vibrations the transducer picks up from damaged or out of balance propellers, from unwanted reflections off deep keels, and especially from cavitation on high speed hulls.

Cavitation results in a film of air bubbles on the face of the transducer that acts as an "acoustic mirror" and blocks sound passage. The "mirror" effect derives from the great disparity between the 1100 feet per second velocity of sound in air and the much higher 4800 feet per second velocity of sound in water. A slightly similar effect occurs with light when a pencil stuck halfway into a glass of water appears to be bent.

Air bubbles are seldom a problem with low speed displacement hulls, except perhaps directly after launching, and this may easily be cured by first wetting the face of the transducer with detergent. On planing hulls, the reason for air bubbles on the transducer may be its incorrect location, its proximity to a hull strake, or the lack of a proper fairing block to smooth the flow of water. The guidance for transducer placement is that it must at all times be in full contact with water; this may place it far aft on a high speed hull and sometimes even abaft the transom on a bracket. Protuberances that disturb water flow should not be directly ahead of the transducer. (See Figure 17–6.)

Fitting the transducer in a hole through the hull is the recommended installation and the one that gives

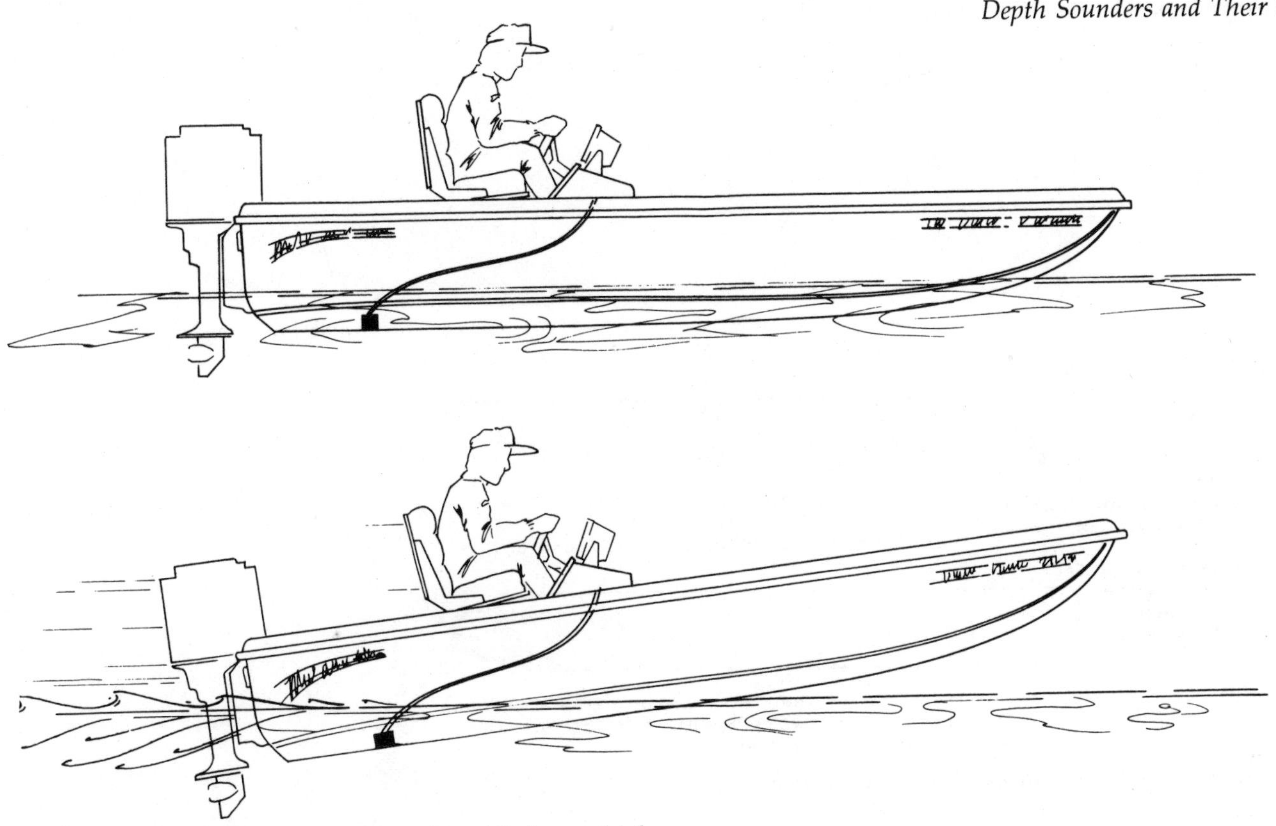

Figure 17-6. Transducer location may be critical on a high speed hull because it must maintain contact with solid water at all attitudes of the boat. (Lowrance Electronics Mfg. Corp.)

optimum results. Fitting the transducer onto the inside of the hull, without drilling through, also is possible, but the usual penalty is degraded performance. For temporary use while fishing, the transducer may simply be hung over the side.

The ideal posture of the transducer is vertical with its face horizontal so that the sound burst and its echo travel vertically down and up. A wide beam sound cone permits some slight variation from this without danger of losing the echo. Manufacturers indicate this tolerance by stating that the transducer may be mounted on hulls with small deadrise without correcting the resultant small deviation from the vertical. Where necessary, the transducer is restored to a vertical position with angle blocks such as the ones shown in Figure 17–7.

One further caution, in addition to those already mentioned, should be observed when selecting the site for the transducer. The location must be far enough athwartship off the center line to prevent the sound beam from hitting the keel. Since manufacturer's specifications list the beam angle of the transducer, it should not be difficult to comply with this. Reflections

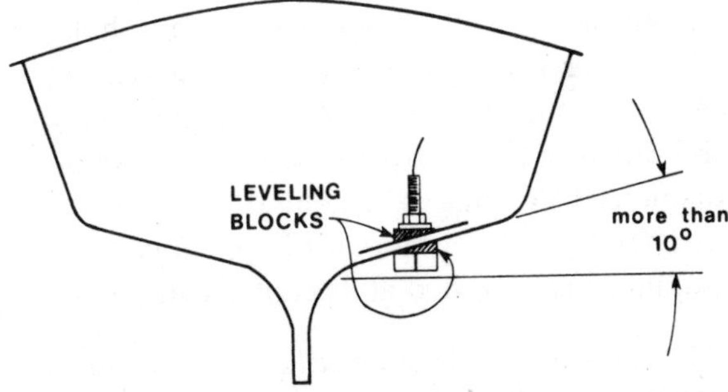

Figure 17-7. The deadrise of the hull determines whether or not leveling blocks must be used to install the transducer. A small angle off vertical for the transducer is permissible but not recommended.

from the keel produce confusing readouts. A long fairing block, like the one shown in Figure 17–8, is a worthwhile addition to assure smooth water passage over the transducer.

Bedding compound is a basic ingredient in a watertight through-hull transducer installation. It is applied liberally before the clamp nut is tightened; the squeezed-out excess caused by nut tightening then is wiped away. The nut is turned home with care when

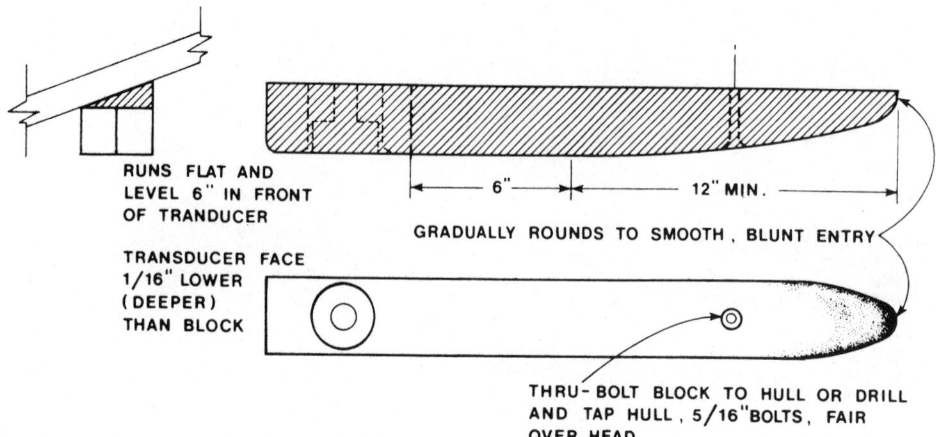

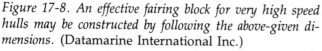

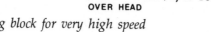

Figure 17-8. An effective fairing block for very high speed hulls may be constructed by following the above-given dimensions. (Datamarine International Inc.)

wood angle blocks are used or on a wooden hull because of the possibility of swelling when immersed.

A water box that permits the transducer to be mounted inside the hull without drilling any holes is shown in Figure 17–9. (Some manufacturers of depth sounders warn that a water box installation reduces the rated range of the instrument.) A short section of large diameter PVC pipe is cemented to the hull at a location chosen in accordance with previous text. A matching end cap is drilled at its center with a hole to take the transducer stem. The lower end of the pipe is cut to hull contour so that the transducer may remain vertical. The pipe is filled with water or with oil. The transducer is thus in contact with the sea water through the intervening hull. Only thin solid hulls will permit this manner of installation; hulls cored with cellular material will not work, nor will those with airspaces between double planking. It may become necessary to adjust the height of the transducer in the water box for best results.

Figure 17-9. It is possible to avoid drilling through the hull when installing a depth sounder transducer by using a water box instead. A short piece of large PVC pipe and a PVC cap for it are arranged as shown in the drawing. Most manufacturers do not approve this type of installation because it degrades performance. See text.

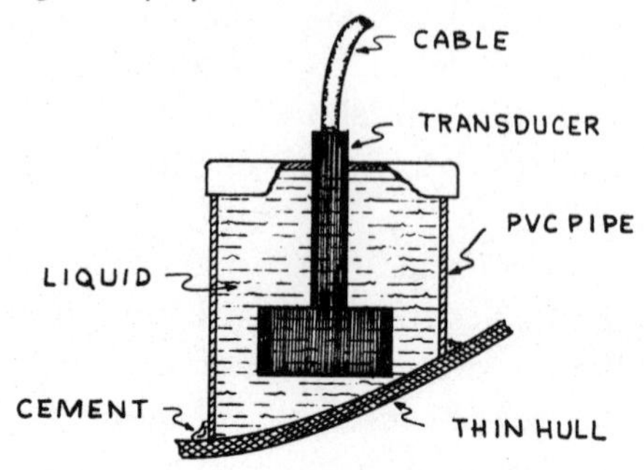

A variation of the water box technique employs a sea chest for the transducer. A short length of large diameter pipe is fitted into a hole in the hull and is capped to make it water-tight when the transducer is fastened within. The purpose here is to avoid any protuberance on the outside of a racing hull. Frankly, a flush-mounting transducer accomplishes this much more easily.

A sailboat heeled in a strong wind could swing the beam of its transducer so far to one side that the echo would be lost and no readout would be obtained. Two transducers, one on each side, are the cure for this condition. A gravity switch connects the low side transducer to the depth sounder. (A drawing of such an installation is shown in Figure 17–10.) As in all instances where the sonic beam does not travel verti-

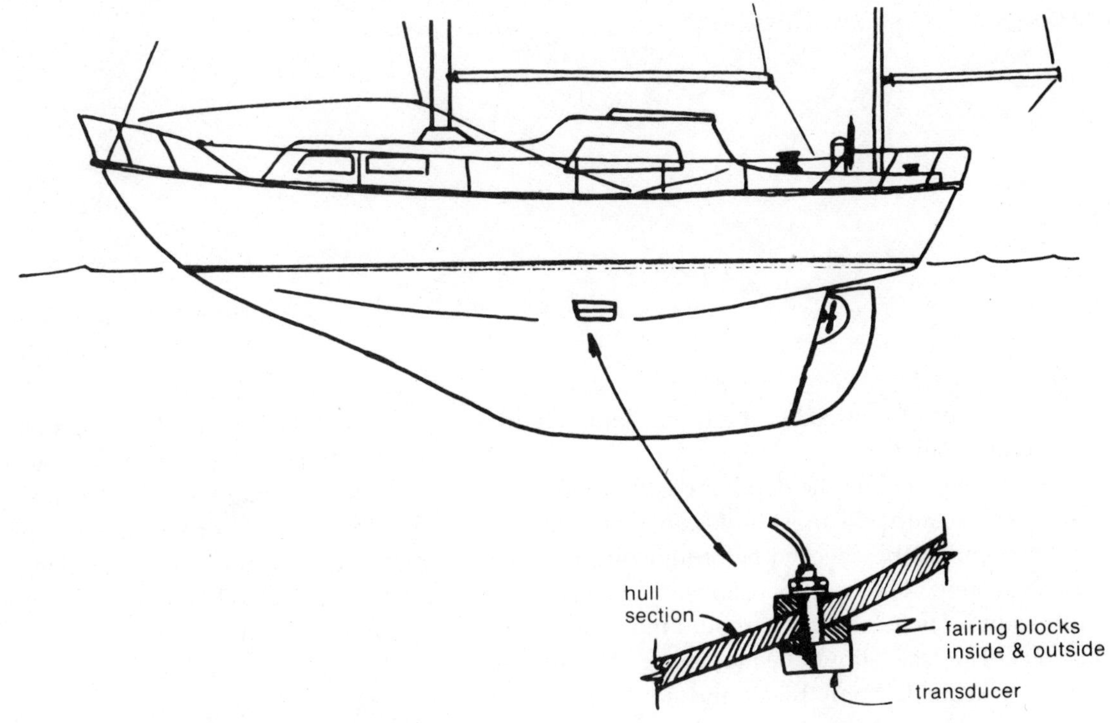

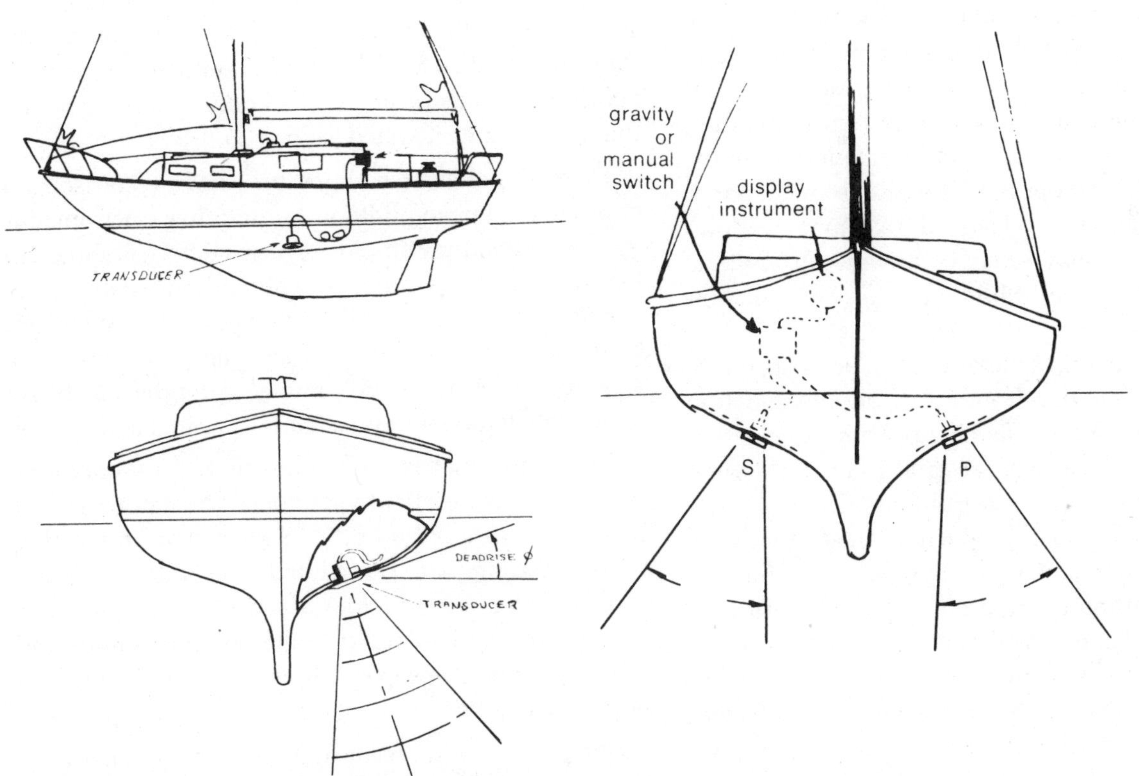

Figure 17-10. One way to overcome the problem of heeling of sailboats is to install two transducers as shown above. The transducers are switched to the indicating unit either manually or automatically as the heel changes. (Datamarine International Inc.)

cally, a small error is introduced into the depth reading because of geometry.

A word on installing the depth indicator on the console near the compass—most units can affect the compass; this should be checked before making the location permanent. The compass is observed closely and depth sounder is turned on, it should indicate a depth of 20 feet. The tests should be momentary because the transducer is loaded only by air and not by the dense water it was designed to drive.

The beauty of the digital indication of depth is that mistaken readings are almost impossible. But fish and underwater conditions are not shown, so it is the old story of having cake and eating it. The flasher depth sounder shows up just about everything under water that can return an echo—but considerable experience must precede the ability to diagnose conditions from the nature of the flashes. The graphic readout from the recording depth sounder is a bit easier to understand. The ground contour is clearly drawn and shadows emanating from it may safely be assumed to indicate vegetation or even trees. Short markings between bottom contours and surface are fish.

With the foregoing general explanations in mind, the reproductions of depth sounder readouts shown in Figure 17–11 will provide familiarity with actual situations. The series of flasher facsimiles illustrates what the skipper will see on his instrument during an average cruise. The graph reproductions have the same purpose and show similar conditions. By contrasting the two forms of information presentation, a prospective buyer may be able to decide what form of depth sounder fills his need better.

Multiple depth readouts are present in both flasher and graphic presentation and need explaining. A powerful sonic burst from the transducer will hit the bottom, bounce back as an echo, and still have enough energy left to repeat the round trip several times. Each repeating echo will leave a flash on the dial or a mark on the chart indicating a multiple of the depth away from the true depth. Thus, for a true depth of 20 feet, the repetitive echoes will appear at 40 feet, 60 feet, and 80 feet; this multiple relationship is the key to their identity. Multiple echoes may be eliminated by reducing the gain of the depth sounder until only the original remains, but this may also remove desirable weak indications.

It is possible (but not necessarily recommended) to test the functioning of a depth sounder without immersing the transducer in a body of water. The transducer is supported vertically with its face parallel to the floor and exactly 4 feet 7 inches above it. When the parallel to the floor and exactly 4 feet 7 inches above it. When the depth sounder is turned on, it should indicate a depth of 20 feet. The tests should be momentary because the transducer is loaded only by air and not by the dense water it was designed to drive.

Professional service shops may have recourse to a depth simulator in addition to the other electronic instrumentation they employ. (Such a device is shown in Figure 17–12.) The simulator contains adjustable delay circuits that slow down the passage of electric current to simulate the time taken by an echo to return from a given depth. A number of simulated depths may be set to test the depth sounder at various points in its range.

The "white line" is a refinement in the presentations of graphic recorder depth sounders; this improvement is aimed at fishermen and is a great help in locating bottom-feeding fish. The white line is a narrow blank space parallel to the bottom that accentuates the blips of fish formerly blended into the bottom contour and lost. The white line separation is produced by internal electronic circuitry.

The graph on the recording depth sounder is literally "burned" into the paper. The voltage applied to the stylus becomes high enough to char and mark the paper each time an echo is received. The resulting fine lines blend into a continuous graph as the paper slowly moves by.

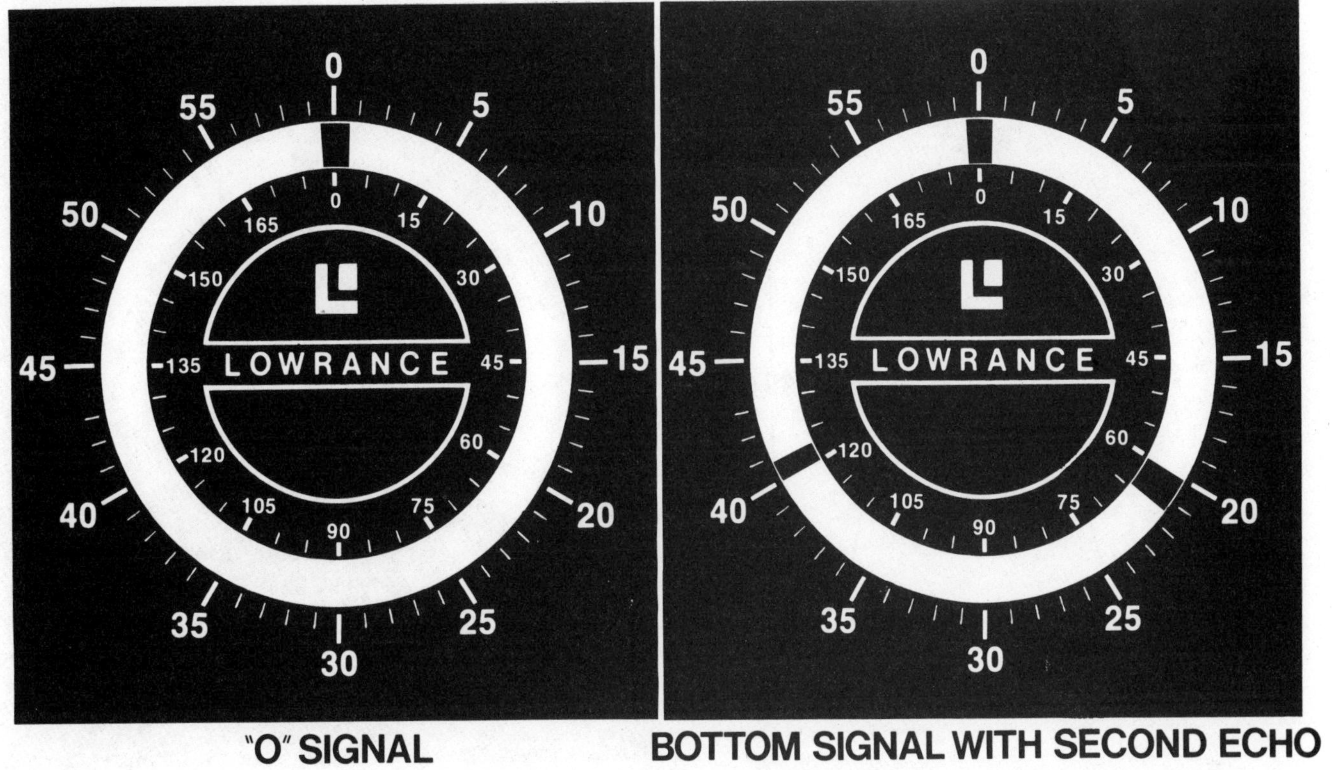

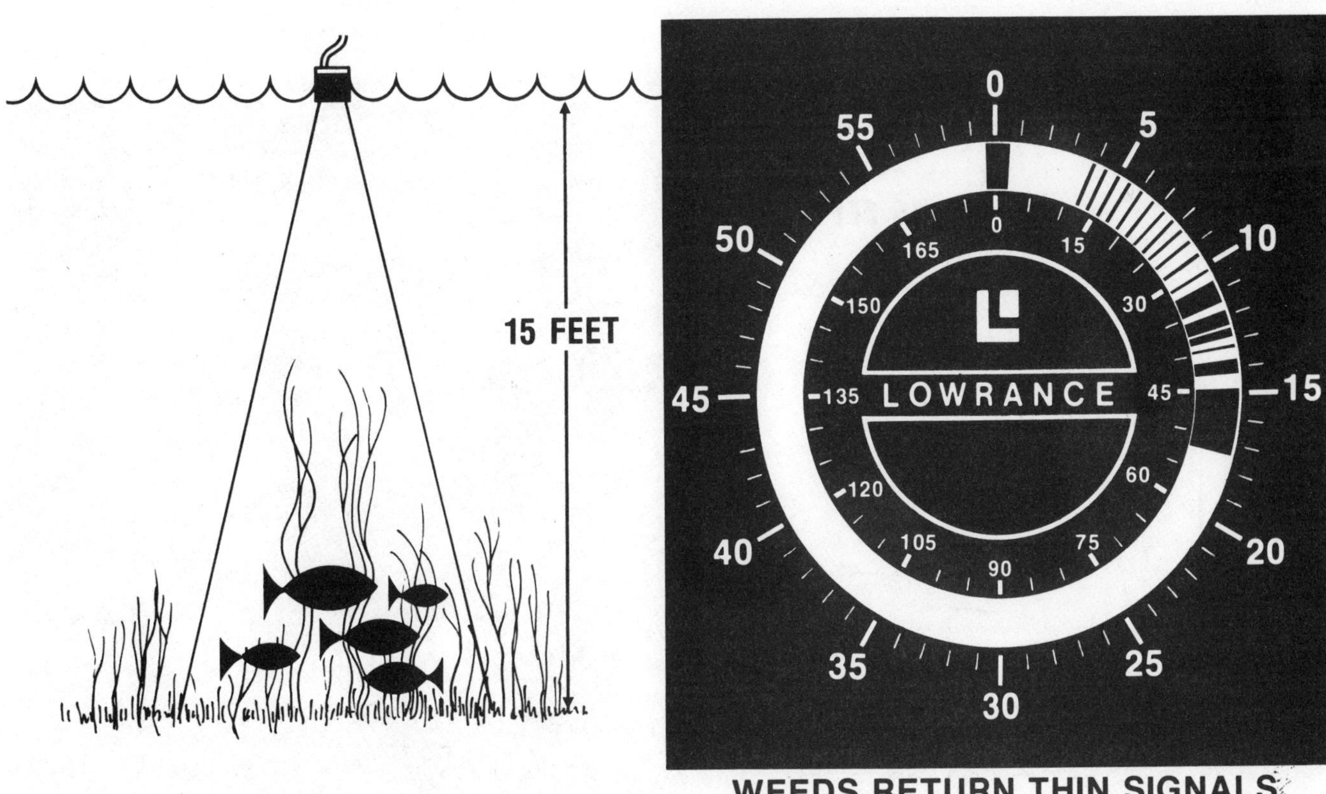

Figure 17-11. The various "readings" seen on the dial of a flasher type depth sounder are reproduced above with an explanation for each. (Lowrance Electronics Mfg. Corp.)

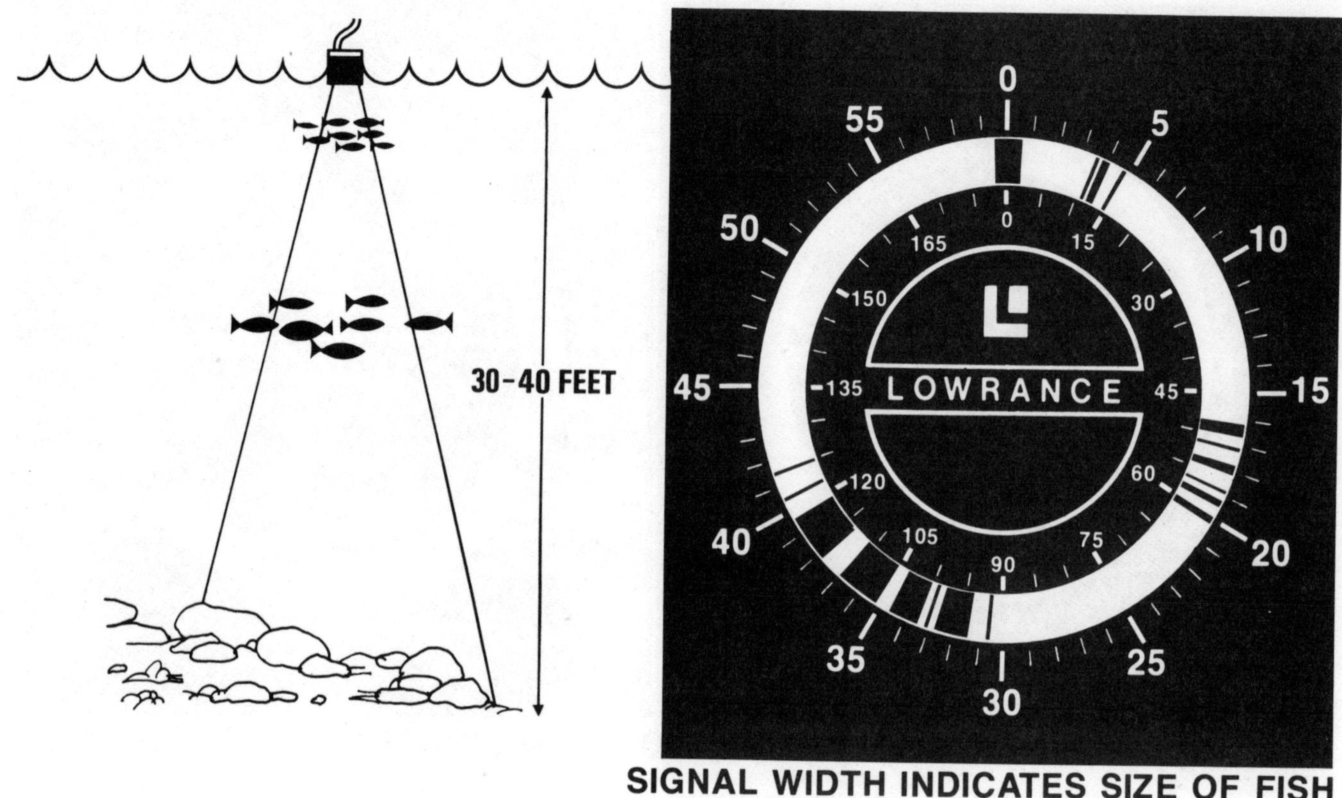

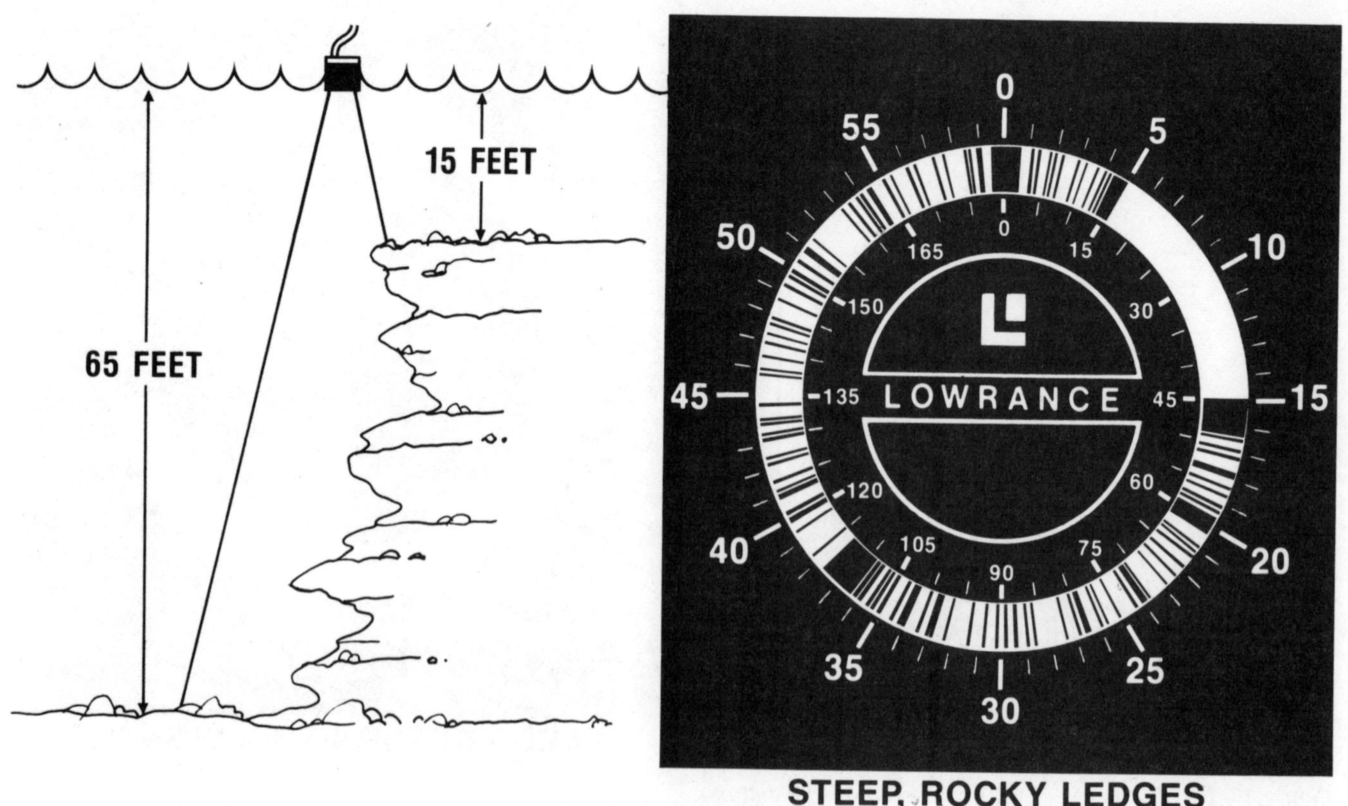

Figure 7–11 (cont'd.)

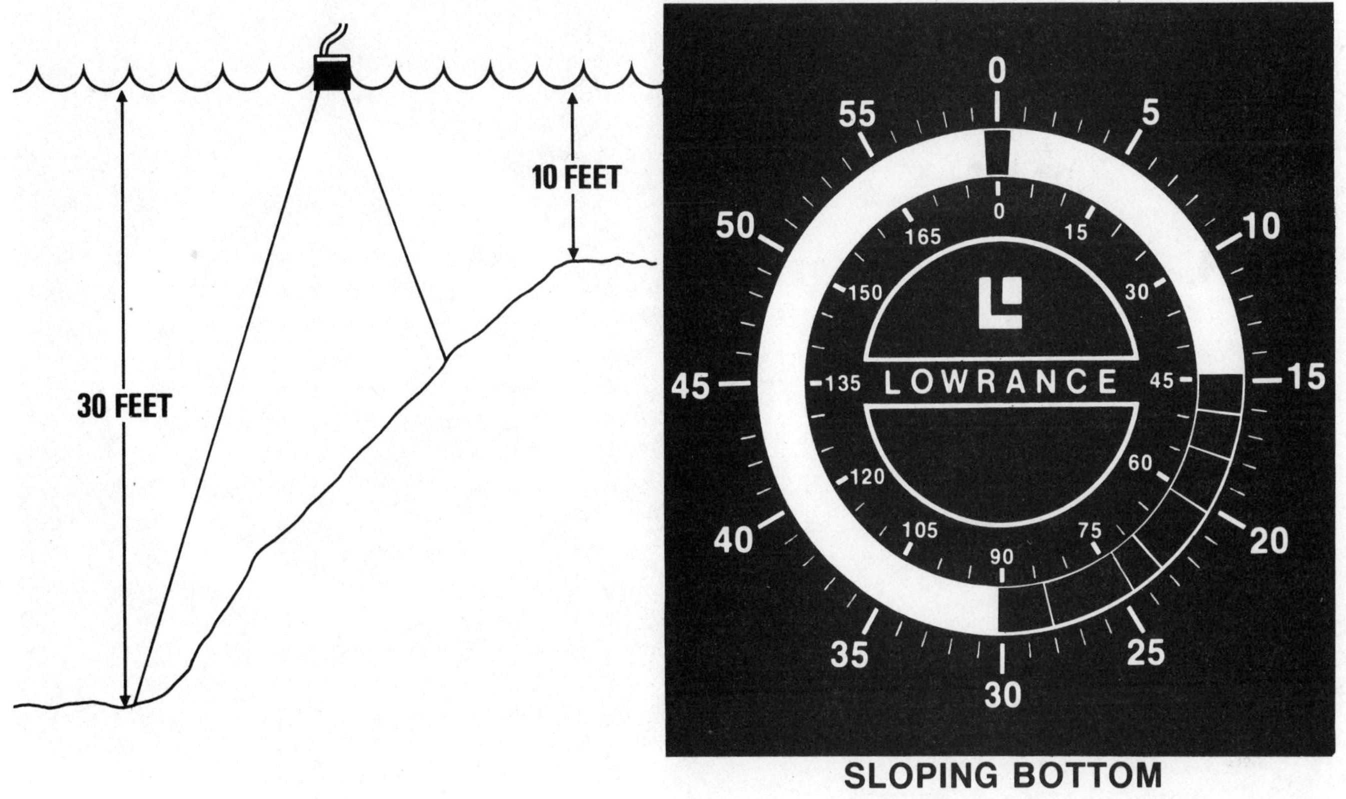

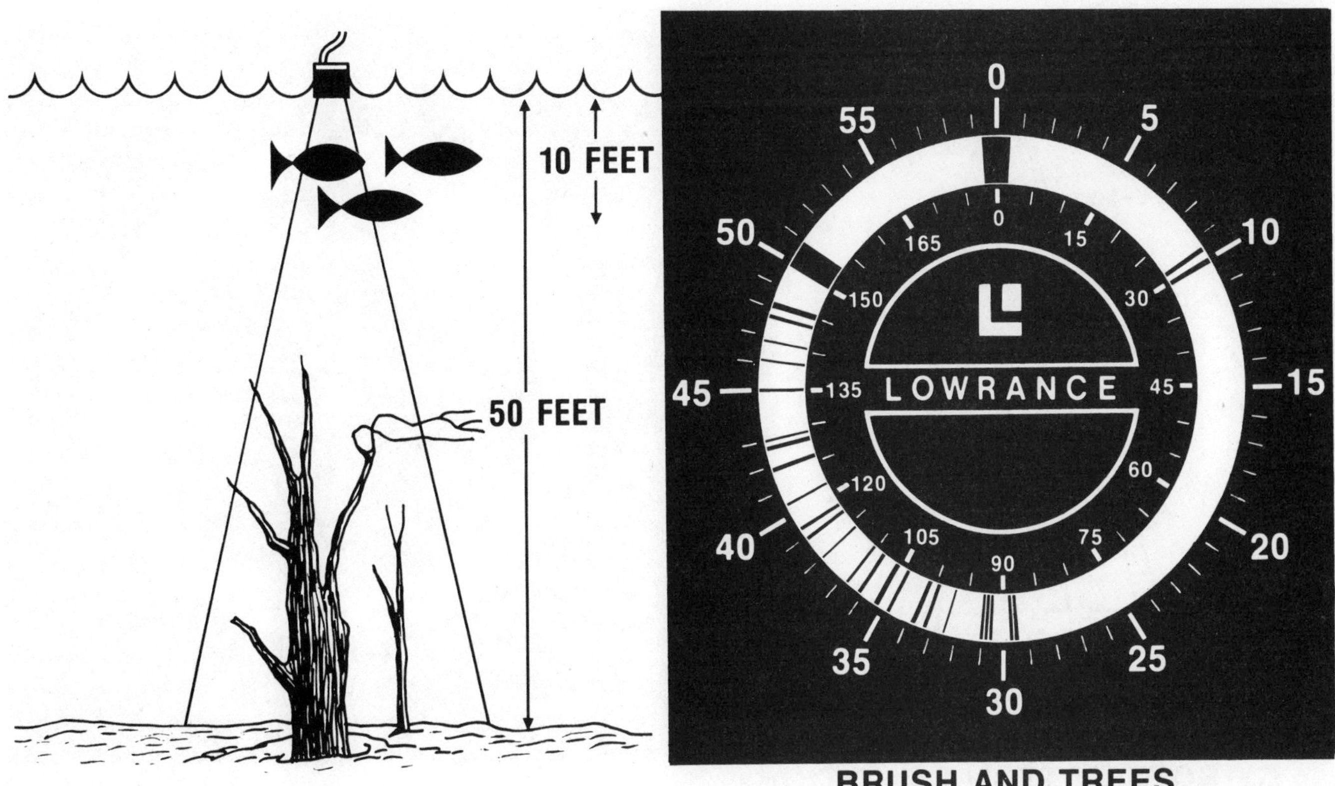

Figure 7–11 (cont'd.)

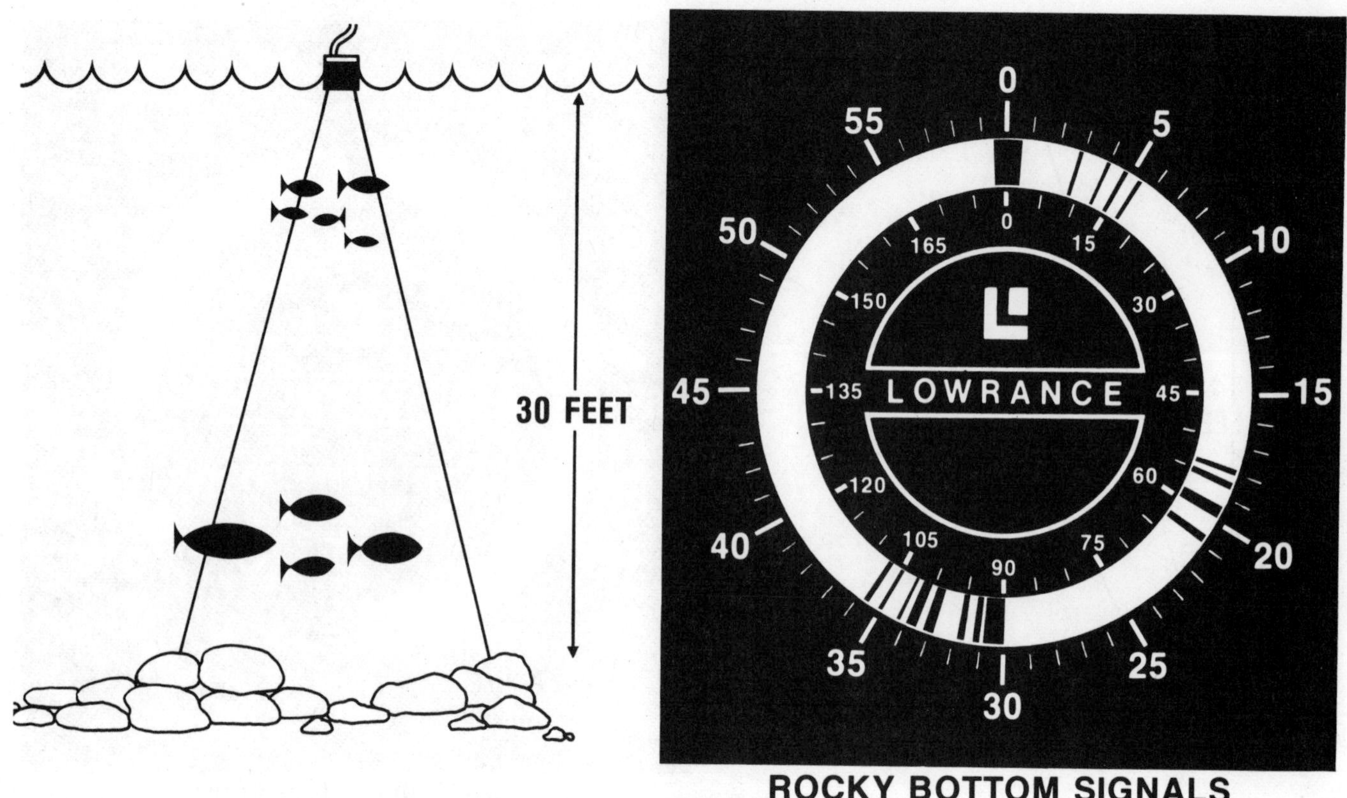

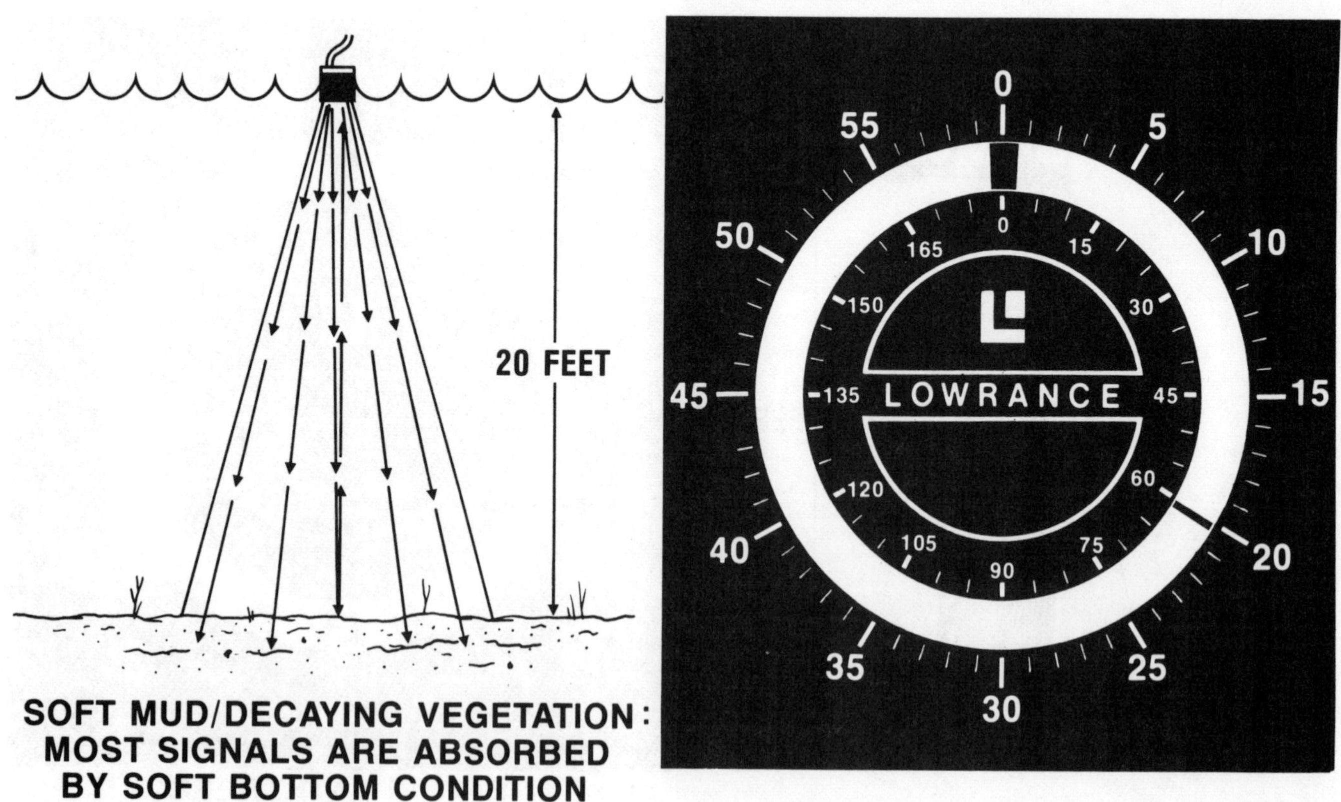

Figure 7–11 (cont'd.)

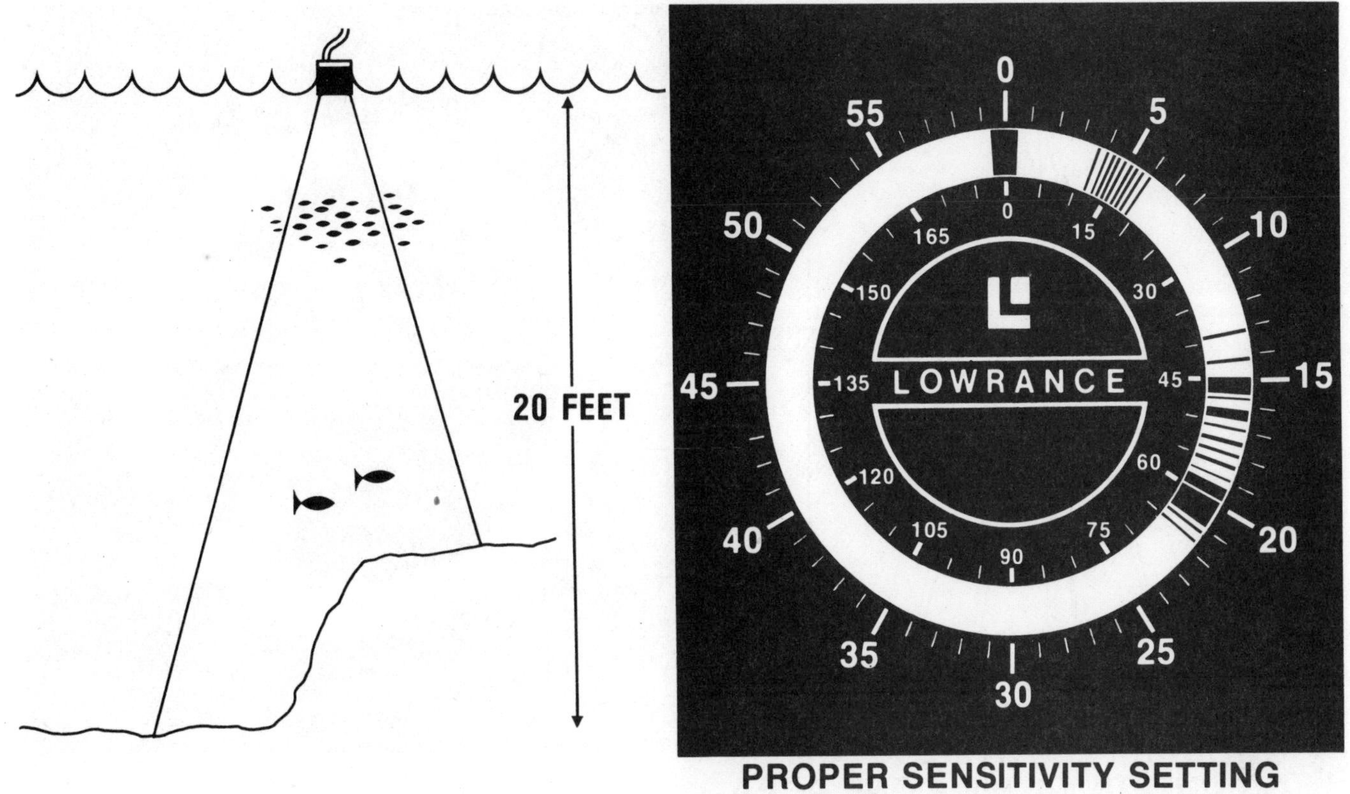

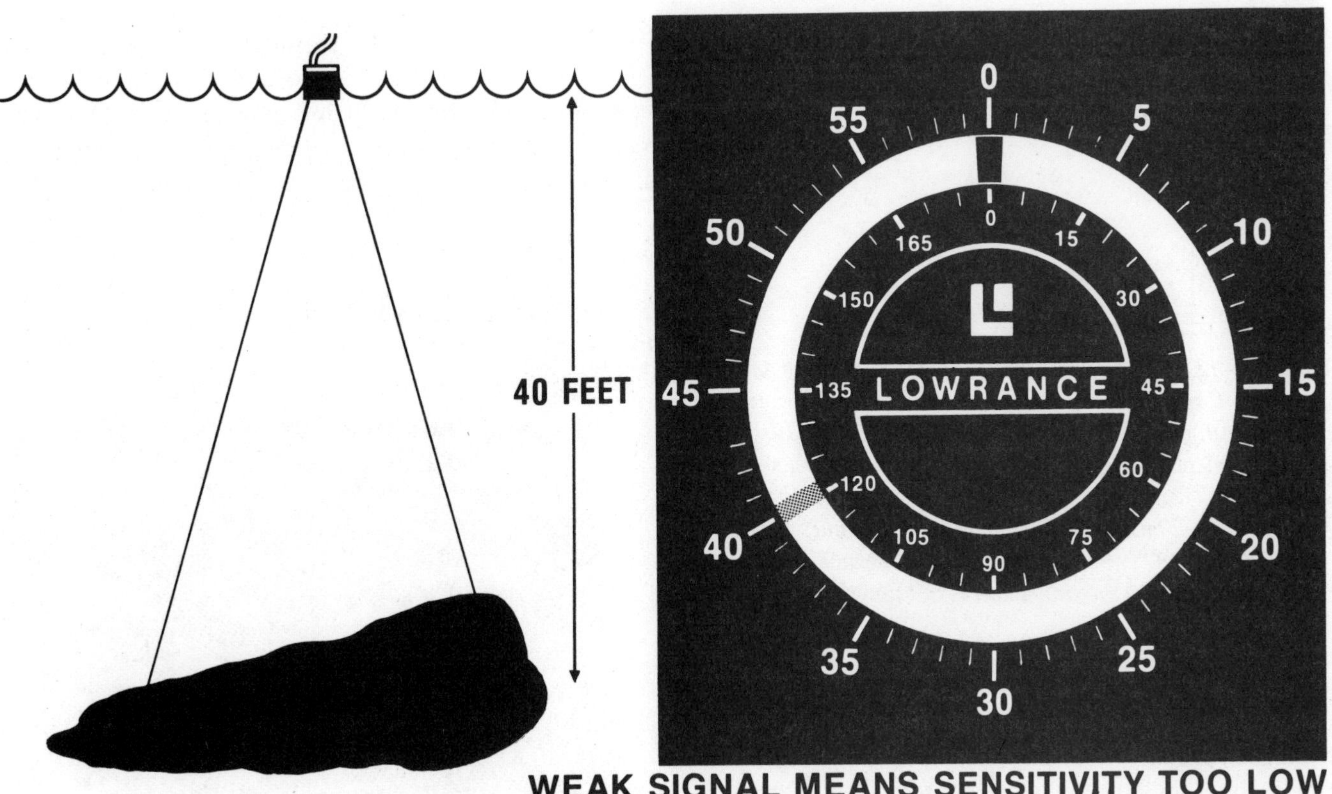

Figure 7–11 (cont'd.)

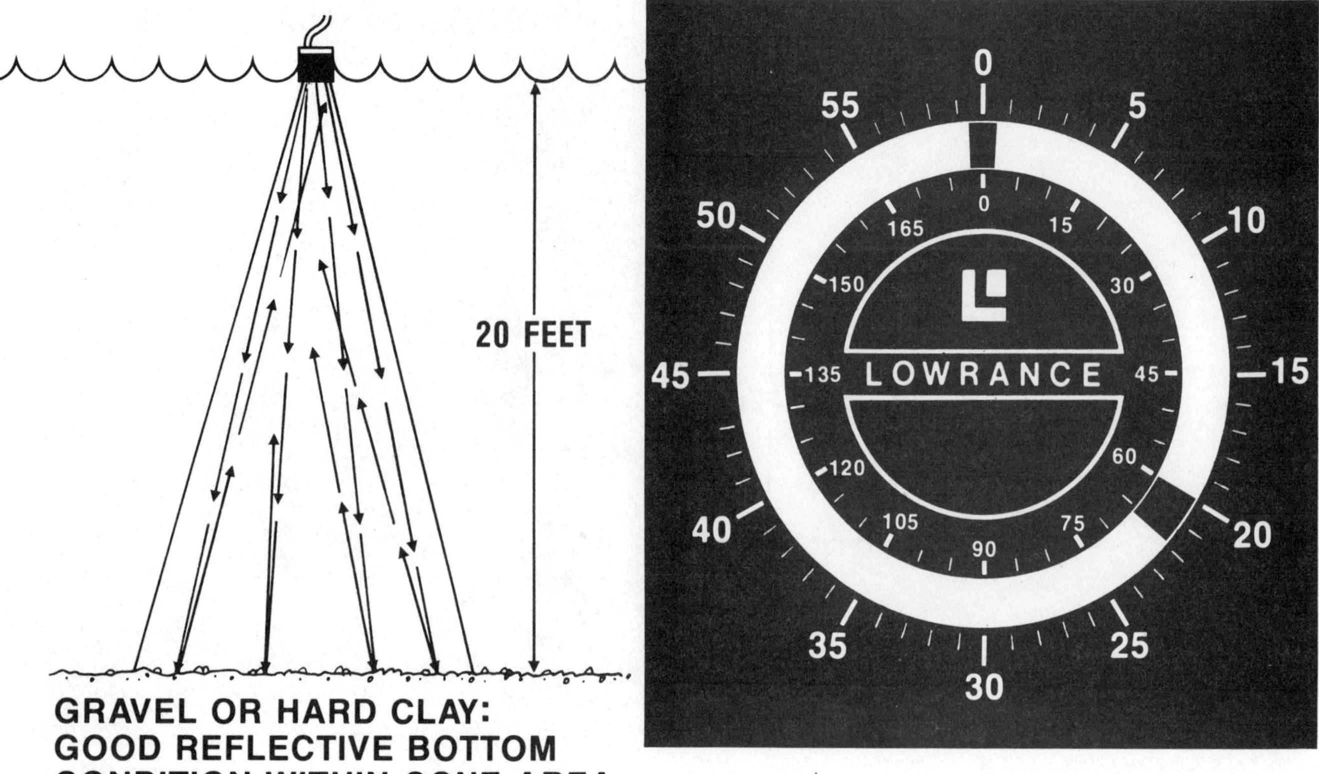

Figure 7–11 (cont'd.)

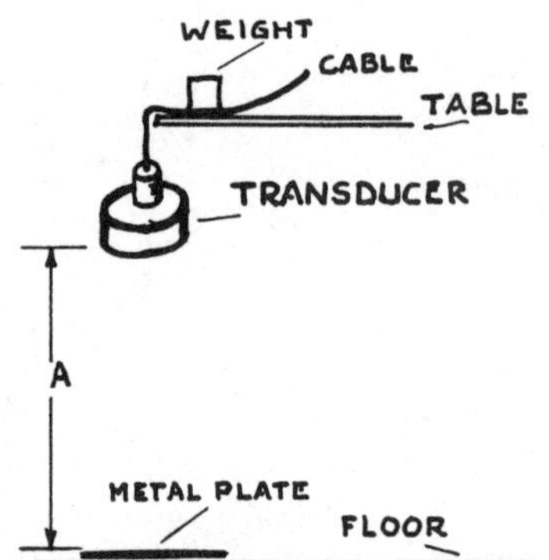

Figure 17-12. A simple method of calibrating a depth sounder without special equipment is shown above. The transducer is hung over the edge of a table or shelf as illustrated and a metal plate is placed on the floor directly below. The distance "A" is measured accurately in feet. With the depth sounder turned on, its reading should be 4.3 times the measurement for "A." (Some manufacturers do not approve of this method because the transducer is not functioning against its normal water load.)

The super-deluxe depth sounder is a so-called "sonar" that searches underwater in a full circle around the boat. The forward search is especially valuable to the skipper in strange waters because it warns of hazards directly ahead and allows time for avoidance. The readout of the sonar is on a cathode ray tube, very similar to the display of a radar unit. Whereas a depth sounder announces trouble only after the boat is in it, the sonar provides early warning. (See Figure 17–13.)

The best depth sounder, like any man-made device, may occasionally pack up and quit. Many of the lesser causative problems are amenable to a step-by-step, common sense approach. A depth sounder totally "dead" when the switch is turned on should trigger a look at the fuse, at the wiring, and even at the state of

Figure 17-13. The Sonar is the super-deluxe depth sounder, in fact, it is more than a depth sounder because it can "look" ahead and give warning. (Wesmar Marine Systems)

NOISE SOURCE	GENERAL TYPE NOISE	SPEED RPM	INDICATOR FLASHES	SUGGESTED CURE
Power Line	Electrical Even or Random	—	Even or Random	Run separate line to the battery. Do not share a common line. Do not share a circuit breaker, AC or DC.
Gasoline Engine	Ignition Even	All Speeds	Evenly Spaced	Install resistor plugs, low noise or shielded ignition wire. Install 1 mfd capacitor to battery side of coil.
Diesel Engine	Acoustic Random	Perhaps Highest	Randomly Spaced	Transducer installed too close to the engine bed. This will be a rare situation.
Generator	Electrical Random	Regulator Cut-in	Randomly Spaced	Remove drive belt to prove. Ground generator frame to engine. Install 1 mfd from armature terminal to generator frame.
Alternator	Electrical Random	Regulator Cut-in	Randomly Spaced	Remove drive belt to prove. Ground alternator frame to engine. Add Sprague 48P100 filter in series with the output and at the alternator terminal.
Tachometer	Electrical Even or Random	All Speeds	Even or Random	Install .01 disc capacitors from each lead to engine ground.
DC Motor Pump	Electrical Random	Normal Speeds	Randomly Spaced	Add 1 mfd capacitor from each lead to the motor (pump) frame.
DC Motor Blower	Electrical Random	Normal Speeds	Randomly Spaced	Add 1 mfd capacitor from each lead to the motor (blower) frame.
DC Motor Windshield	Electrical Random	Normal Speeds	Randomly Spaced	Add 1 mfd capacitor from each lead to the motor (wiper) frame.
DC Motor Refrigerator	Electrical Random	Normal Speeds	Randomly Spaced	Add 1 mfd capacitor from each lead to the motor (compressor) frame.
Propeller Shaft	Electrical Even or Random	Higher Speeds	Even or Random	Run heavy duty brush on main shaft and ground to engine. (Polish shaft with steel wool).
Propeller	Acoustic Even or Random	Higher Speeds	Even or Random	Bent blade. Transducer too close to the propeller. Reflection from extra hard shallow bottom.
Hull (Transducer) Movement	Acoustic Random	Higher Speeds	Randomly Spaced	Check transducer location. Must be in solid water. Check fairing block for shape and finish. Check for hull protrusions ahead of the transducer.
Power Plant	SEE ABOVE FOR GASOLINE OR DIESEL AND ALTERNATOR NOISE SUPPRESSION RECOMMENDATIONS.			
Transducer	Electrical Random	All Speeds	Random	Adjust noise balance capacitor for minimum observed noise

Figure 17-14. This table will help determine what outside influences are bothering the depth sounder. (Ross Laboratories, Inc.)

the battery. Has the transducer cable connection come loose?

If everything appears normal when the switch is turned on but no depth indications appear, the problem is either inside the instrument or externally in the transducer portion. If disconnecting the transducer cable restores the zero depth indication, the trouble is in the connector, the cable, or in the transducer itself. The usual check the technicians use is to substitute a known good transducer. The trouble lies in the unit if the cable disconnection has no effect.

One common problem is electrical "noise" that makes itself known with a multitude of spurious indications in addition to the depth. This is not a depth sounder problem, but merely the depth sounder announcing that it is being harassed by outside electrical disturbances. Ignition is a prime suspect, as are generators, alternators, voltage regulators, and motors. The culprit is found by turning each one on and off separately while the depth sounder is in operation. Manufacturers supply filters that can cure these annoyances. (The table in Figure 17–14 should help.)

18. *Direction Finders and Their Use*

Electronic direction finding is possible only because of the unique reception characteristics of a loop antenna. Place this antenna edgewise to the direction of an incoming signal and it is a fairly good receptor; turn it a quarter circle so that it is broad on to the arriving wave and reception drops almost to zero. In effect, the loop antenna seeks out the direction in which the sending station is located. Simple chart manipulation can turn such information into a reasonable exact fix of a boat's position.

Some early radio broadcast receivers were fitted with loop antennas as a substitute for the then customary long wire "aerials" that festooned the roofs of houses. Listeners soon found that changing from one station to another also meant swinging the loop antenna to achieve best reception.

The typical loop antenna was a square or round frame approximately 18 inches across and wound with wire. It was rotatably connected with the receiver either by means of pigtails or by brushes that maintained wiping contact with collector rings. The drawings in Figure 18–1 explain the loop's angles of signal acceptance and rejection.

At (a) in the sketches, the plane of the loop is at right angles to the arriving radio wave. Both sides of the loop are "cut" simultaneously by the wave. Under the laws of electromagnetic phenomena that govern such action, voltages are generated in both sides that are equal in intensity and identical in polarity. Because of their opposing situation the voltages cancel and are not fed to the receiver. The loop is in its "null" position.

Figure 18-1. When the loop is broadside to the oncoming radio wave, (a) the response is a null. When the radio wave arrives in the plane of the loop and "cuts" through first one side and then the other, (b) the response is maximum.

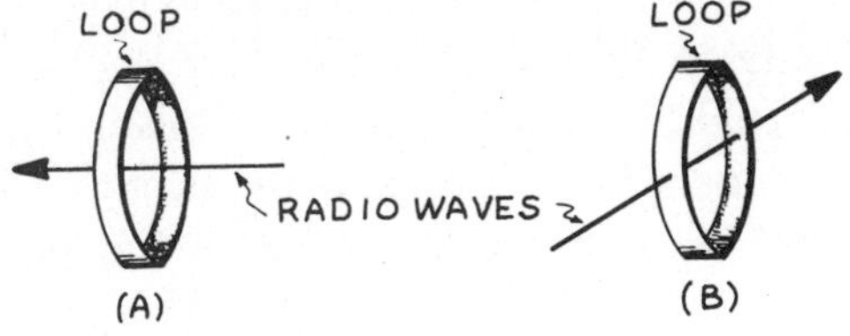

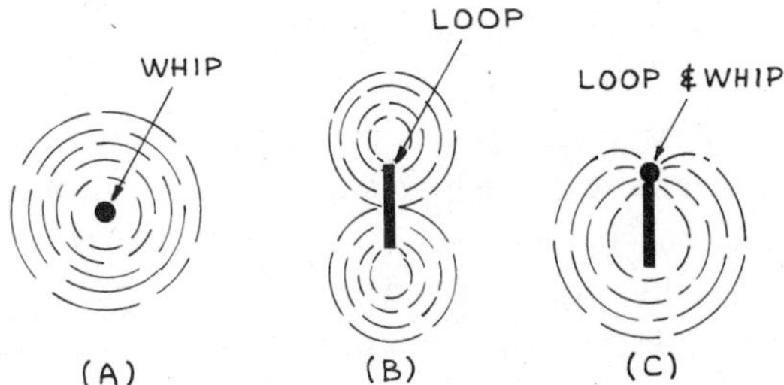

Figure 18-2. Combining a whip antenna with the loop antenna overcomes the ambiguity of the loop alone. At (a) is the omni-directional pattern of a whip antenna. At (b) the two-lobed pattern of the loop is shown. Combining the two results in the unidirectional pattern of (c).

At (b) the loop has been turned 90 degrees and now its plane is in line with the arriving radio wave. The two sides are cut one after the other by the wave, the voltages become additive, and the total is fed to the receiver. The loop is in its orientation of maximum sensitivity to signals from that particular direction.

There is one small drawback to this otherwise fortuitous behavior of the loop antenna—it is subject to ambiguity. The response described would be identical if the signal arrived from the exactly opposite direction. Although in most situations this ambiguity may be solved by common sense, it could be critical to a skipper enveloped in thick fog.

Luckily, this ambiguity can be overcome with very little technical fuss. The addition of a small whip antenna does it. The voltage from the whip is fed to the receiver coincidentally with the voltage from the loop. The scheme works because the polarities are such that the voltages cancel on one side to deepen the desired null and add on the other side to remove the unwanted null. (The drawings in Figure 18–2 explain this.)

At (a) the effective reception pattern of a whip is shown. The view is that seen from directly above the tip of the whip and it reveals a circular disk with the whip at its center. Translated into words, the whip is omni-directional; it receives signals from all points within the circumference with equal effectiveness.

A similar view of the loop antenna at (b) emphasizes its directional characteristic. The two small disks depict the sensitivity from the sides and show the lack of reception at the top and bottom, the two nulls. Superimposing the drawing a upon the drawing b results in drawing c and illustrates the effect of the ensuing addition and cancellation of voltages. The upper half, where the voltages cancel, now has a sharpened null. The lower half has lost its ambiguity because of voltage addition.

Many large vessels with permanently installed radio direction finders mount a loop antenna on top of the wheelhouse. Usually this is a "shielded" loop. The loop wire is encased within a metal tube ring that shields it from many electrical disturbances. If the tube were a complete ring, the contained loop would of course lose its effectiveness. To prevent this the ring is cut and separated at its top center by an insulator. The shielded loop is rotated from below where its position is continouosly shown on an azimuth dial. (See Figure 18–3.)

The rotation of a loop antenna may also be simulated electrically with the loop itself remaining stationary. This is accomplished with a form of loop antenna

Figure 18-3. This shielded loop is found atop many wheelhouses of large ships; it is rotated from below when taking a bearing. Note the insulated split. (See text)

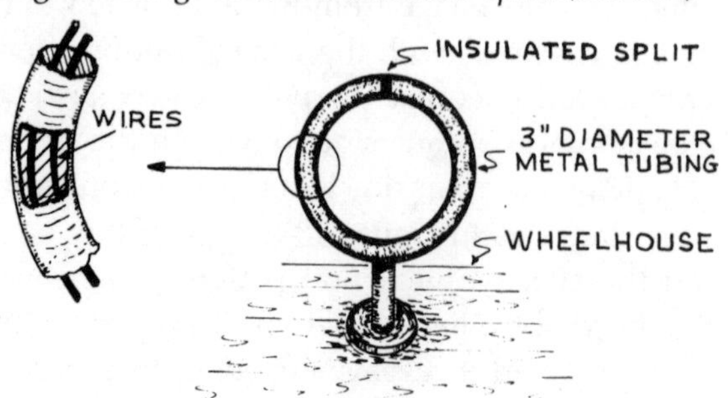

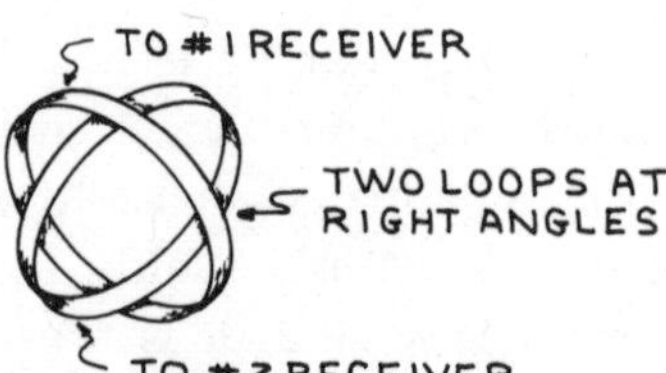

Figure 18-4. A single loop must be rotated in order to obtain directive readings. Bellini and Tosi eliminated this problem with the loop construction named after them. Two loops are mounted together at right angles to each other. The receiving equipment combines the outputs of both loops to arrive at a directional indication.

evolved many years ago by two scientists, Bellini and Tosi. Two loops are crossed at right angles as shown in Figure 18–4. The voltages generated in the loops by the passing radio waves go to four coils oriented as shown in the receiver indicator. Note that the two loop antennas are in independent circuits.

The arrangement of four coils at right angles to each other with a rotatable coil at their center is known as a "goniometer" and this principle of operation is much used in automatic direction finders. When the central coil is turned to a null position, it will point in the direction of the incoming signal exactly as though it were the actual receiving loop. As before, ambiguity is prevented by the addition of a whip antenna that often is placed at the center of the Bellini-Tosi structure like the vertical axis of a sphere.

The Adcock antenna is still another form of receptor to be found on large commercial radio direction finders. The two loops already discussed are replaced by four whips in the Adcock type. The whips are placed at the four corners of a square and are connected to the receiver as shown in Figure 18–5. An advantage of the Adcock antenna over the conventional loop is its lower susceptibility to the "night effect" errors that occur when reflected and polarized sky waves and ground waves combine. The horizontal portions of the Adcock are balanced to be insensitive to horizontally polarized radio waves.

Figure 18–5. This directionally sensitive antenna for a VHF receiver allows the simultaneous showing of the direction from which the message is arriving. (Regency Electronics)

The large loops under discussion may be considered to have a "core" of only air. This limits their effectiveness because the magnetic permeability of air is merely unity; this is the reason for their required large size. If the core were of a material of high magnetic permeability, the loop could be much smaller without losing its capability to intercept radio waves. Ferrite is such a material.

Ferrite is a ferrous-ceramic composition whose magnetic permeability is thousands of times greater than that of air. A ferrite core no thicker than a pencil and half as long, with wire wound upon it, is an extremely sensitive antenna and it retains the null characteristics of a loop. This is the antenna employed in modern portable direction finders. (Wire-wound ferrite rods called "loopsticks" are the antennas in transistor pocket radios; this explains why these little receivers must be oriented to the broadcasting station for best reception. See Figure 18–6.)

The receivers of radio direction finders use the superheterodyne circuit described in Chapter 16. The sensitivity of the receiver is listed by each manufacturer as so many "microvolts per meter." This is a measurement of the weakest radio signal to which the receiver will respond adequately and, obviously, the smaller the number the better.

Modern direction finders tune to several frequency bands, but early models could receive only the beacon band—and the latter may still be the purist's preference because beacons are the best radio aids. Most commercial portable ratio direction finders are capable of tuning to the broadcast band extending from 540 to 1,650 kilohertz, to the beacon band from 150 to 420 kilohertz, and to the marine band extending from 1,600 to 4,600 kilohertz. Some models may add citizen's band (CB) and VHF reception. However, CB and VHF are of questionable value and the use of broadcasting stations for direction finding can be confusing because transmitters may not be at the geographical location listed for the call letters.

Figure 18-6. The output of a large air-core loop may be duplicated with a small so-called "loopstick." The winding of a loopstick is upon a core of ferric powder. This is the type of antenna used in modern radio direction finders.

The receivers of portable radio direction finders are designed to process amplitude modulated signals and, unless a "beat frequency oscillator" (BFO) is included in their circuit, they are insensitive to single sideband (SSB) and continuous wave (CW) transmissions. As its name implies, this local oscillator beats with the incoming signal and generates an audible tone. The pitch of this tone may be varied to suit the listener by slightly tuning the beat frequency oscillator. The BFO is also a help in finding weak stations.

The superheterodyne circuit has one unwanted characteristic that must be guarded against in the design—it is susceptible to the reception of "images." An image, in radio parlance, is an unwanted interfering station whose frequency differs from the desired signal by twice the intermediate frequency of the receiver. A simple numerical example makes this clear.

Assume that the radio receiver has an intermediate frequency of 500 kilohertz and that a desired signal of 2,000 kilohertz is being heard. To accomplish this, the receiver's oscillator is running at 1,500 kilohertz because that results in the required difference frequency of 500 kilohertz which the intermediate frequency amplifier accepts. But a simultaneously incoming signal of 1,000 kilohertz also produces the 500 kilohertz difference frequency and so will be heard as interference. This latter is the image. By similar arithmetic it can be shown that had the oscillator been running at 2,500 kilohertz the same desired signal would be received, but the image would now be at 3,000 kilohertz. The image always is removed from the desired signal by twice the intermediate frequency.

As stated earlier, the cure lies in the design. Sufficient tuned circuits must be placed at the front end of the receiver to prevent the would-be image from reaching the mixer where the trouble begins. The ability to keep the image from interfering is called "image rejection" and is rated in dBs. The greater the number of dBs of image rejection listed in the specifications, the better.

The reading of a direction finder is always taken from the null for auditory reasons both technical and human. Technically, as the previous diagrams will attest, the curve of response to the incoming signal is much sharper on the null side with a pronounced dip at the critical position. This accentuates the movement of a meter needle if such be the indicator used. Furthermore, when a loudspeaker is the "indicator," the characteristics of the human ear fit in nicely. The ear responds to the absence of sound, or to its sudden diminishment more readily than to a slight change in sound level.

Theory assumes that radio waves travel out in straight lines from the transmitter. In actuality, various natural conditions bend or refract the waves from their direct course and the angle at which they arrive may not necessarily be the reciprocal of the exact direction to the transmitter. This influence on the waves of the environment through which they pass is called "quadrantal error." If the waves crossed land and sea "refractive error" may have taken place; this is caused by the differences of land and sea in the ability to reflect and conduct. The changes in the ionic conditions of space that take place at sunrise and sunset may cause "polarization error" as the waves seesaw between horizontal and vertical polarization.

The errors in radio wave travel just enumerated are chargeable to nature and beyond human correction. One final error, however, is indigenous to the ship on which the radio direction finder is installed. This is the reflection and reradiation, of the incoming waves caused by mental masts, metal rigging and other metal surfaces within the "view" of the direction finder. These may be measured and recorded in a deviation table that is used to correct direction finder readings. (A sample of such a table is shown in Figure 18–7.) Note that this deviation table is valid only so long as the metal surfaces are not moved, removed, nor added to, and the direction finder remains in the same relationship to them. The deviation table for the direction

Figure 18-7. Just as nearby iron objects cause deviation in compass readings, so may metal rigging and other metallic surfaces cause deviation in radio direction finder bearings. Once these deviations are found (see text) a deviation table similar to that shown in Figure 18-7 should be made for future referral.

Direction Finder Bearing	Deviation Correction
0	+2
15	+1
30	−3
45	0
60	+4
etc.	+2
	etc.

finder is compiled by "swinging ship" in a similar manner as done for the compass, but this time disregarding magnetic headings and recording only relative angles of reception.

Radio direction finders also come in "junior" form as hand-held models; a representative unit is shown in Figure 18–8. These hand-held units are generally lower in price, but their main advantage is that they may be taken out on the deck away from local deviatory influences that affect not only the radio but the compass as well. The hand-held RDF includes a compass that may be viewed directly for bearings or through a line-of-sight prism for greater ease and accuracy. Tuning is with a knob, as with a radio, although one eliminates the variable tuning and selects its targets with a pretuned module for each station. The null is generally determined audibly either with a loudspeaker or with stethoscope-type earphones, but some units have meters.

Most full-size radio direction finders tune their receiver with a knob that moves a pointer along a frequency scale. This is the fastest and easiest method of sweeping across a band. Digital tuning also is available at the more deluxe level. The desired frequency is selected by punching the required buttons on a ten digit panel: synthesizing (see Chapter 7) makes every frequency within the designed span instantly available and indicated digitally.

The almost universal source of power for portable direction finders is the 12 volt ship's storage battery. The exception is the hand-held units that contain their own batteries, either of the flashlight type or of the 9 volt transistor style, although some portables provide for connection to the boat's 12 volt lines. In all cases, the power consumption is minimal and battery life is long.

Figure 18-8. This hand-held radio direction finder is aimed for a null from the desired station. The direction in degrees then is read from the card of the incorporated magnetic compass. One advantage of the hand-held unit is that it may be taken to a location on the ship least bothered by interference from metal stays and masts. (Epsco Inc.)

The basic difference between manual and automatic radio direction finders boils down to who turns the loop. The operator obviously is the loop turner for the manual models while a small electric motor takes over in the automatics. In both cases the first order of business is to latch on to the radio transmitter to be used. When the received signals are weak, for reasons of distance or natural barriers, tuning them in can be a delicate procedure not easily done in rough seas. For this reason automatic acquisition of a desired signal is a valuable feature of more sophisticated, more expensive models; the frequency is punched in, the machine does the rest.

Radiobeacon positions and the exact locations of broadcasting station antennas are shown on marine charts and are the primary targets of radio direction finders. The transmitting frequency is indicated for each broadcaster together with its official call letters. The information for radiobeacons includes the frequency, the morse code identification letter or letters, and the numerical position in the sequencing. (Full details of all radiobeacons are contained in "Radio Aids to Navigation," a government publication that lists all East coast and Great Lakes aids in book number 117A and West coast aids in book number 117B.)

Sequencing is necessary to conserve radio spectrum "space" and affects all but primary radiobeacons. These primary units transmit continuously. The others are tied into a sequencing pattern that gives groups of six beacons a total of six minutes, each one to be on the air for one minute. The order of appearance of each beacon in this sequence is denoted by a Roman numeral from I to VI. A typical radiobeacon is shown in Figure 18–9 together with the symbol that represents it on a chart. The morse code alphabet listed in Figure 18–10 will aid in identifying charted beacons. The detailed descriptions found in the government publication include the name of the radiobeacon, its latitude, its longitude, its range, its frequency, its morse code signal, and the hours of its operation.

Figure 18-9. The chart symbol for a radio beacon is a magneta circle with a central dot. The photograph is of a typical radio beacon.

Those radio direction finders that are installed fixedly should be well away from the ship's compass because of the strong permanent magnets in their meters and speakers. Owner's manuals usually state the minimum safe separation. There are several ways of aligning the azimuth curser or degree scale for different methods of operation, but the simplest is to set the diameter running from 0° to 180° parallel to the keel, with 0° to the bow. This will reduce the null readings to easily handled relative bearings. Such alignment, of course, does not apply to hand-held direction finders that have their own compass and are minus the curser.

Radio direction finders serve the skipper either to establish a "fix" or to enable him to "home" on a target. The creation of the fix requires at least two and

Figure 18-10. The Morse code consists of dots and dashes with the dashes several times as long as the dots. The code can be of great help in certain situations and every skipper would do well to learn it.

A	.—	N	—.
B	—...	O	———
C	—.—.	P	.——.
D	—..	Q	——.—
E	.	R	.—.
F	..—.	S	...
G	——.	T	—
H	....	U	..—
I	..	V	...—
J	.———	W	.——
K	—.—	X	—..—
L	.—..	Y	—.——
M	——	Z	——..

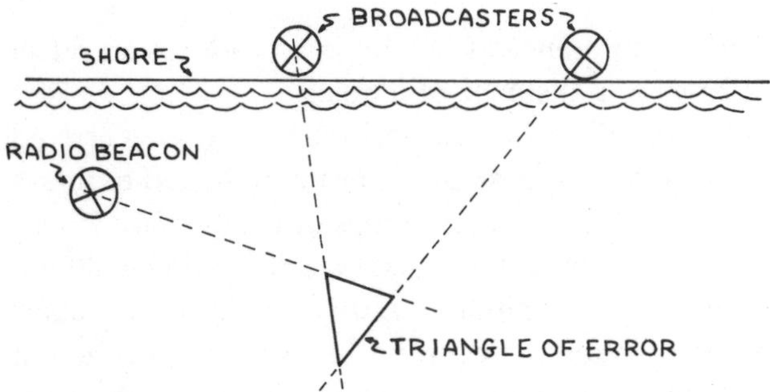

Figure 18-11. The accuracy of the fix from a radio direction finder is greatly improved if three targets are used to obtain three lines of position. The vessel's position then is within a small triangle of error.

preferably three beacons or other transmitters if radio is to be the sole medium. Only the one transmitter at the desired location is neeced for homing.

To accomplish homing, the target is tuned in nulled, and the bearing noted. The ship is now swung to this heading to bring the null directly on the bow. This is not the piece of cake that it sounds, because judgment and vigilance are required to avert accident. Careful inspection of the chart should verify that no shoals or other dangers intervene on the homing line. Perhaps an actual happening will best illustrate another danger in homing: In 1934 a vessel using the Nantucket Lightship as a homing target ran it down and sank it.

A so-called "line of position" is the basis for all straightforward marine position determination, in other words, for making a fix. The radio direction finder yields such a line of position. The information contained in a line of position is limited to the fact that the ship is on that line; no cue is given as to the point on that line. However, by geometric reasoning, if two independent lines of position are crossed, the ship must be at the intersection since that is the only possible way for the ship to be simultaneously on both.

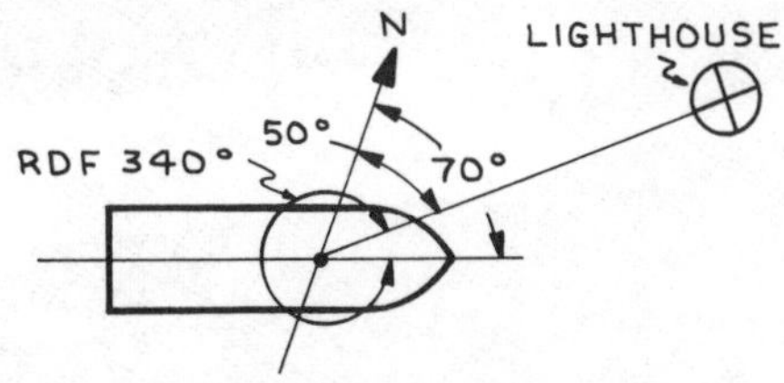

Figure 18-12. In the drawing above, the ship is on a course of 70° when it sights the lighthouse on a bearing of 340° relative to that. The line of sight is plotted at 50° arrived at as shown and explained in the text.

Theoretically, the intersection is the exact fix, but the inherent errors make that conclusion doubtful. In practice, three lines of position at widely divergent angles usually result in a small included triangle that marks the vessel's probable location. The more careful and accurate the procedure in obtaining the lines of position, the smaller the triangle. (See Figure 18–11.)

With the radio direction finder activated, the target chosen and tuned in, the null is determined either manually or automatically. The directional ambiguity is solved, if necessary, with the sense antenna. Now the direction finder reading on the curser and the compass heading must be noted simultaneously and deviational corrections made to both. The sum of these two corrected readouts produces the chart plotting angle for the line of position.

A numerical example will prove the simplicity of all this. Suppose the corrected direction finder reading to have been 340° and the corrected compass reading to have been 70°. The sum is 410°. Subtracting 360°, the usual practice when the sum is greater than 360, gives 50° as the angle for the line of position. (See Figure 18–12.)

Had the foregoing sighting been made with a hand-held radio direction finder, the procedure would be even simpler. Presumably, the skipper has taken the unit to a point on the ship where deviation is negligible and only variation must be considered, but this cancels out automatically if the magnetic rose on the chart be used for drawing the line of position. Assuming zero variation, sighting the null across the compass of the hand-held unit immediately reveals 50° as the answer.

For all run-of-the-mill piloting, the angle found by radio direction finding may be laid down on the chart directly as the direction of the line of position emanating from the target. The error will be insignificant even though radio waves travel the great circle route while the chart is a mercator projection. But for distances from the transmitter of hundreds of miles the correction had best be made even though small. Table No. 1 in Bowditch provides such correction figures without mathematical juggling and gives full instructions.

Errors in determining the line of position angle, that creep in because of human and other causes, were mentioned in the text. What is their approximate magnitude? A difference of one degree produces an offset of 552 feet at a distance of 6 miles.

Although the fix in the earlier example was arrived at with all lines of position determined by radio, this is not an essential condition. Lines of position are amenable to being mixed. For instance, a fix could be plotted with one line of position found visually, one by radar, and one by radio direction finder. The errors inherent in each method could well balance out when the various types are combined—or they could add, as luck would have it. Incidentally, the individual errors are not added arithmetically, but are summed by a process called "the square root of the sum of the squares."

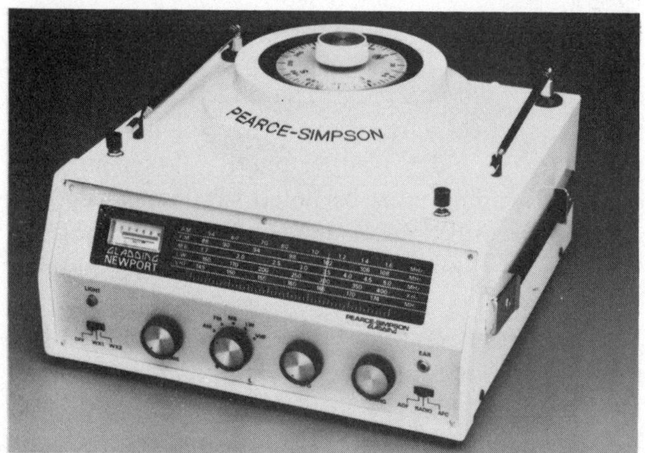

(a)

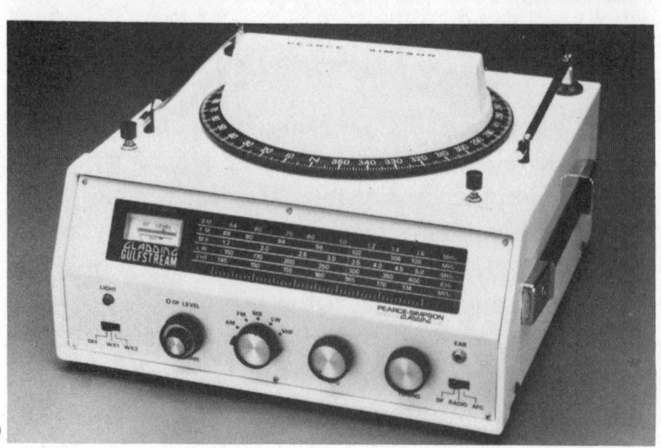

(b)

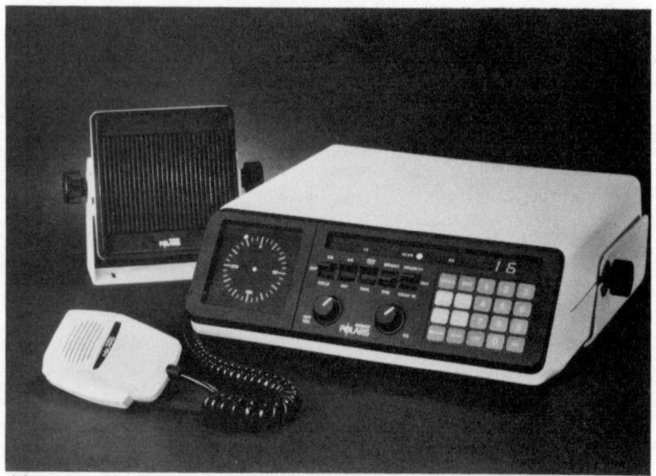
(c)

Figure 18-13. Pictured are: (a) an automatic radio direction finder (ADF), (b) a manual direction finder (RDF), and (c) a novel combination of an automatic direction finder and a VHF transceiver that employs an antenna consisting of four dipoles. (a and b: Pearce-Simpson; c: Regency Electronics)

Modern automatic radio direction finders proclaim their readings by pointer, by radially spaced lights, or even by small television-type electron ray tubes; a digital readout in degrees is planned to hit the market. The pointer is the simplest and easiest to read. The pointer swings over a 360° scale that may itself be rotated for relative positions. The lights are placed around a circle every ten or so degrees and obviously cannot give as accurate a reading. The television-style readout eliminates the motorized pointer. (See Figure 18–13.)

Perhaps the great advantage of automatic radio direction finding over the manual operation is that the human factor is eliminated from bearing determination; the errors are only those of the machine. Manual manipulation adds the little quirks that man is heir to and this degrades whatever accuracy the electronic device innately has. The need for sharpening human skill thus becomes clear—and the best sharpener is practice, plenty of practice.

The ideal practice is in fair weather, within sight of a transmitter, so that accuracy may be checked visually. The procedural habits thus established can be counted on as a reliable guide in bad weather.

19. *Loran and Its Use*

Ghosts of oldtime navigators, still clutching their sextants, chronometers, and sight reduction tables, must look with envy at modern skippers who merely copy two numbers from an electronic gadget, go to the chart, and say with complete assurance, "This is where we are." Such is the wonder of Loran. Such is the fantastic ability of radio waves to pinpoint a location at sea within less than a few hundred feet.

Loran is one of the few beneficial spinoffs from the barbarity of war. In its original form as Loran-A, it did yeoman service in position determination until recent years. Then Loran-A metamorphosed into and has been supplanted by the much more accurate and more easily operable Loran-C currently in use. It truly fulfills the description "*lo*ng *ra*nge *n*avigation" implied by the acronym "Loran."

Loran-C signals are now available for position determination over 16,000,000 square miles of the Great Lakes and the oceans contiguous to the United States, the so-called "coastal confluence zone." This coverage presently is provided by eight Loran-C groups, each consisting of one master and several secondaries. Planned additions of both American and foreign groups will make coverage worldwide. (The "master/slave" terminology of Loran-A has become "master/secondary" for Loran-C.) Each group is identified by its "group repetition interval" (GRI), the time in microseconds between the start of each repeated master transmission. The secondaries are allotted the necessary portion of the GRI for their own transmissions. (A microsecond is one-millionth of one second.)

Masters and secondaries of Loran-C transmit on 100 kilohertz in contrast to the very much higher almost 2 megahertz of Loran-A. This low frequency (and the corresponding extremely long wavelength) has a more reliable ground wave, greater distance coverage, and is less affected by atmospheric and other interference.

The 100 kilohertz transmitting frequency imposes theoretical restrictions on the maximum length of the group repetition interval within which the master and all its secondaries must finish one complete transmission; thus the GRI cannot be longer than 99,990 microseconds, nor shorter than 10 microseconds. In commercial practice, only the span between 40,000 and 99,990 is used with intervals of ten microseconds. Since the last digit always is zero, it is dropped. This makes the GRI of any Loran-C group a four digit number between 4,000 and 9,999, and the group is identified by that number on charts. (A table of GRI values is given in Figure 19–1.)

The simplicity of Loran-C operation is a great boon to pleasure boat skippers, especially after the complexity of Loran-A. The expertise required to use Loran-C is just about zero. Once the Loran-C group that best serves the boat's area is determined, its GRI is keyed into the receiver by push button or dial. From then on the rushing electrons do the rest without outside aid. Within minutes two six-digit numbers are displayed. These are "time differences" (TDs) and apply to correspondingly numbered curves on the chart, each a "line of position" (LOP). As is normal in the art of marine position finding, where two or more LOPs cross becomes a "fix."

Figure 19-1. The Group Repetition Interval (GRI) of present and future Loran C chains is given in the table.

Group Repetition Interval (GRI)	*Coverage Area*
4990	Central Pacific
5970	Southeast Asia
5990	Canadian West Coast
7930	North Atlantic
7960	Gulf of Alaska
7970	Norwegian Sea
7980	Gulf of Mexico
7990	Mediterranean
9930	U.S. East Coast
9940	U.S. West Coast
9960	U.S. Northeast
9970	Northwest Pacific
9990	North Pacific

(a)

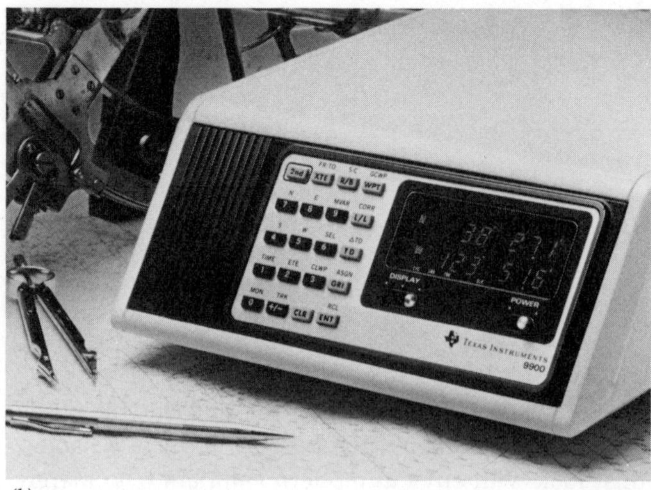

(b)

Figure 19-2. (a) This early Loran receiver, circa World War II, with its many knobs and controls, took time and skill to operate. The modern Loran C receiver (b), gives a dramatic comparison of the old and the new. (Texas Instruments)

The fantastic functioning of a Loran-C receiver is all the more remarkable when its small size is considered. It is about one-tenth the bulk of an early Loran-A receiver, as the photos in Figure 19–2 prove. The contrast between the many knobs on the Loran-A and the few controls on the Loran-C also tells a story. Each of these knobs required experienced manipulation before the end result, a series of pips on the cathode ray tube, could be deciphered into a TD applicable to a chart. Think of all this complex procedure, with its many opportunities for error, in comparison to the simple reading of two numbers from the front panel of a Loran-C. How navigators in World War II bombers could successfully use Loran-A under combat conditions remains a mystery.

Simple position determination is only one in the bag of tricks built into a modern deluxe Loran-C receiver. All the secondaries of a master can be computed simultaneously and each time difference displayed to provide additional lines of position to verify fixes. Destinations may be programmed into the receiver and a warning will alert the skipper as each is reached. Course to steer to a desired point is computed. And the time differences may automatically be converted to latitude and longitude so that a fix may be set down on a standard chart without superimposed Loran curves. It boggles the mind.

The accuracy of the Loran-C system is achieved and maintained with atomic clocks. All transmitters in a group, master as well as secondaries, are governed by their own caesium frequency standards whose innate error may be in the order of one second in a year. Secondaries in a group thus maintain their timed position independently and not, as in Loran-A, by a signal from another. The accuracy of each group is checked continuously by a monitoring station that either corrects any disparity or else causes all receivers to "blink" as a warning to navigators. The monitoring station is known as a SAM, system area monitor.

Nature provides the basis for Loran operation by holding the speed of travel of radio waves to a fixed 300,000,000 meters per second. With speed known, the inclusion of elapsed time results in distance. The entire Loran concept depends on the difference in time of arrival at the vessel of two radio pulses sent out at predetermined intervals from two known locations. This same time difference will be found all along a curve, known mathematically as a hyperbola, that becomes a line of position with a GRI number on specially prepared charts.

The technical functioning of a Loran-C group is so nearly perfect that any subsequent errors in geographic position of the fix nearly always are traceable to causes other than radio—for instance, discrepancies in the chart. Timing of the transmitted pulses is maintained to within two one-hundredths of one microsecond! This is an interval of time so microscopic that it cannot be compared to any familiar happening. If, for any reason, timing "slumps" to two-tenths of one microsecond, the blinking alarm goes out automatically.

The typical Loran-C group or chain consists of one master, M, and several secondaries individually identified by the letters W, X, Y, and Z. The time difference

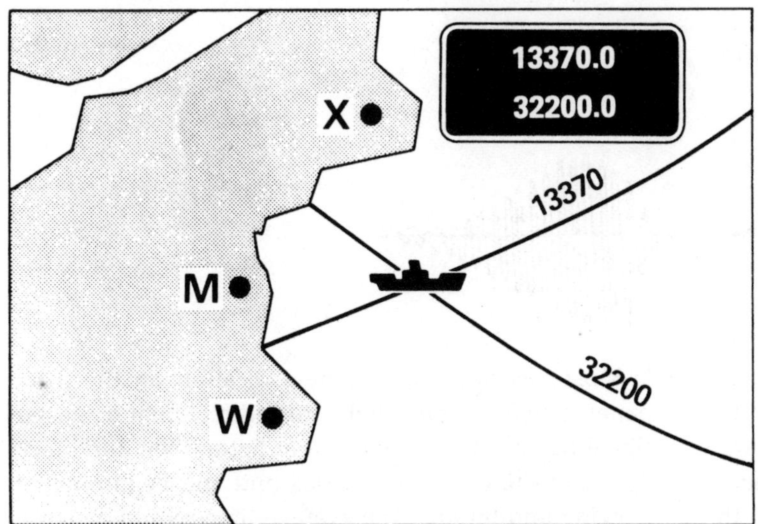

Figure 19-3. The dark insert in the above diagram is a reproduction of the digital readout in a given Loran C receiver. Since the vessel is simultaneously on the 13370.0 TD curve and on the 32200.0 TD curve, it must be at the intersection of these two curves. The intersection is the fix at that particular time. (U.S. Coast Guard)

between the arrival of pulses from the master M and from a secondary, for instance, from secondary Y, may be written as TDMY, but more generally is stated as simply TDY with the master understood. The groups and their various transmitters are spread out and located to ensure that a vessel anywhere in the service area receives at least two time differences that may be referred to charted TDs. This fulfills the need for at least two lines of position in order to designate a fix. In most cases, the vessel will receive three TDs and will be able to refine the accuracy of the fix. (See Figure 19–3.)

The increased accuracy of Loran-C over Loran-A is brought about by an improvement in the method of time difference determination. Loran-A measured the difference in arrival time of whole pulses of radio waves; Loran-C refines this into basing its measurements on individual cycles or phases of the pulses. The effect may be compared to looking at something directly or through a powerful magnifier.

The knowledge gained from a quick look into the nitty gritty of the Loran system will hasten familiarity with Loran receivers and their use in navigation. It is common knowledge that Loran is a hyperbolic system and that lines of position derived from it are hyperbolas (although for short distances they may be considered to be straight lines). Mathematically, a hyperbolia is the curve generated when a cone is cut at an angle by a plane as shown in Figure 19–4. The hyperbola is pertinent to Loran because it contains all points that are the same difference in distance from two fixed locations. This too is explained in the drawing. In the example shown, the distance from M to A is three units and from S to A is five units; the difference is two units. The distance from M to B is ten units and from S to B is twelve units, again with a difference of two units. A and B and all points with this same two unit difference are on this "two" hyperbola. For Loran this hyperbola would be tagged with the microsecond equivalent of two units.

Since all Loran-C transmissions are on 100 kilohertz, the question of interference may arise. Although the Loran-C groups are placed at widely spaced locations, some overlapping of signals results under certain at-

Figure 19-4. A hyperbola is generated when a cone is cut at an angle by a plane. The point A is 3 units from M and 5 units from S—a difference of 2. This point B is 10 units from M and 12 units from S—also a difference of 2. A and B lie on the hyperbola TD that contains all points with a difference of 2.

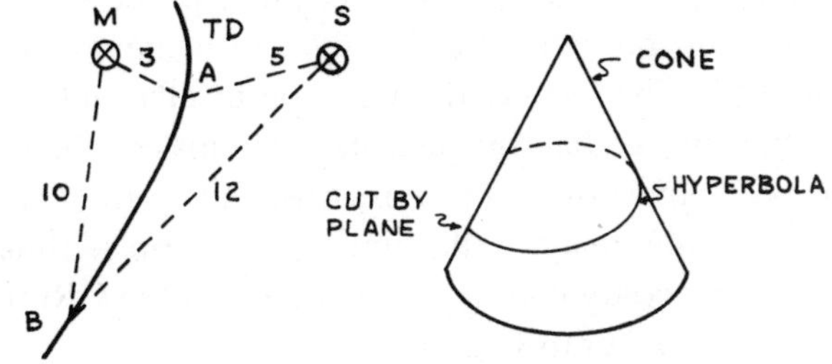

Figure 19-5. The pulses received by the Loran C receiver as seen on an oscilloscope. Note how the time differences are measured from a chosen cycle in each pulse.

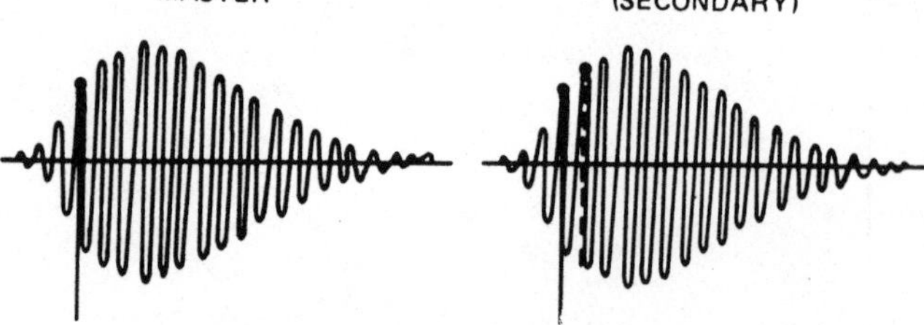

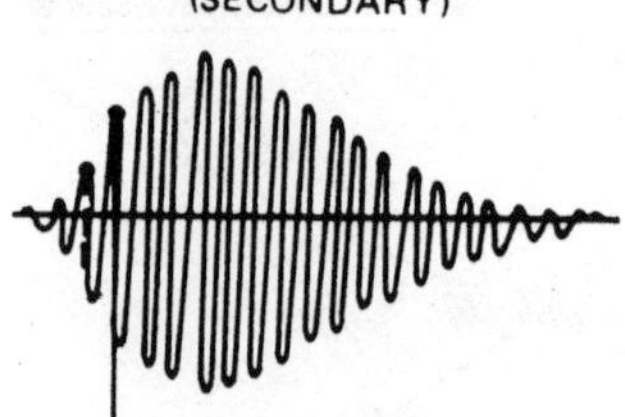

mospheric conditions; why don't they garble each other as two broadcasting stations would under the same circumstances? The secret lies in the different GRIs, the different group repetition intervals. When a certain GRI is punched into a Loran-C receiver, the electronic circuitry accepts only that one for processing and rejects all others. Although all the keys for a brand of lock are similar, the lock opens only for the key with the exact required contour.

If a Loran-C pulse were examined on an oscilloscope, it would provide the picture shown in Figure 19–5. Only the first three cycles of this pulse are of interest to the receiver and highly sophisticated integrated circuitry seeks them out and locks onto them. Timing for determining the time difference, or TD, begins as the third cycle of the master pulse ends at the zero line; this is marked in the drawing. As mentioned earlier, this improvement over Loran-A timing is the reason for the greater accuracy and repeatability of Loran-C.

The composition of a single complete Loran-C transmission is diagrammed in Figure 19–6 and comprises one group repetition interval. The master pulses are followed in sequence by the pulses from secondaries X, Y, and Z. The master transmission consists of nine pulses, the first eight separated from each other by 1,000 microseconds plus a space of 2,000 microseconds between the eighth and ninth. All secondary transmissions consist of eight pulses spaced 1,000 microseconds apart. There may be variations in the amplitudes of pulses from master and secondaries because of the difference in distances travelled by the radio waves.

The longer spacing of the extra pulse in the master transmission provides the clue by which the receiver identifies it. The secondaries wait specifically set times after the master has finished its transmission before they begin their own transmissions; this is the "secondary coding delay." The sum of all the pulse times and the secondary coding delays becomes the GRI for that particular Loran-C group.

Most receivers have some indication on the front panel, usually a pilot light, to indicate that a master signal has been acquired. Other lights then announce the acquisition of secondaries and the beginning of tracking and timing. The average time required before the TD numerals are flashed out for the skipper is less than two minutes.

The ability of the receiver to function at great distances from the group transmitters is dependent upon the sensitivity of the receiver's circuits and the ratio of the signal strength to the electrical noise in the area, the S/N in technical lingo. Under very poor S/N conditions, the automatic Loran-C receiver is better able to pick the signal out of the morass than manual operation.

The electrical noises that degrade the S/N condition come from many sources, such as atmospheric discharges, distant radio transmitters of all types, and especially electrical and electronic equipment aboard. Some of this noise gets into the receiver because of pickup by the Loran antenna; some arrives via the battery power lines. Onboard noises are best eliminated at the source, although this may not be easy to do. The worst offenders are radar, inverters, fluorescent lights, strobe flashers, TV sets, electronic tachometers, and the generators and alternators on propulsion engines. Ignition systems of gasoline engines, that bugaboo of shipboard radio transceivers, could be particularly

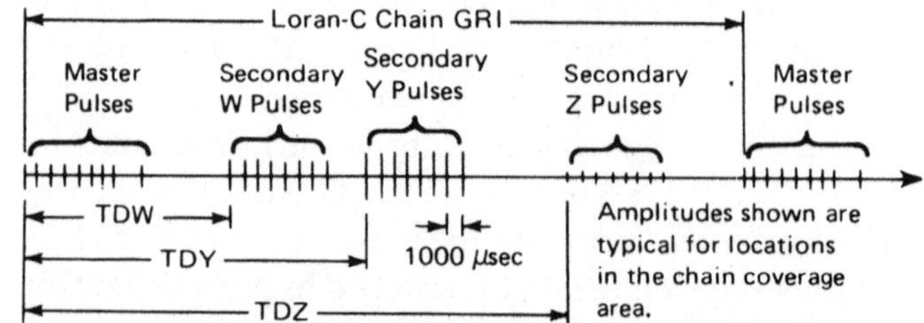

Figure 19-6. The "anatomy" of a Loran C pulse is diagrammed above. Note that the Group Repetition Interval is long enough to accommodate the transmission from the master station plus the succeeding transmissions from the secondaries. (TD is the time difference.)

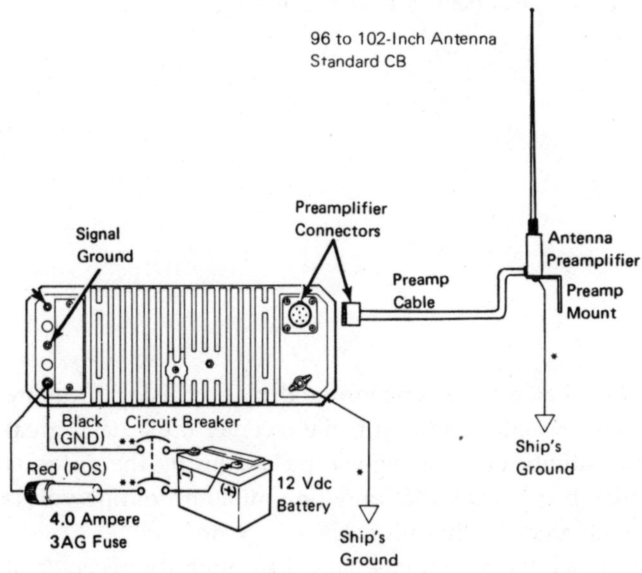

Figure 19-7. The preamplifier is directly at the antenna for maximum effectiveness in keeping out local electrical noise and maintaining a high signal to noise ratio. All other installation connections also are shown in the drawing above. (Texas Instruments)

troublesome because of the extremely low working frequency of Loran-C.

The standard elimination treatment for onboard noises is to shunt the offending device with a large value capacitor of the proper type; these are available in electronic stores, often packaged for specific purposes. The metal cases and chassis of all electrical equipment should be grounded. This operation is largely a matter of "cut and try," unit by unit, until results are achieved.

The antidote for interference that reaches the receiver by way of the antenna system is the "notch filter." (See Chapter 7.) A notch filter is so named because it removes a notch, or narrow segment, of frequency from the spectrum to which it is connected. The good Loran-C receiver has internal notch filters and provision for additional external notch filters if they be needed. One filter tunes just below the 100 kilohertz operating frequency, the other tunes just above, and they are manipulated until the offending signal is removed. Obviously, neither notch filter is tuned to 100 kilohertz because that would remove the signals with which the Loran-C functions. Some receivers include a meter that shows the effectiveness of the notch filter.

Most Loran-C receivers include an antenna coupler as part of the antenna system. The coupler contains a preamplifier that boosts the incoming signal before it is sent down the cable to the receiver. The preamplifier is tuned for maximum response when the unit is installed. The antenna, a standard whip, is connected directly to the coupler without intervening cable in order to reduce both loss and pickup of interference. The electric power needed by the coupler is brought to it by extra conductors in the cable to the receiver. Figure 19–7 shows a typical installation.

Although VHF-FM receivers that work way up in the megahertz portion of the spectrum do not require a ground, a good ground becomes more and more important as the frequencies become lower, and at the Loran-C frequency of 100 kilohertz the ground is as critical as the antenna. All makers of Loran-C sets stress this, and with good reason. At very low radio frequencies the antenna and the ground are equal parts of the antenna system and both must be good in order to achieve good reception. The ground wire from the Loran-C receiver should go directly to the ship's ground and should consist of wire no smaller than #10 AWG (preferably it should be a copper strap).

A line connecting the master and one secondary is known as the "base line" for that pair, and the portion of this line extending beyond the secondary is the "base line extension." The accuracy of position determination varies inversely with the distance from the base line. The error may be as little as 50 feet on the base line and increase gradually to perhaps several miles on the extreme 1,200 mile fringes of the coverage area. (This is an average; manufacturers may claim more or less.) However, the foregoing does *not* apply to the base line extension. This area behind a secondary produces uncertain TD readings and is unsuitable for navigation. Readouts that cause LOPs to cross here should be discarded in favor of others that do not.

A modern Loran-C is essentially a radio receiver plus a minicomputer programmed to carry out the operations necessary for getting the designed results. The receiver is fixed tuned to 100 kilohertz, the universal Loran-C operating frequency. This receiver must be able to steadily maintain its tuning despite fluctuations in battery voltage and the extreme changes in temperature that may be encountered by a cruising pleasure boat. Equally important, the receiver must have great sensitivity because signals picked up by the antenna may be so weak that only one-millionth of one volt is generated for the receiver to work on.

Once the signal has passed through the receiver, it goes into the realm of the computer, and what happens to it there is dependent upon the sophistication of the particular Loran-C. It may be processed simply for time-difference readouts. More likely, it will hatch out additional goodies such as latitude/longitude conversion and the other features mentioned earlier. Manufacturers even offer course plotters and direct connection to autopilots. Highly complex integrated circuits make all this possible.

Loran transmissions arrive at the vessel by ground wave, by sky wave, and sometimes by both. The preferred messenger is the ground wave that, as its name indicates, follows the ground in its direct travel; thus its relationship to time is straightforward. The sky wave makes its way by bouncing back and forth between earth and ionosphere (sky) and reaches the antenna after one or several hops. Obviously, its travel time must be longer than that of a ground wave from the same transmitter and so corrections must be made.

A standard eight-foot whip antenna is recommended for Loran-C. It should be of the "straight through" stainless steel or fiberglass type, not one that is capacitor coupled or contains a loading coil. Various mounts are commercially available for securing the antenna to horizontal, vertical, and inclined surfaces. Some antenna couplers also act as the mount for the antenna.

The location of the Loran-C antenna is a critical factor in the overall success of the installation—as it is with any antenna in any radio service. The chosen site should give the antenna the greatest possible height and a clear "view" in all directions. Proximity to metal guys, masts, booms, other metal surfaces, and especially to other antennas must be avoided. All these metal objects can shield the Loran-C antenna from its necessary signals—and may even reradiate spurious signals. (A typical Loran-C antenna installation is shown in Figure 19–8. Note that if compromise must be made between antenna height and clear all-around exposure, sacrifice of height is recommended.)

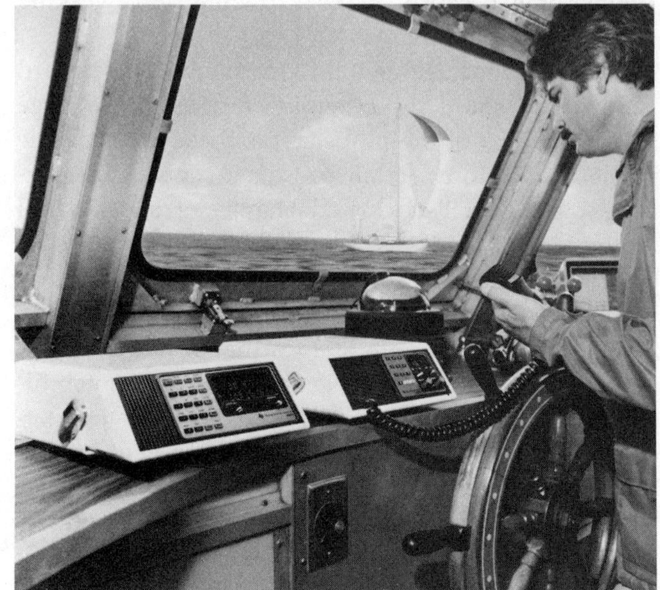

Figure 19-8. This compact Loran C receiver is installed at the steering station where the helmsman can see it conveniently. (Texas Instruments)

Loran installation does not legally require a licensed technician, and manufacturers' instructions are clear enough to enable a handy skipper to do the job. However, it cannot be denied that an experienced Loran installer may be able to coax that last bit of performance out of the receiver.

Some sophisticated Loran-C receivers are programmed to test their own internal workings and report discrepancies. A simple overall test the skipper may perform at any time is to take readings while tied to his home pier whose exact location on the chart he knows. Plotted and actual locations of course should coincide. Any further meaningful testing requires expensive laboratory equipment and is beyond the ability of most skippers.

The usual Loran-C position determining procedure is to get two readouts from the receiver and plot them on the chart. But Loran lines of position, like all LOPs, may be intermixed with those obtained by other means. The Loran LOP, for example, may be crossed with a radio direction finder LOP in order to establish a fix.

The readouts from the Loran-C rarely will fall directly on TD lines printed on the chart; they will have some value between two TD lines and interpolation will be necessary to locate them accurately. A linear interpolator is printed on the Loran chart for this purpose. (See Figure 19–9.)

The linear interpolator permits graphical solution of what otherwise would be a mathematical problem in proportion. Only a divider is required. The divider is set to the space between the two TDs in question and this is transferred to the linear interpolator at the point equivalent to the microsecond difference between the two TDs. At this location the divider is reset to the readout difference by means of the calibrated lines. The divider now can mark the proper spacing of the desired LOP between the two TDs.

Figure 19-9. The linear interpolator, printed on the chart, permits accurate interpolation between two TD curves with only a divider.

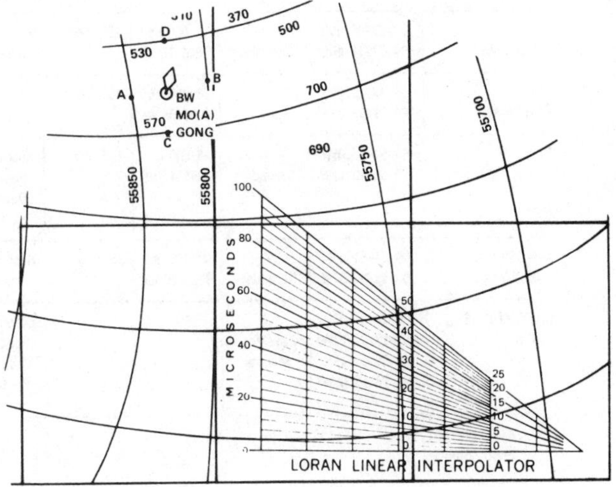

The Loran-C has a few minor quirks that should be borne in mind. When the boat is at the pier ready to leave for a long cruise, the Loran receiver is in an area of good signal strength and should get down to business quickly. In addition, its readings may easily be confirmed because the correct ones are known. Coming home into the fringes of the coverage area, the conditions are opposite. Signals are weak and the S/N ratio may be poor. The Loran-C needs more time to settle down and its operation should be watched closely.

A skipper who remains on the same coast of the United States hardly needs to touch his Loran-C except to turn it on at the start of a voyage. Only on his first run with the new receiver will he have to concern himself with selecting the GRI of the appropriate Loran-C group and tuning out local and ambient interference. Once everything works properly it will continue to do so, barring the occasional little upsets that electronics is heir to.

A goodly portion of the world's seas is already covered with Loran-C transmissions. Only the ultimate passagemaker may find himself reverted to his sextant, chronometer, and sight reduction tables (although it is fun to use these even in a covered area and compare results).

A listing of worldwide Loran-C groups is given in Figure 19–10. Many low-frequency radio transmitters located around the world are potential sources of interference for Loran-C receivers. Those most commonly encountered are listed in Figure 19–11 together with detailed readings of the interfering signal strength at various locations. The notch filters, mentioned earlier in this chapter, that are part of all commercial Loran-C receivers, are able to overcome the interference in most instances.

Figure 19-10. These charts give complete data for all the chains in the worldwide Loran C system. Latitude and longitude are given for the master and for all the secondaries in each chain.

LORAN-C DATA SHEET
MEDITERRANEAN SEA CHAIN—RATE 7990 (old rate SL1 (79,900 usec))

STATION	COORDINATES LATITUDE & LONGITUDE	STATION FUNCTION	CODING DELAY & BASELINE LENGTH	RADIATED PEAK POWER	REMARKS
SIMERI CHICHI, ITALY	38-52-20.23N 16-43-06.39E	MASTER		250 KW	Exercises Operational Control of Chain.
LAMPEDUSA, ITALY	35-31-20.80N 12-31-29.96E	X Secondary	11,000 us 1755.98 us	400 KW	
KARGABARUN, TURKEY	40-58-20.22N 27-52-01.07E	Y Secondary	29,000 us 3273.23 us	250 KW	
ESTARTIT, SPAIN	42-03-36.15N 03-12-15.46E	Z Secondary	47,000 us 3999.76 us	250 KW	
RHODES, GREECE	36-25-20.66N 28-09-31.92E	System Monitor			Control for X & Y.
SARDINIA, ITALY	39-10-51.26N 09-09-35.02E	System Monitor			Control for Z.

LORAN-C DATA SHEET
NORTHWEST PACIFIC CHAIN—RATE 9970 (old rate SS3 (99,700 usec))

STATION	COORDINATES LATITUDE & LONGITUDE	STATION FUNCTION	CODING DELAY & BASELINE LENGTH	RADIATED PEAK POWER	REMARKS
IWO JIMA, BONIN IS.	24-48-04.22N 141-19-29.44E	MASTER		3.0 MW	
MARCUS IS.	24-17-07.79N 153-58-53.72E	W Secondary	11,000 us 4284.11 us	3.0 MW	
HOKKAIDO, JAPAN	42-44-37.08N 143-43-10.50E	X Secondary	30,000 us 6685.12 us	400 KW	
GESASHI, OKINAWA, JAPAN	26-36-24.79N 128-08-55.99E	Y Secondary	55,000 us 4463.24 us	400 KW	
YAP CAROLINE IS.	09-32-45.84N 138-09-55.05E	Z Secondary	75,000 us 5746.79 us	3.0 MW	
SAIPAN, MARIANA IS.	15-07-47.07N 145-41-37.62E	System Monitor			Controls W & Z.
KAMI SEYA, JAPAN	NOTE 1.	System Monitor			Controls X & Y. Time Service Monitor

NOTE 1. Monitor station and/or antenna physically relocated. Positions given on old Data Sheets no longer valid. System control established using correlated numbers.

LORAN-C DATA SHEET
U.S. EAST COAST CHAIN—RATE 9930 (old rate SS7 (99,300 usec))

STATION	COORDINATES LATITUDE & LONGITUDE	STATION FUNCTION	CODING DELAY & BASELINE LENGTH	RADIATED PEAK POWER	REMARKS
CAROLINA BEACH, NORTH CAROLINA	34-03-46.50N 77-54-47.29W	MASTER		1.0 MW	Exercises operational control of chain.
JUPITER, FLORIDA	27-01-58.85N 80-06-53.59W	W Secondary	11,000 us 2695.51 us	400 KW	
CAPE RACE, NEWFOUNDLAND	46-46-31.88N 53-10-29.16W	X Secondary	28,000 us 8389.57 us	2.0 MW	Host nation manned. Double-rated to NORLANΓ chain (7930)
NANTUCKET, MASSACHUSETTS	41-15-12.29N 69-58-39.10W	Y Secondary	49,000 us 3541.33 us	400 KW	
DANA, INDIANA	39-51-08.30N 87-29-12.75W	Z Secondary	65,000 us 3560.73 us	400 KW	
Electronics Engineering Center, Wildwood, N.J.	38-56-58.59N 74-52-01.94W	T Secondary	82,000 us 2026.19 us	200 to 400 KW	Experimental station. Not normally on air.
BERMUDA, U.K.	32-15-53.18N 64-52-34.27W	System Monitor			Control for W,X &Y.
EGLIN AFB, FLORIDA	NOTE 1.	System Monitor			Control for Z.

NOTE 1. Monitor station and/or antenna physically relocated. Positions given on old Data Sheets no longer valid. System control established using correlated numbers.

LORAN-C DATA SHEET
NORWEGIAN SEA CHAIN—RATE 7970 (old rate SL3 (79,700 usec))

STATION	COORDINATES LATITUDE & LONGITUDE	STATION FUNCTION	CODING DELAY & BASELINE LENGTH	RADIATED PEAQ POWER	REMARKS
EJDE, FAROE ISLANDS	62-17-59.64N 07-04-26.55W	MASTER		400 KW	Host nation manned. Double-rated to NORLANT (7930)
BO, NORWAY	68-38-06.55N 14-27-48.46E	X Secondary	11,000 us 4048.16 us	250 KW	Host nation manned.
SYLT, GERMANY	54-48-29.24N 08-17-36.82E	W Secondary	26,000 us 4065.69 us	400 KW	
SANDUR, ICELAND	64-54-26.07N 23-55-20.41W	Y Secondary	46,000 us 2944.47 us	1.5 MW	Host nation manned. Double-rated to NORLANT(7930)
JAN MAYEN, NORWAY	70-54-51.63N 08-43-56.57W	Z Secondary	60,000 us 3216.20 us	250 KW	Host nation manned.
SHETLAND IS., U.K.	60-26-25.27N 01-18-05.22W[1] 60-26-17.49N 01-18-19.08W[2]	System Monitor			Exercises operational control of chain. Control for W, X, Y, Z.

[1] North antenna.
[2] South antenna.

Figure 19–10 (cont'd.)

LORAN-C DATA SHEET
NORTH PACIFIC CHAIN—RATE 5930 (old rate SH7 (59,300 usec))

STATION	COORDINATES LATITUDE & LONGITUDE	STATION FUNCTION	CODING DELAY & BASELINE LENGTH	RADIATED PEAK POWER	REMARKS
ST. PAUL PRIBILOFF IS., ALASKA	57–09–12.10N 170–15–07.44W	MASTER		400 KW	Controls X,Y,Z. Exercises operational control of chain.
ATTU, ALASKA	52–49–44.40N 173–10–49.40E	X Secondary	11,000 us 3875.17 us	400 KW	
PORT CLARENCE ALASKA	65–14–40.35N 166–53–12.95W	Y Secondary	28,000 us 3068.97 us	1.8 MW	
SITKINAK, ALASKA	56–32–19.71N 154–07–46.32W	Z Secondary	42,000 us 3284.83 us	400 KW	

LORAN-C DATA SHEET
CENTRAL PACIFIC CHAIN—RATE 4990 (old rate S1 (49,900 usec))

STATION	COORDINATES LATITUDE & LONGITUDE	STATION FUNCTION	CODING DELAY & BASELINE LENGTH	RADIATED PEAK POWER	REMARKS
JOHNSTON IS.	16–44–43.85N 169–30–31.63W	MASTER		300 KW	Exercises operational control of chain.
UPOLO PT., HAWAII	20–14–50.24N 155–53–08.78W	X Secondary	11,000 us 4972.38 us	300 KW	
KURE., MIDWAY IS.	28–23–41.11N 178–17–29.83W	Y Secondary	29,000 us 5253.08 us	300 KW	
FRENCH FRIGATE SHOALS	23–52–05.23N 166–17–19.60W	System Monitor			Controls X & Y

LORAN-C DATA SHEET
NORTH ATLANTIC CHAIN—RATE 7930 (old rate SL7 (79,300 usec))

STATION	COORDINATES LATITUDE & LONGITUDE	STATION FUNCTION	CODING DELAY & BASELINE LENGTH*	RADIATED PEAK POWER	REMARKS
ANGISSOQ, GREENLAND	59–59–17.19N 45–10–27.47W	MASTER		1.0 MW	Host nation manned.
SANDUR, ICELAND	64–54–26.07N 23–55–20.41W	W Secondary	11,000 us 4068.07 us	1.5 MW	Host nation manned. Double-rated to Norwegian Sea Chain (7970)
EJDE, FAROE ISLANDS	62–17–59.64N 07–04–26.55W	X Secondary	21,000 us 6803.77 us	400 KW	Host nation manned. Double-rated to Norwegian Sea Chain (7970)
CAPE RACE, NEWFOUNDLAND	46–46–31.88N 53–10–29.16W	Z Secondary	43,000 us 5212.24 us	2.0 MW	Host nation manned. Double-rated to U.S. East Coast Chain. (9930)
KEFLAVIK, ICELAND	NOTE 1.	System Monitor			Control for W & X. Exercises operational control of NORLANT chain.
ST. ANTHONY, NEWFOUNDLAND	NOTE 1.	System Monitor			Host nation manned. Control for Z.

NOTE 1. Monitor station and/or antenna physically relocated. Positions given on old Data Sheets no longer valid. System control established using correlated numbers.

LORAN-C DATA SHEET
SOUTHEAST ASIA CHAIN—RATE 5970 (old rate SH3 (59,700 usec))

STATION	COORDINATES LATITUDE & LONGITUDE	STATION FUNCTION	CODING DELAY & BASELINE LENGTH	RADIATED PEAK POWER	REMARKS
SATTAHIP, THAILAND	12–37–06.91N 100–57–36.58E	MASTER		400 KW	Exercises operational control of chain.
LAMPANG, THAILAND	18–19–34.19N 99–22–44.31E	X Secondary	11,000 us 2183.11 us	400 KW	
CON SON, RVN	08–43–20.18N 106–37–57.39E	Y Secondary	27,000 us 2522.07 us	400 KW	Contractor operated through DOD funding.
TAN MY, RVN	16–32–43.13N 107–38–35.39E	Z Secondary	41,000 us 2807.28 us	400 KW	Contractor operated through DOD funding.
UDORN, THAILAND	17–22–44.20N 102–47–12.40E	System Monitor			Controls X,Y,Z.

Figure 19-12. Loran C coverage in various areas is shown by these diagrams. A—U.S. West Coast; B—U.S. Southeast; C—U.S. Northeast; D—, Great Lakes. (U.S. Coast Guard) *on page 161*

Figure 19-11. The radio transmitters most commonly heard as interference on Loran C receivers are listed above together with their signal strengths at various locations. (U.S. Coast Guard)

Frequency	Transmitter Location (source of interference)	Authorized Power (kw)	Observed Field Strength (dB/uV/m) Cap Elizabeth ME	Sandy Hook NJ	EECEN Wildwood NJ	Mayport FL	Eglin FL	New Orleans LA	Plumbrock OH
70.387	Newfoundland	1.2			31.5				
71.142	Nova Scotia	2.7	43.5		40.8				
71.437	Quebec	1.2			35.7				
73.6	Nova Scotia	250	74.0	65.4	68.0	51.6	39.5	39.8	35.0
76.4	—	—							79.0
77.5	—	—			32.5			40.0	
84.465	Newfoundland	1.2			33.1				46.0
84.75	Ontario	0.9							
85.37	Nova Scotia	2.4	49.6		40.0	38.9			
85.43	Nova Scotia	2.4		44.2					
88.0 (±.85)	Annapolis	50		86.1					
112.3	Ontario	3.0			52.0			35.3	56.0
112.5	Halifax	15.0	66.0			40.7			
113.2	Ottawa	3.0	43.5	49.5	52.3	46.1			51.0
113.827	Nova Scotia	2.4					32.3		
114.3	Quebec	1.2	36.1					37.4	
115.3	Halifax	250		48.0					
116.0	—	—	40(est)						
116.8	—	—	36.0						
117.157	Quebec	1.2		46.2					
119.85	Norfolk	2.0	30(est)			42.4			
125.8	Manitoba	1.0		33.9					
128.25	New York	—							
128.25	Newport, RI	—	52.2	66.5	30.6				
128.25	Nonolulu	—					30.7		
131.1	Ontario	3.0	39.7	48.0	37.7				
133.15	Halifax	15	58.1		38.5				
134.9	Annapolis	100		77.7					
137.7	—	—						39.5	58.0
139.8 (±85)	Newport, RI	20							
139.8 (±85)	Norfolk	100		72.4	79.9	57.7	45.2	43.9	64.0
143.6	Nova Scotia	40	56.8		36.3				
145.4	—	—							
148.5	Ontario	3.0		57.3	48.3				

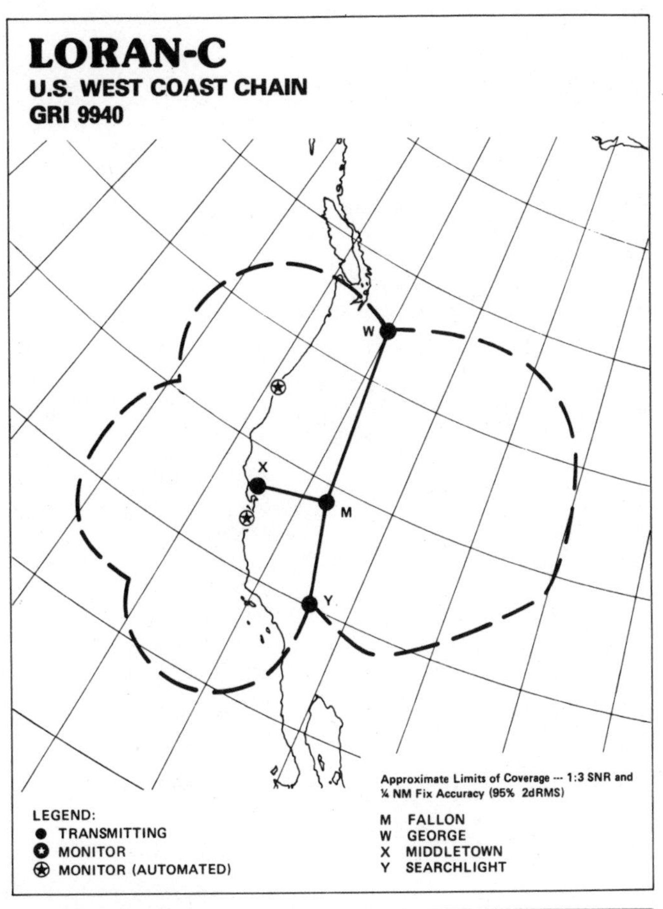
LORAN-C
U.S. WEST COAST CHAIN
GRI 9940
W
X
M
Y
Approximate Limits of Coverage — 1:3 SNR and ¼ NM Fix Accuracy (95% 2dRMS)
LEGEND:
TRANSMITTING
MONITOR
MONITOR (AUTOMATED)
M FALLON
W GEORGE
X MIDDLETOWN
Y SEARCHLIGHT

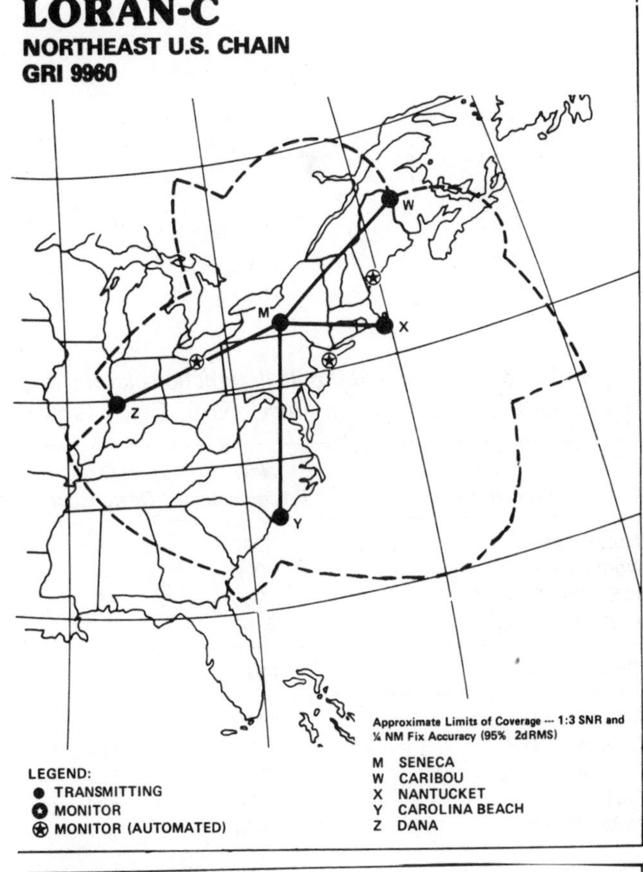
LORAN-C
NORTHEAST U.S. CHAIN
GRI 9960
W
M
X
Z
Y
Approximate Limits of Coverage — 1:3 SNR and ¼ NM Fix Accuracy (95% 2dRMS)
LEGEND:
TRANSMITTING
MONITOR
MONITOR (AUTOMATED)
M SENECA
W CARIBOU
X NANTUCKET
Y CAROLINA BEACH
Z DANA

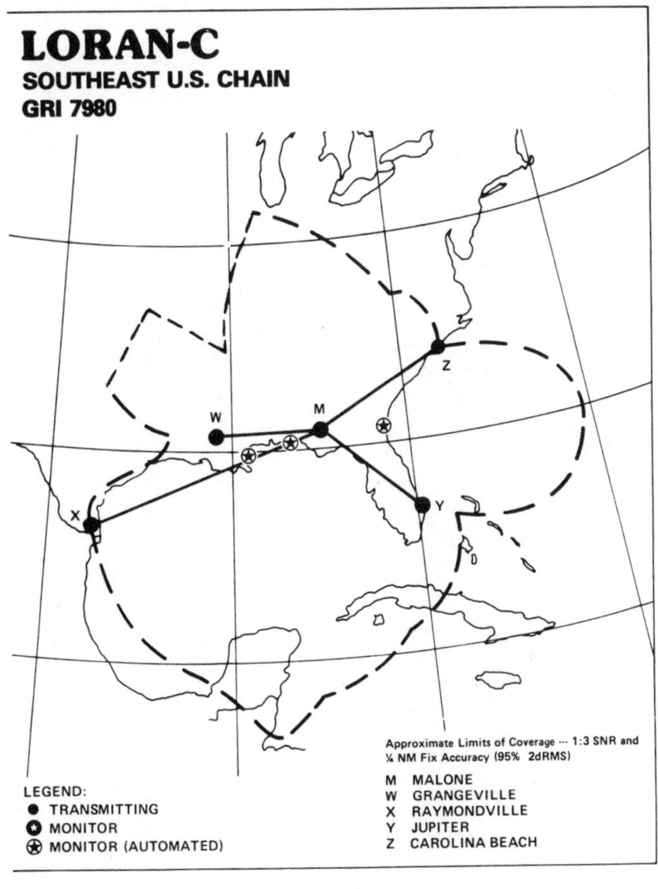
LORAN-C
SOUTHEAST U.S. CHAIN
GRI 7980
Z
W
M
X
Y
Approximate Limits of Coverage — 1:3 SNR and ¼ NM Fix Accuracy (95% 2dRMS)
LEGEND:
TRANSMITTING
MONITOR
MONITOR (AUTOMATED)
M MALONE
W GRANGEVILLE
X RAYMONDVILLE
Y JUPITER
Z CAROLINA BEACH

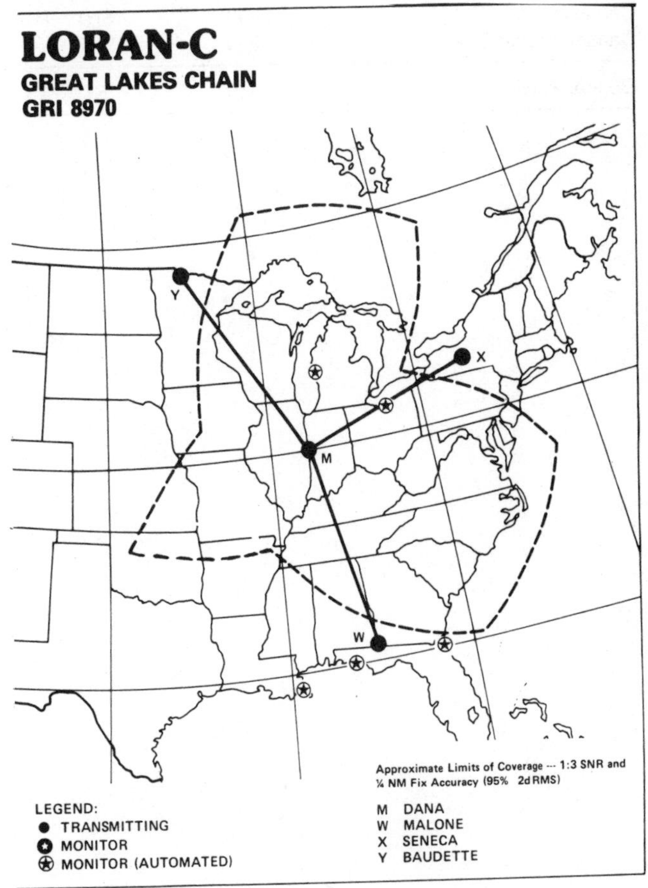
LORAN-C
GREAT LAKES CHAIN
GRI 8970
Y
X
M
W
Approximate Limits of Coverage — 1:3 SNR and ¼ NM Fix Accuracy (95% 2dRMS)
LEGEND:
TRANSMITTING
MONITOR
MONITOR (AUTOMATED)
M DANA
W MALONE
X SENECA
Y BAUDETTE

Figure 19-13. The new and the old designations for the existing Loran C chains are given above. (U.S. Coast Guard)

Chain Name	New Designation	Old Designation
North Pacific	9990	SS1
Northwest Pacific	9970	SS3
U.S. East Coast	9960	SS4
U.S. West Coast	9940	SS6
Great Lakes	8970	NONE
Mediterranean Sea	7990	SL1
Southwest U.S.	7980	SL2
Norwegian Sea	7970	SL3
Gulf of Alaska	7960	SL4
North Atlantic	7930	SL7
Canadian West Coast	5990	SH1
Canadian East Coast	5930	SH7
Central Pacific	4990	S1

20. *Radar and Its Use*

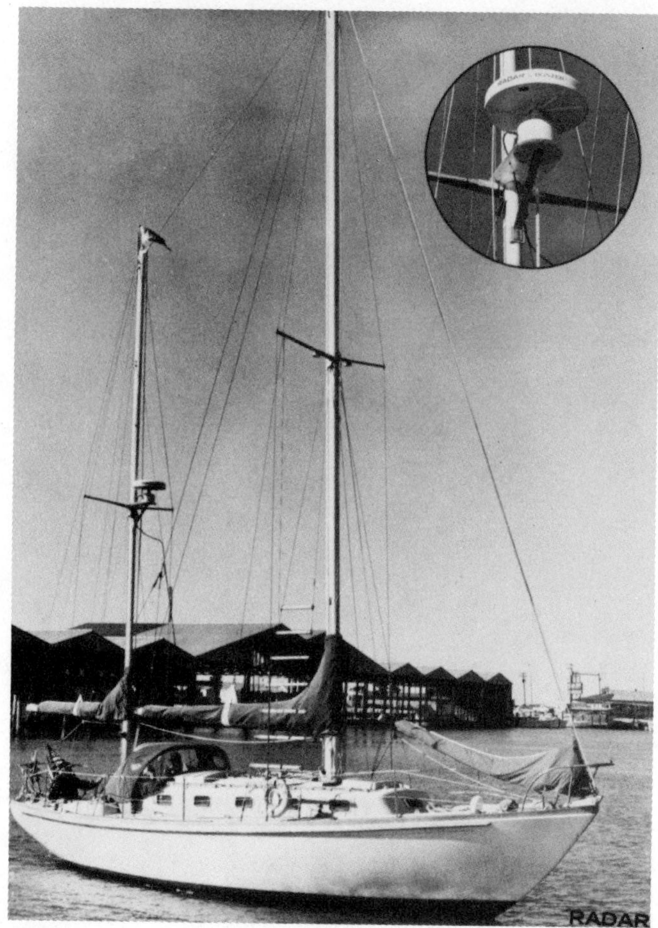

Figure 20-1. Radar is finding a place on constantly increasing numbers of sailboats and powerboats. Sailboats offer the advantage of height for the antenna unit. (Bonzer Inc.)

Mariners often blew a whistle or rang a bell and timed the return of the echo in order to estimate their distance off a cliff-lined shore. Their medium was sound waves that travel at only 1,100 feet per second.

Radar also times the return of an echo, but here the medium is radio waves that travel at 186,000 *miles* per second. So, while the mariners waited many seconds for their echoes, the radar echo is back with its information in a few *millionths* of *one* second. For all practical purposes, the return of the radar echo is instantaneous.

Timing the second echo was done simply with a watch or often only by counting. Timing the radar echo is infinitely more difficult because no known mechanical device is able to do this; the job must be left to modern electronics. Fortunately, sophisticated circuitry can divide one second into a million parts and then go right on dividing that into tenths and hundredths of one-millionth of one second. Such microscopic divisions time the travel of radar echoes very accurately. In fact, this fantastic timing ability may be said to have made radar suitable for practical navigational use.

Radar is an electronic "eye" that can "see" through fog and is not inhibited by darkness. It can measure with great accuracy the distance to the object "viewed." Specifically, radar is a system of *ra*dio *d*etection *a*nd *r*anging developed during World War II by the Americans, by the British, and perhaps also by the Germans. It has become an almost common piece of equipment on large vessels and small, commercial and pleasure boats. The suspicion even arises that on many small craft the radar antenna on its high perch serves the double purpose of also being a status symbol. Figure 20–1 shows a typical marine radar installation.

A radar system consists of a rotatable antenna, a radio transmitter, a radio receiver, and an indicator with a cathode ray tube similar to the well-known TV picture tube. Pulsing and synchronizing circuits interconnect these units and provide means for rapidly and continuously switching the system between transmitting and receiving. What the radar "sees" is depicted in shadowgraph form on the tube; the fine detail of television is totally absent. Rough gobs of light called "pips" represent the targeted objects and it takes considerable experience to identify the pips correctly.

If the searchlight atop a wheelhouse were rotated continuously in full circles, it would approximate the action of radar, except that it uses visible light waves to project its energy instead of invisible radio waves. This difference in medium, the contrast between microscopically short light waves and tremendously longer radio waves, accounts for the difference in how a target is seen by searchlight and by radar. (A typical view on a radar scope is shown in Figure 20–2.)

Just as it takes considerable imagination to see the various human and animal forms the Greeks attributed to the star constellations, so it becomes a task mostly for memory to connect the pips with the objects that cause them. Memory and the consequent ability to unravel pips are developed by long practice in clear day-

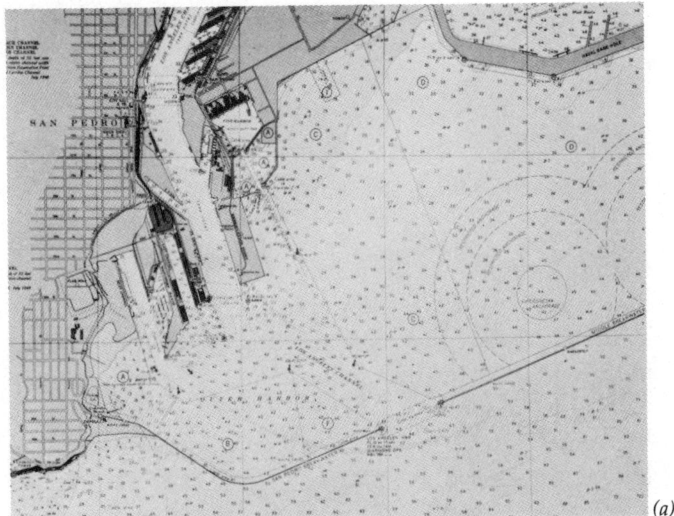

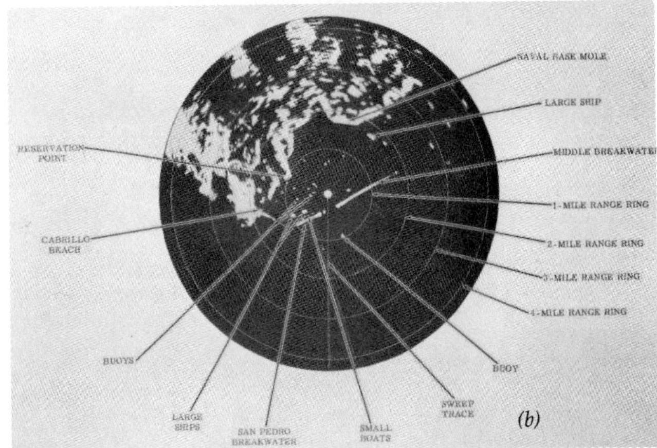

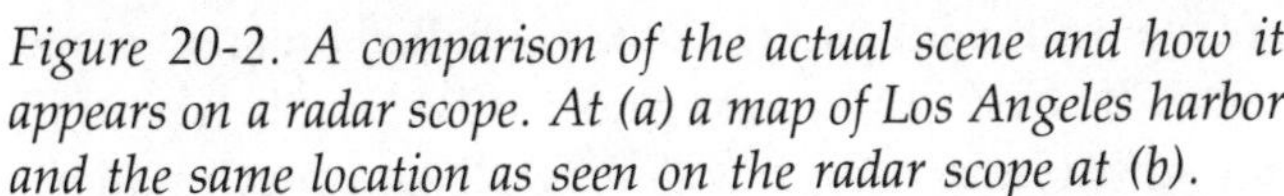
Figure 20-2. A comparison of the actual scene and how it appears on a radar scope. At (a) a map of Los Angeles harbor and the same location as seen on the radar scope at (b).

light when target and pip may be seen simultaneously. Such practice also builds the confidence a skipper needs later when he must rely on radar at night or in fog.

The sequence of operations of a radar system begins with the generation of a pulse of radio energy by the transmitter and its projection into space by the antenna. The antenna immediately thereafter is switched from the transmitter to the receiver to await the return of the radio echo that is then amplified into an equivalent electrical pulse. This pulse, fed to the indicator unit, causes the little burst of light that is the pip on the radar screen. Synchronizing circuits keep the angular position of the pip identical with the angular position of the antenna, thus establishing correct azimuth. The entire sequence is repeated as many as thousands of times per second while the antenna rotates and scans the area.

Radar makes use of extremely high radio frequencies, far above those commonly devoted to communication. For convenience, these are grouped into "bands" that extend into the thousands of megahertz. The exact radar wavelength in centimeters for any operating frequency in megahertz may be found easily by the following arithmatic shorthand: Divide 30,000 by the frequency. Thus a popular radar that functions at 9,410 megahertz is found to be on a wavelength of 3.188 centimeters and, by referring to the table, this is in the X band.

The shorter the wavelength of the projected radio waves, the smaller the distant object that can cause a pip to appear on the radar scope and, vice versa, the longer the wavelength, the larger the minimum size of a target that will reflect an echo. For example, a radar beam consisting of two or three centimeter waves will bounce back from rain drops and from snow flakes and these weather conditions will be indicated on the radar screen. Yet beams of ten centimeter and longer waves will not be obstructed by these precipitations, will not show rain or snow on the scope, and will pass right through the storm to go on to distant targets.

The fact that the conditions just enumerated exist gives rise to a form of radar interference called "clutter." As one would expect from the name, clutter fills the radar scope with indiscriminate little pips that confuse the main targets. Rain and snow are the most common sources of clutter on short wave radars, such as most of those designed for use on yachts. Clutter from the sea is also common; the echoes are returned from the flat surfaces presented by wind-driven waves as shown in Figure 20–3. Sea clutter usually is stronger on the windward side (this becomes self-explanatory from the drawing).

There is an inherent difference in the natures of the

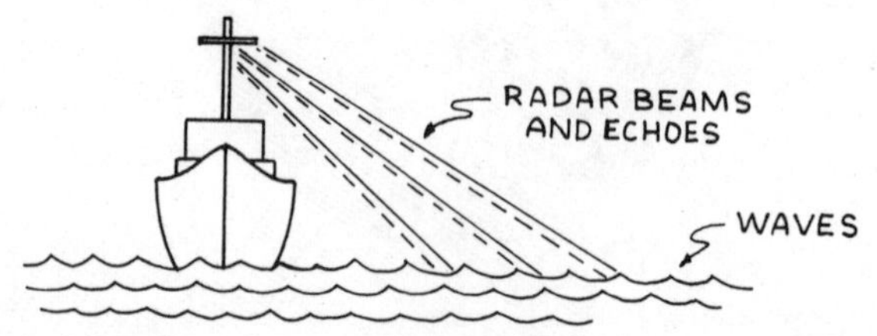

Figure 20-3. Sea clutter on the radar scope is caused by radar beams that strike waves and generate return echoes.

echo returns from weather clutter and from sea clutter; most radar sets take advantage of this technical dissimilarity to provide two separate controls for alleviation of this nuisance. One control, the "fast time constant" (FTC), reduces the clutter produced by precipitation. The other control, the "short time constant" (STC), subdues sea clutter. Both controls must be used judiciously because they may also wipe out desired targets while they are doing away with undesired clutter.

Manufacturers' literature presents long lists of technical specifications to describe their radars; understanding these should help a skipper to make the best choice for this vessel. Among the many parameters specified are: Operating frequency, peak power, pulse length, pulse repetition rate, ranges, display size, antenna type, and current requirement. Physically, modern radars are built into two units. the antenna assembly and the indicator cabinet, the two connected by cable.

Perhaps the most misunderstood of all the specifications is the one giving the range of the equipment. The skipper of a 42 foot cruiser opts for a 32 mile radar because he thinks it would be great always to be able to "see" that far in all directions. Will the installed set give him a 32 mile range? Of course not. Why? Because the distance to the horizon from the 20 foot high perch of the radar antenna is only about six miles—and radar waves travel line of sight. His working distance will be a bit more than six miles because refraction caused by atmospheric conditions will bend his radar beam down slightly beyond the horizon while the distant target may rise above it. (See Figure 20–4.)

In actual use, this skipper with the 32 mile equipment will discover that reliance on the shorter ranges of his instrument is his best bet. True, a landfall with high cliffs or high buildings will appear on the scope at longer ranges and this may be helpful to an ocean passagemaker. One further but minor point may be stated in favor of the long range radar installed on an average yacht: If it be of the very short wavelength type and has sufficient power, it will show distant precipitation.

Manufacturers express the strength of the output beams of their radars as "peak effective power" (PEP), and the unit generally used is kilowatts. The radars offered commercially for pleasure boats range from approximately 3 to 5 kilowatts PEP. If Ohm's law, that states 1 watt is equal to 1 volt times 1 ampere, be borne in mind, the arithmatic result of 5 kilowatts from a 12 volt battery becomes a technical impossibility. The radar would draw a minimum of 417 amperes even at 100 percent efficiency—patently absurd. Actually, a typical 5 kilowatt radar draws only 8 amperes from its 12 volt battery. What is the explanation?

The answer to the apparent anomaly lies in the factor of time and in the definition of peak effective power. The team "power" is meaningless until the duration of its existence is added. In this radar the full 5 kilowatts is generated and projected for less than one-millionth of one second at a time. The actual power is only a few watts.

Figure 20-4. The very short radio waves sent out by radars travel like light, in a straight line, although there is a slight bending around the curvature of the earth that increases detection range.

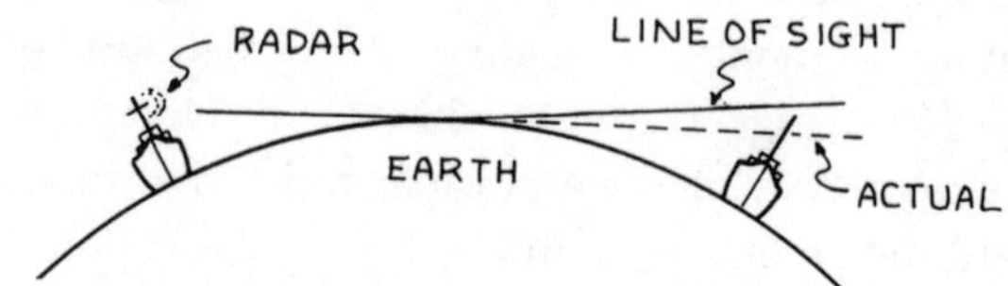

Consider the beam projected by a radar to be like a slice of pie placed vertically with its point at the antenna. The width horizontally and the height vertically are expressed in degrees as shown in Figure 20–5. Horizontal width should be as narrow as possible for best resolution of the targets. The narrow beam can distinguish between two objects close together while a wider one would "illuminate" both together at some instant and consequently would show them as one on the scope. A typical horizontal beam width is approximately two degrees.

Figure 20-5. The radar beam is shaped like a piece of pie on edge with its apex at the antenna. Width and height are specified in degrees.

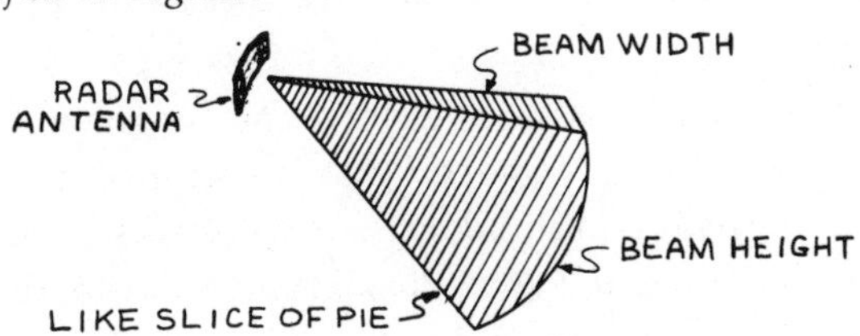

In contrast, the vertical beam should be as wide as possible. The reason for this is to maintain contact with a target while the boat is rolling in a sea. Vertical beam widths on yacht radars approximate twenty degrees.

Since the radar beam is specified in degrees, its actual width naturally becomes greater the farther it is from the antenna. At five miles from the antenna, the two degree beam mentioned above is approximately 310 yards wide and any two targets at this range must be separated more than this amount in order to show up individually. Another unwanted distortion also enters the picture because of beam width: Targets appear wider than they are. This occurs because the wide beam "illuminates" the target longer than would the ideal but unattainable small spot.

The distorted width of a target could amount to several degrees and must be accounted for if accurate bearings are to be taken for line of position (LOP) readings. The usual practice is to estimate the middle of the pip and to use that point for measurement with the cursor on the indicator scope. Incidentally, a moving target will increase the width distortion of the pip when it is crossing the bearing of the observing vessel.

The pulse repetition rate (PRR) has a direct effect on the distance capability of a radar set. The faster the PRR, the shorter the range over which the unit can function. This is easily explainable by considering that the echoes can be accepted by the antenna only during the pauses in transmission; the shorter these pauses, the less time for the beam to reach the target and for the echo to return (in other words, less distance of travel). However, a compromise is made here because enough beam hits must be made on the target to insure strong pips on the scope. On many radars the PRR and the pulse length change with changes in the range.

The diameter of a round brush determines the smallest dot that can be painted on the canvas. In this sense, the electron spot of the cathode ray tube also paints a picture, and fine detail may be had only with a spot that approaches a point in size. Controls that govern voltages in the picture tube circuits, and consequently spot size, are provided on the panels of most radars.

The duration of each pulse in the train of pulses that make up the pulse repetition rate determines how much radio energy hits the target—and how much can be reflected back. Longer pulses therefore make longer ranges possible. Because the projected energy declines as the square of the distance (like light), the need for a longer pulse to cover greater range adequately can be understood. Pulse lengths vary from a few hundredths of one millionth of one second to one tenth of one millionth of one second. Pulse repetition rates generally lie between 1,000 and 2,000 per second.

It can readily be seen that the rate of rotation of the antenna also affects the amount of "illumination" the target receives. Twenty revolutions per minute seems to be an average for the radars available to the yachtsman. Of course it is imperative that the sweep line on the radar screen rotate in absolute angular synchronism with the antenna; this is accomplished automatically by so-called "selsyn" motors or by various forms of synchronizing circuits.

The antennas used in early big ship radars were sections of parabolic reflectors, as pictured in Figure 20–6. The action in these is identical with what takes place in a searchlight, except that the upper and lower sections of what would be light reflectors are removed in order to get the pie-shaped radio beam. The output of the transmitter is located where the bulb of a searchlight would be.

Modern radars, especially those designed for yachts, replace the reflector with a slotted waveguide antenna. The beam-forming action here is entirely different from what has just been described. A "waveguide" is a hollow metal tube of circular or rectangular cross section through which radio energy of extremely high frequency may be sent with minimal loss. By taking a short length of such a waveguide and slotting it lengthwise, the electric and magnetic fields existing within it are shaped into the slice-of-pie configuration required for radar use. (A slotted waveguide antenna is shown in Figure 20–7.) A more complete discussion of this antenna is found in Chapter 24 together with a further explanation of waveguides.

Figure 20-6. Earlier high-powered radars had parabolic reflectors like this. The top and bottom of the parabola are cut away in order to shape the beam.

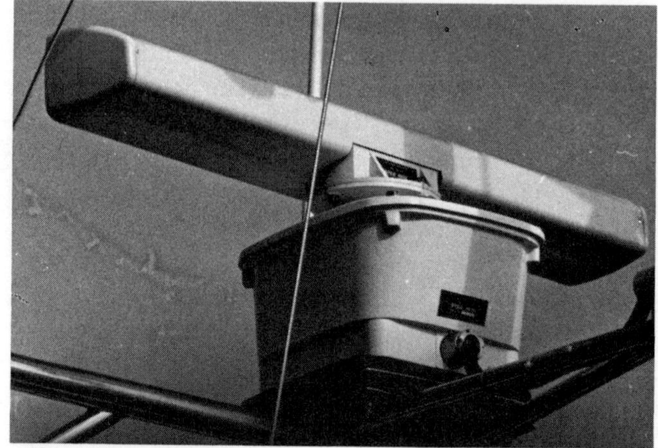

Figure 20-7. Modern radars use the more efficient and simpler slotted wave guide antenna, often enclosed in a plastic dome.

In present-day commercial designs of radars the operating frequency of thousands of megahertz is retained within the antenna housing structure. All conversions to a lower, more easily handled frequency take place therein. This supersedes the older method of transmitting the high radar frequency down to the indicator unit via a waveguide. Thus only a multi-wire cable is needed to connect the unit on the mast with the indicator by the helmsman—and a cable is much simpler to install than a waveguide.

The extremely high yacht radar frequencies are generated with the aid of resonant cavities instead of with coils and capacitors as is done in the transmitters and receivers described in earlier chapters. Minute variations in the true volumes of these cavities, even ambient changes, can offset the operating frequency sufficiently to affect performance. To counteract this, many radars have a tuning control on the front panel and usually also a pilot light to indicate resonance. A tiny twist of this control may often make a marked difference in the clarity and strength of target pips.

The inner face of the cathode ray tube screen is coated with phosphors that become fluorescent and emit light when struck by the electron beam. The scanning beam is such a line of light. It rotates from the center like the hour hand of a clock gone berserk and duplicates the antenna in speed and position. The phosphors have persistence and the glow of a pip remains even after the beam has passed, thus giving the impression of a continuous picture. (This beneficent glow may become confusing when quickly shifting from one range to another because for some seconds the old picture interweaves with the new.)

Incidentally, the view presented on the radar screen is called a "plan position indication" (PPI) because it resembles what would be seen from an airplane directly overhead (in other words, in the language of drafting, it is a plan view). The presentation on yacht radars is in "relative motion" but may be in "true motion" on more sophisticated equipment. With the relative motion picture everything moves when the ship does because the relative distances to targets constantly change; one's own ship is always at the center. Connecting the radar to the gyro compass on big ships makes true motion possible. The azimuth angle of any target from the ship may be measured by turning the cursor and reading the graduations.

The cursor normally is left with its zero mark directly at the top, at the index. Thus, one's own ship, at the center of the screen, is on a *radar course* of 000 degrees—and this does not necessarily have any relationship to the true course or the compass course the ship is travelling. With the cursor in this position, all targets are read on the degree scale with their azimuth *relative to the ship's course*. This target reading may then be converted to true or compass, if desired, by adding it or subtracting it from the ship's true or compass course.

An alternate method is to set the cursor to the true or compass course the ship is following. Then all azimuth readings of targets will be directly either true or compass. The "heading flasher," provided on most radars, can now prove helpful; it instantaneously brightens the scanning line when the antenna is pointing dead ahead and offers a constant reference.

Many yacht radars are used only cursorily for an occasional glance ahead to spot a buoy or a bridge abutment in bad weather. But the daytime, fair weather uses of radar are many, and encompass such piloting operations as finding lines of position and measuring distances. Concentric range circles on the screen are a fixed distance apart for every range to which the radar is set. Knowing the distance between two circles makes it easy mentally to interpolate and determine how far the target is from the ship. When this distance is desired exactly, right down to small fractions of a mile or even yards, an accessory called a "variable range marker" may be added to the radar. Such a device is shown in Figure 20–8; its distance readout is digital.

Figure 20-8. An accessory called a "variable range marker" gives a digital readout of distance to a target. (Intermarine Electronics, Inc.)

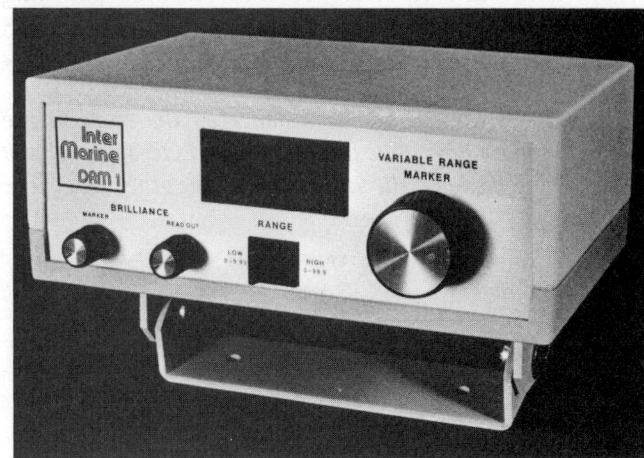

Lines of position may be derived from radar readings of range, azimuth, or a combination of both, and these often may be combined into fixes. Figure 20–9 shows the simplest LOP, in this case a circular one, obtained with one reading of the range of a distant lighthouse. The ship is somewhere on that circle. If a fortuitous second range can be obtained simultaneously, then the intersection of the two circles provides a fix, as also shown.

In Figure 20–10 two radar bearings have been made on charted objects and the resulting straight lines of position cross to form a fix. It should again be borne in mind that lines of position may be intermixed to get a fix. By whatever means they are established, LOPs are all compatible to solve the puzzle of the ship's location.

Radar of course achieves its greatest usefulness in the avoidance of collisions. And yet the anomaly remains: Some of the worst collisions of modern times have been between radar-equipped ships. Perhaps the obvious inference from this is that radar, wonderful as it is, cannot supplant human brains and is impotent to overcome human incompetence. Even the Rules of the Road have mandatory instructions concerning the use of radar.

Avoiding collisions entails more than frantic action at the moment when two ships obviously are about to crash into each other. Avoidance requires planning long before collision is imminent, and for this radar in competent hands is unexcelled. How much lead time is

Figure 20-9. Measuring the distance on the scope to the target provides the radius for the circular LOP on the chart. The bearing angle on the scope provides the other LOP. The crossing is a fix.

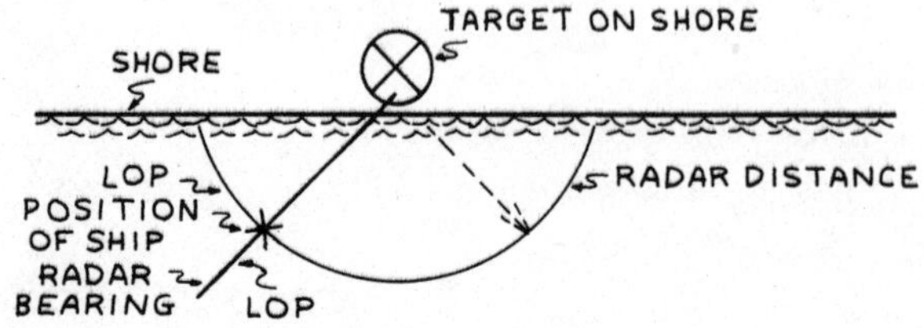

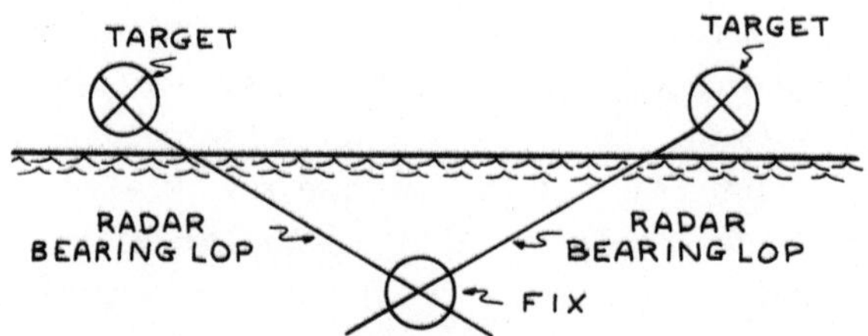

Figure 20-10. Bearings from two convenient targets measured on the radar scope provide a fix when transferred to the chart.

adequate depends upon the sizes and speeds of the ships in question. The larger and faster the vessel, the more time it requires to change course or to stop; big boats cannot be handled like outboards.

Plotting is the "tool" the skipper uses in connection with his radar to predict what will happen when his vessel and another are on converging courses. By taking several azimuth and range readings of the other boat, and plotting his findings, he can reach a clear decision whether to take evasive action or continue "as she goes." Plotting may look complicated but it is not.

Plotting is ideally done directly on the face of the radar screen with a grease pencil. But "ideally" in this case refers to big ship radars with large screens positioned conveniently for drawing lines on them. Pleasure boat radars do not permit this; they have neither the screen size nor the handy location. Yacht skippers have recourse to "maneuvering boards"; these are pads of paper with preprinted polar coordinates and distance scales, like the one shown in Figure 20–11. Frankly, it is more convenient and far easier to use the maneuvering board. (Maneuvering "boards" are available in pads as HO-2665-10 from the government printing office.)

The operations in plotting are straightforward. A skipper notes that a ship some distance off to starboard appears to be on a converging course with his own vessel. A look at his radar reveals that this ship is 4 miles distant and at an azimuth angle of 95°. He notes this down together with the exact time. Five minutes

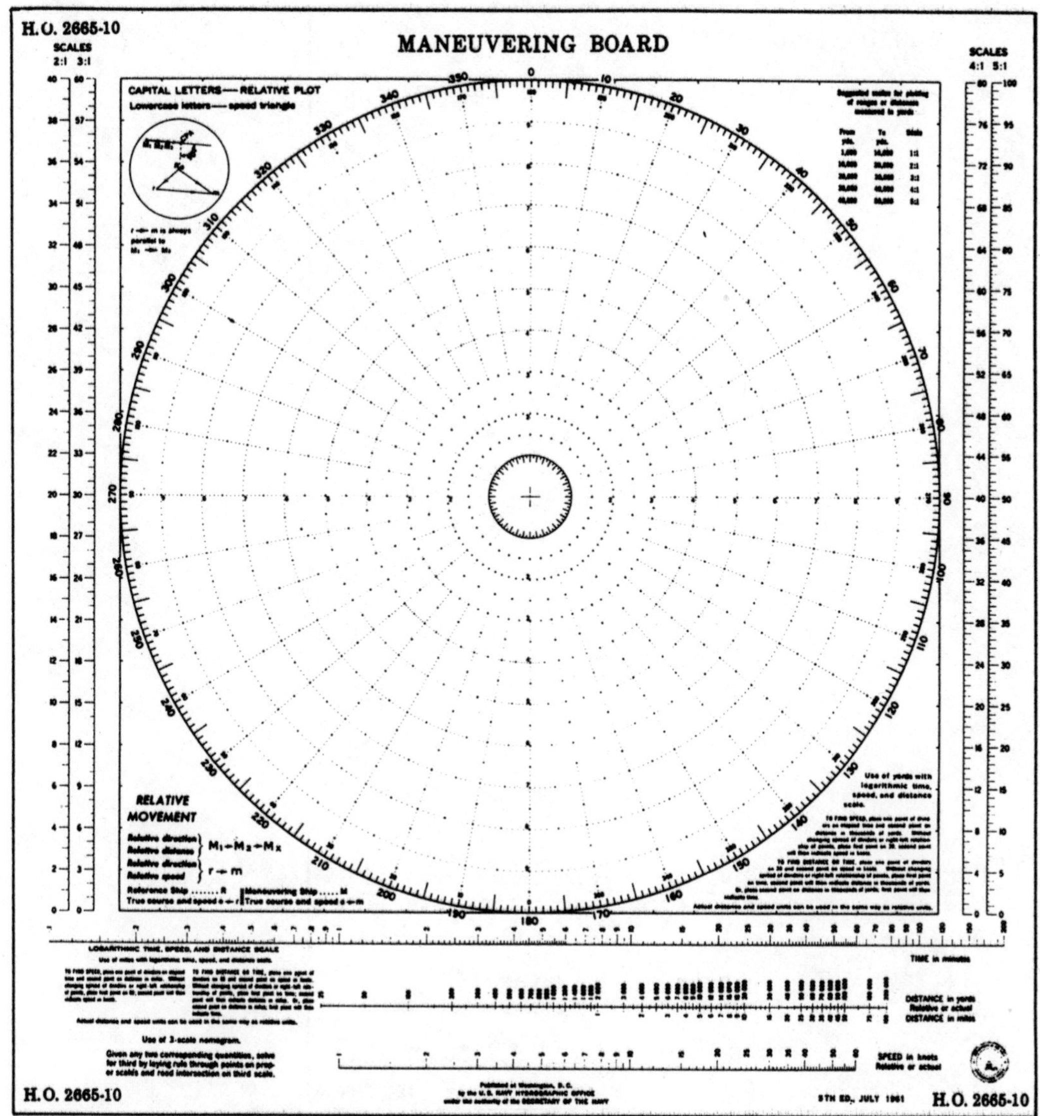

Figure 20-11. Maneuvering "boards" like this one are available in pads under the government number HO-2665-10.

later the radar readings are 90° and 3 miles and this too is noted. Another 5 minutes is allowed to pass before he checks his radar again and now the ship is 2 miles away at an angle of 80°. The skipper now needs to know the answers to the following questions: What are the direction of relative movement and the speed of relative movement of the distant ship? At what time will it cross his own course? Will this entail danger of collision? What will be the "closest point of approach" (CPA) of the two vessels if both continue exactly as they are going?

All of these questions are answered easily by plotting the information gained from the radar on a maneuvering board as shown in Figure 20–12. The only additional data the skipper requires is the speed of his own vessel—and this he may obtain from a knotmeter, from an engine rpm meter, or simply from his knowledge of his boat. Whether the intervals between radar readings should be five minutes, more, or less depends upon local conditions and common sense. The idea is to get plotting points that are far enough apart to provide a clear diagram without ambiguity.

Although there are several methods of making radar plots, the one explained in Figure 20–12 is quick and is the favorite of many cruising yachtsmen. However, one early warning danger signal that supersedes any plot must always be borne in mind: If the azimuth angle, the bearing, of the distant ship remains the same for several successive sightings and the distance is decreasing, a collision will result unless an abrupt course change is made quickly. The use of the scales at the bottom of the maneuvering board to solve time-

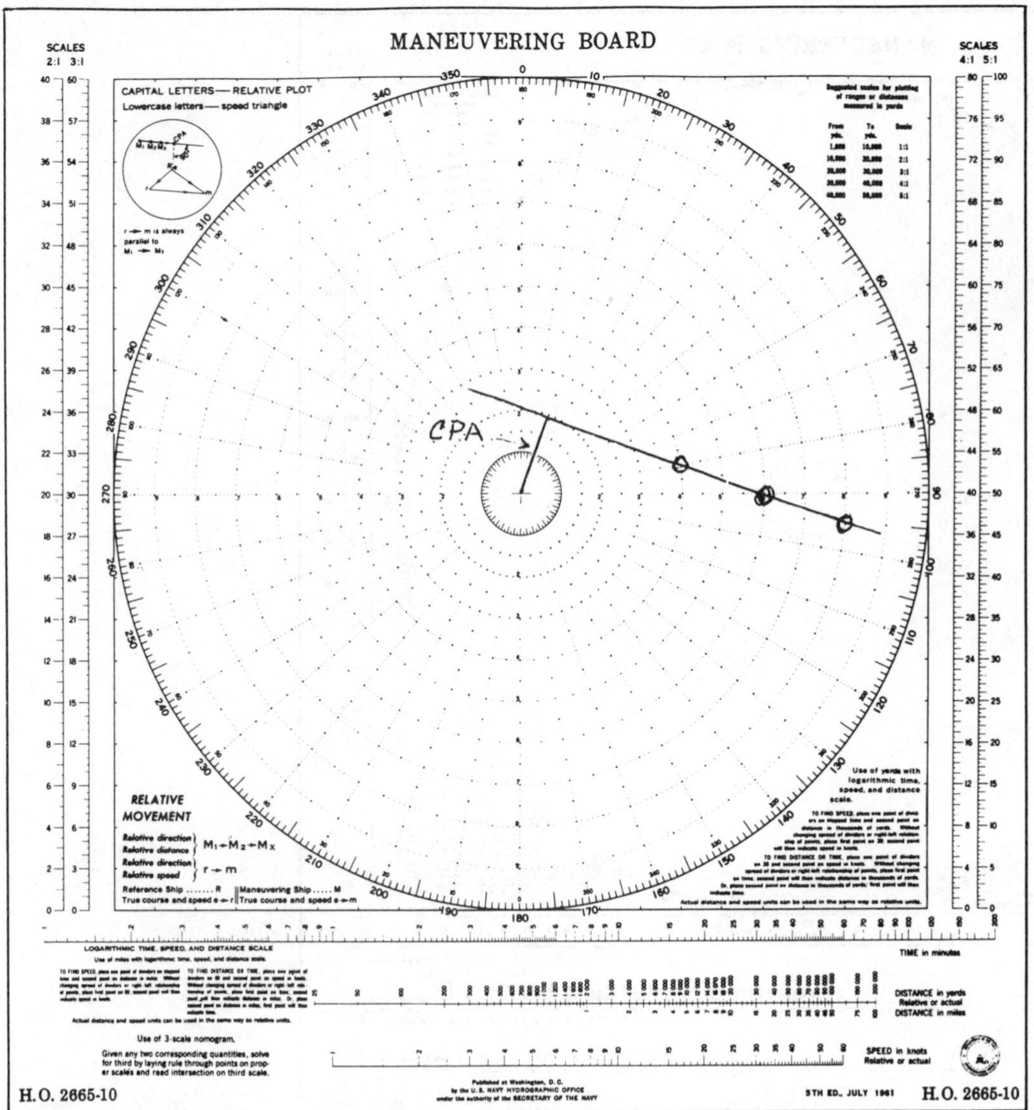

Figure 20-12. The figures in the text have been plotted on this maneuvering board at half scale. The closest point of approach (CPA) is seen to be one mile and at that time the other vessel will have a relative bearing of 20°. Measuring with dividers shows that the CPA will occur approximately 18 minutes after first sighting and that the relative speed of the other ship is approximately 12-3/4 mph.

speed-distance problems also is explained in the diagram.

An extremely useful radar accessory is shown in Figure 20–13. It is a self-contained electronic solid state device that maintains continuous watch of the radar picture and sounds an alarm if any target is found within a danger zone. The target could be another ship, a navigational aid, or a coastline that has become closer than the distance the skipper previously has set.

This warning device finds its greatest usefulness on vessels equipped with an autopilot and on which the helm often is left unattended. Digital readouts on the unit give the distance to the target causing the alarm and also the bearing in degrees. To avoid false alarms, circuits within the device evaluate the situation during at least three revolutions of the antenna before going into the danger mode.

A few cautions on installing the radar: Since the extremely high radio frequency employed by radar is a line-of-sight phenomenon, the antenna should be installed as high over the water as possible. This provides the greatest distance to the horizon, and the

Figure 20-13. This instrument monitors the radar scope continuously. It sounds an alarm whenever the radar detects a target within a previously selected safety zone. (Radar Devices, Inc.)

exact amount of distance for any height may be found from the table in Bowditch. The antenna support must be strong enough to carry the weight of the unit, which may reach fifty pounds, and also absorb the effect of windage.

Radar antennas on pleasure boats often are found directly in front of the flying bridge. This may prove a hazard to personnel because whoever is steering there is exposed to the radar beam. The effect on the person could be like that from a small diathermy machine and is best avoided. The situation is akin to the cautions about radio leaks from microwave ovens.

It is advisable to have the entire installation made by a technician with a radar license because such a person is legally required anyway for tuning and adjustment.

Many of the older radars, especially the more powerful ones, relied on magnetrons and klystrons to generate power and maintain frequency. The magnetron is actually a specialized diode under the influence of a powerful permanent magnet (see Figure 20–14). The klystron (see Figure 20–15) is a cavity resonator. The volume of the cavity is minutely adjustable to achieve a desired frequency.

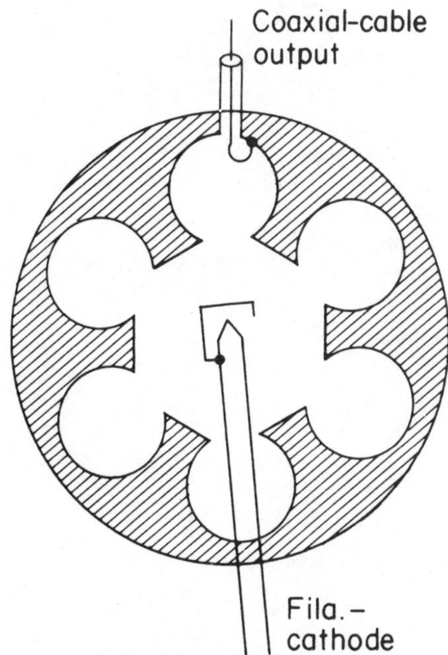

Figure 20–14. This is the anode of the magnetron that generates that ultra-high frequency of many radars. The size of the cavities determines the output frequency of the device.

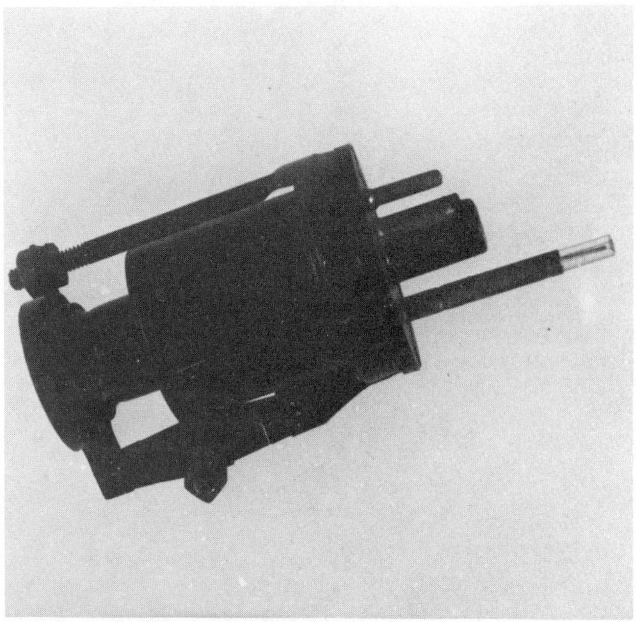

Figure 20-15. In many radars, the klyston, shown above, is the local oscillator that sets the radar frequency.

21. *Autopilots and Their Use*

Despite their name, autopilots do not "pilot"—they merely steer the course for which the skipper has set them. But this clarification of terms in no way denigrates them. A good autopilot easily is worth a couple of human helmsmen, and it does not eat nor does it sleep. Continuous steering is a monotonous job. On a long cruise an autopilot may make the difference between tedious work and pleasant recreation.

But note well that steering and piloting are two different things and that piloting remains the province of the skipper. It is up to him to maintain a watch to assure that the course is clear because the autopilot is "blind" and can blithely run into danger without a qualm. How constant that watch must be is determined by the density of traffic, by ambient conditions, and above all else by plain common sense.

All autopilots, whether simple or sophisticated, expensive or low in cost, consist essentially of three functional parts: A sensor unit, a control unit, and a power unit. These three usually are entirely separate but in one commercial design all three are in one housing. However, separate or not, functioning of all is identical.

The sensor unit aligns itself with some permanent directional phenomenon, usually the earth's magnetism, and monitors the relationship between this and the course desired to be steered. Any deviations from the course are converted into electrical signals that vary in form in different makes of autopilots.

The deviation signals are passed to the control unit that acts somewhat like a computer and analyzes what is happening. Is the course change small or large, is it a gentle falling off or an abrupt swerve? Are the changes random and intermittent, as in a rough sea, or does there seem to be a constant push, as from wind or current? The output from the control unit is a directive to the power unit to use force of the intensity and direction required to correct the unwanted deviation.

The power unit is the muscle that responds to the brain and moves the rudder. It derives its force through the medium of a motor or a servo motor that runs on battery current. Since the actual angular change of the rudder is small and comparatively slow, a miniature, high speed motor can be geared down to do the work with minimal battery drain.

Of course, each commercial autopilot embodies some unique feature of design that makes it superior in the eyes of its maker—and this gives the skipper a choice in purchasing. Price usually also enters into that choice. Autopilots are broadly classed into their suitability for powerboats or sailboats, wheel steering or tiller stearing. All autopilots on the market are good and will steer a boat, but the real separation of the superb ones from the just plain ordinary comes when conditions are bad and boats are cranky—the so-called separation of the men from the boys. The photos in Figure 21-1 show some autopilots suitable for yachts.

An autopilot must be compatible with the steering system inherent to the boat on which it is to be installed. Broadly, these systems are either mechanical or hydraulic, and each may take several forms. Under certain conditions, it is possible to add a mechanical type of autopilot to an hydraulic steering system; in such a case the autopilot simply turns the helm and leaves the hydraulic lines intact.

Choosing the correct autopilot for a specific boat entails certain factors that are often overlooked or else not given their proper importance. Every vessel has its own peculiarities when it comes to steering. Some take more muscle on the steering wheel, some take less. Some require the wheel to move only a spoke or two for a given turn, while others need a generous swing. Then again, some boats will not hold a straight course with the rudder amidships; the rudder must be offset a few degrees. All these items affect a proper wedding of autopilot to boat.

First, exactly how much force is needed to be exerted on the steering wheel in order to move the rudder? This force is expressed technically in "pound-inches"—and autopilot manufacturers likewise rate

Figure 21–1. Three styles of autopilot are shown below. (a) A unit for small sailboats that attaches to the tiller. (TillerMaster) *(b) A sailboat unit that operates the steering wheel directly.* (First Mate) *Three units of a standard autopilot are shown: the helmsman's control, (d) the remote control and the power source (c)* (Cetec Benmar) *(d) This automatic pilot adapts to the steering wheel of a sailboat (f)* (TillerMaster)

(a)

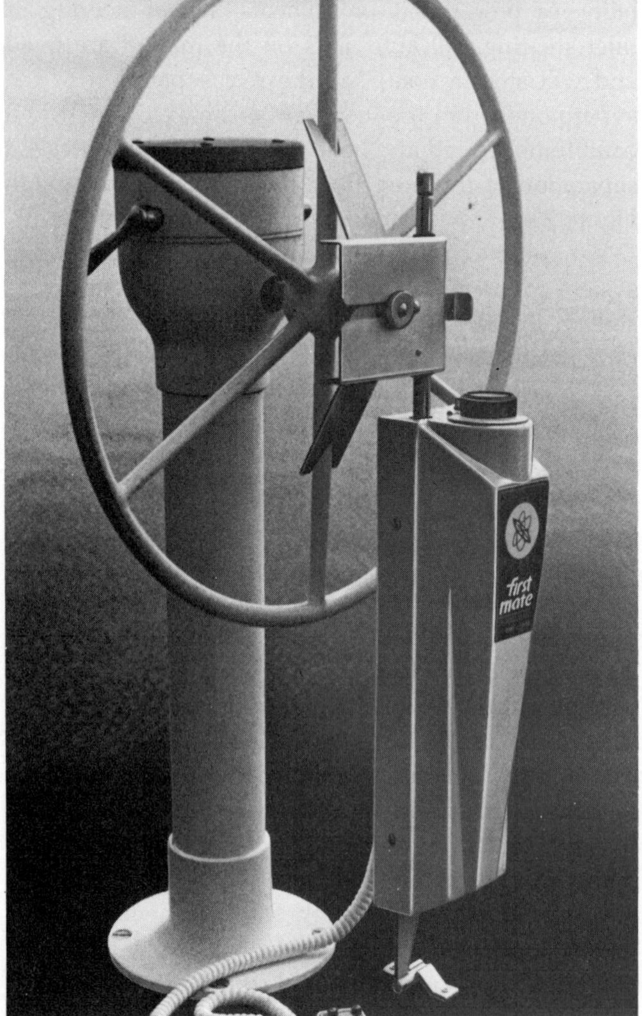

(b)

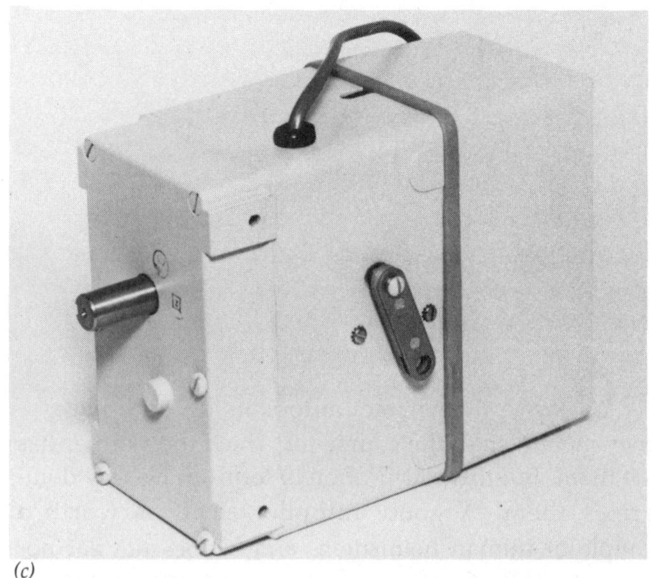
(c)

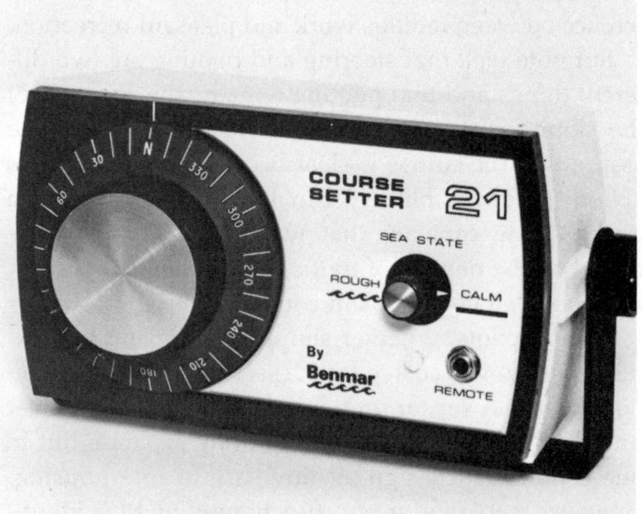

(d)

(e)

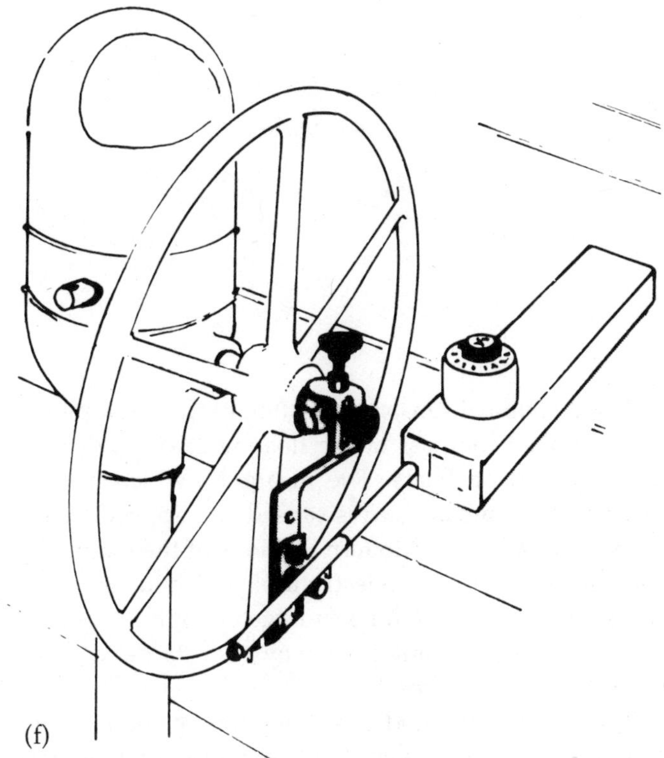

(f)
Figure 21–1 (cont'd.)

the abilities of their units in pound-inches to enable a correct choice. (A pound-inch is exerted when a twisting force of one pound is applied at a radius of 1 inch. Thus a 10 pound force at a radius of 10 inches produces 100 pound-inches—as would also happen when 25 pounds acts at a radius of 4 inches, or any similar combination whose product is 100.)

The measurement of the actual pound-inch requirement of a boat is a simple process and is illustrated in Figure 21–2. The only tools used are a fish weighing scale calibrated in pounds and a ruler. The scale is pulled tangentially until the steering wheel turns and then the scale reading is noted and multiplied by the radius of application in inches. The force will vary with the speed of the vessel, being least when standing at the pier and most when moving fast. The maximum figure is the one on which to base the choice of an autopilot. Also entering into the choice is the number of turns of the wheel that must be made to move the rudder from the port stop to the starboard stop, a swing called "hard over to hard over."

Figure 21-2. A common fish scale is rigged as shown in order to determine the number of pound-inches of torque the autopilot must supply to steer any given boat.

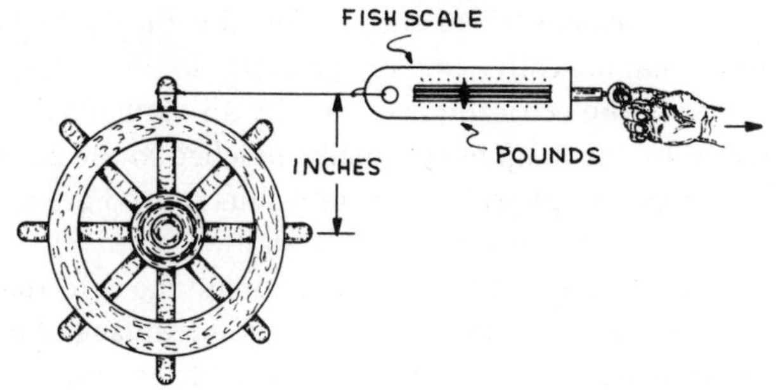

In addition to being offered a selection from various gradations of force and power, the skipper may also choose from three styles of performance: Hunting, non-hunting with a "deadband," and non-hunting with proportional rate action. The hunting type is the original and oldest, the least sophisticated, the lowest in cost, and it puts the heaviest load on the battery and on the steering system. The reason for this latter fact is that this type of autopilot is constantly in motion, constantly applying rudder correction in this direction or that to counteract the slightest yaw of the vessel. Since these applications are not modulated but always full force, oversteering results, and this must be corrected. The ship follows a course that is actually serpentine, with the undulations minimal with some hunting autopilots and pronounced with others.

The ideal toward which autopilot designers strive is to build a machine that duplicates the actions of an experienced helmsman. Such a man does not steer a boat with swift bursts of wheel turning. He anticipates and then tempers his response to the rate of falling off course as well as the amount. If, for instance, he gives left rudder to overcome a swing to starboard, he will apply right rudder at just the moment needed to stop the return swing directly on course without overshooting; seamen call this "meeting her."

The hunting autopilot obviously does not perform like an experienced helmsman. The non-hunting deadband machine mirrors the ideal a bit more closely while the non-hunting proportionate rate unit achieves a fairly good duplication. Quite naturally, these latter two autopilots are the more expensive and the most expensive, respectively.

The deadband is a small segment about the desired course within which the autopilot remains inactive. If the vessel does not fall off enough to get outside the deadband, no action is taken. On most units this deadband is adjustable from perhaps two degrees to as many as ten degrees. The skipper makes the adjustment in order to keep the ship more comfortable in rough sea conditions. What results is an averaging situation in which the errors to one side usually balance those to the other.

The autopilots so far discussed require feedback from the rudder; in order to make the proper response they need to know the exact rudder position at all times. This feedback may be mechanical, for instance with a flexible shaft, or preferably electrical with a potentiometer driven by the rudder post. The electrical output of the potentiometer is an analog of the rudder angle. Even a human helmsman feels more comfortable when an indicator on the console tells him what the rudder is doing.

One advantage that the makers of non-hunting proportional rate autopilots claim is that their units require no feedback and thus they avoid problems inherent in feedback devices. These autopilots sense not only the amount of divergence from the desired course, but also the rate at which it is happening. (Herein lies the similarity to the "experienced helmsman.") This information, in the form of electrical pulses, is integrated and differentiated by computerized circuits in the control unit. The computations occur instantaneously and the output signal to the power unit restricts the rudder movement to the predetermined rate and amount.

The efficacy of an autopilot may be judged by how well it keeps a vessel on a straight line. Perfection would be achieved when the serpentine path mentioned earlier is entirely absent. Since the shortest distance between two points is a straight line, it follows that undesired undulations from it add to distance travelled—and consequently also add to time taken and fuel used.

The directional sensor most often used in autopilots is a magnetic compass of the floating card type commonly found on boats. The means by which movement of the compass card is sensed and applied differ with manufacturers. Modern versions of the magnetic compass that function electronically without moving cards are also used. On boats large enough to sport a gyro compass, the autopilot may take its directional input from that source.

The simplest method of sensing the position of the compass card is with very light cat whisker wires that touch contacts as the vessel swings. The more sophisticated systems avoid physical contact with the compass card, and this is accomplished either with light rays or by electronics. In the "compasses" without cards, the magnetic field of the earth is dealt with directly, again by electronics. Each method has something in its favor.

Light rays can "read" compass card positions. In some cases the card is a transparent disc with certain portions of its surface made non-transparent in carefully controlled patterns. Small incandescent lamps are fixed to the case above the card and light sensitive diodes are stationed below it. Since the card remains pointed North while the case turns with the boat, it can be seen that the amount of light on the diodes varies and therefore their electric current also varies. This varying current, which also differentiates between right and left, is sent on to the control unit of the autopilot to compute what is needed to keep the vessel on course.

The compass card of the sensor unit may also be "read" magnetically. Stationary coils, located below the card and electrically activated by an oscillator, are affected by the magnets normally attached to the card. The changes in phase and current thus caused are the signals that influence the control unit.

These magnetic sensors that eliminate the card that characterizes a compass have the basic advantage of no moving parts. A finely wound coil with a high permeability core is sensitive to the angle at which the earth's magnetic field "cuts" it; this generates the necessary error signal for the control unit to work on. The earth's field may also be called upon to work on a conductor in the so-called "Hall effect" phenomenon.

About a century ago a scientist named Hall discovered that magnetism, applied in certain ways, affected the ability of a conductive material to carry electric current. The discovery remained a laboratory curiosity until modern electronic technology made it possible to amplify the minute Hall effect to useful levels. The Hall effect "compass" monitors the angle at which the earth's magnetic field strikes it by the changes in its electric circuit. (See Figure 21–3 for a graphic explanation of the Hall effect.)

As noted earlier, each method of directional sensing has its good points and often, unfortunately, its bad. Those systems that rely on the intereffect of the compass card magnets on a nearby pickup coil may be sensitive to the tilting of the card that takes place in a bad sea. This places card and magnet at continuously varying distances from each other with corresponding differences in electrical signals generated.

One general improvement in autopilots seems to be industry-wide—the apparent abandonment of electro-mechanical relays to control the motor that moves the rudder. Relays have contacts that are subject to pitting and sticking; they have moving armatures that are vulnerable to the marine environment. Transistors, the "switches without moving parts,"

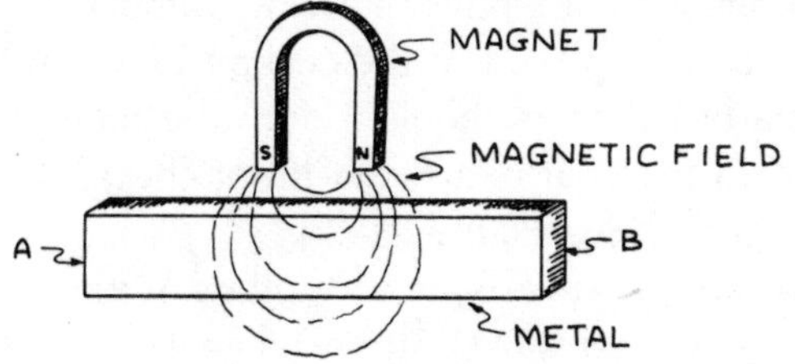

Figure 21-3. If a magnetic field be applied across the metal conductor as shown above, the electrical resistance from A to B will change. This is the Hall effect. When this phenomenon is used to guide autopilots, the magnetic field is that of the earth.

Figure 21-4. This form of autopilot compass has no pivoted card or other moving part and thereby avoids the effects of roll and yaw. (Alpha Marine Systems)

have taken over and theoretically they show no wear in service.

There is some controversy about whether or not the compass in the autopilot sensor unit should or should not be adjusted to compensate for deviation. The answer seems to lie in the piloting method the skipper habitually uses. If he sets his course by ship's compass and then cuts in his autopilot merely to follow it, deviation adjustment really can be overlooked. If, however, the dial on the autopilot is the primary means of course setting, obviously the sensor compass must be adjusted to as near correct as possible.

Speaking of the compass brings up a question relating to those autopilots in which the sensor compass is a separate remote device. (Such a combination is shown in Figure 21–4.) Where should this compass be located for best operation? The answer is a compromise between actual and ideal.

At first blush, it would seem that up near the top of the mast is ideal; there it would be away from deviatory influences. But that location greatly magnifies rolling, yawing, and pitching to the detriment of compass accuracy and compass bearings. Theoretically, low in the hull, at the center of gravity, all boat motions are at their minimum and this spot would seem to be perfect. Actually, the spot seldom is available for use and often is plagued by magnetic neighbors. The final answer is to stay away from the mast and get as close to the center of gravity as ambient conditions permit. (The sketches in Figure 21–5 explain boat motions.)

Figure 21-5. As noted in the text, the rolling and yawing of the boat may affect the compass response of the autopilot. The drawings explain both yaw and roll. (Benmar)

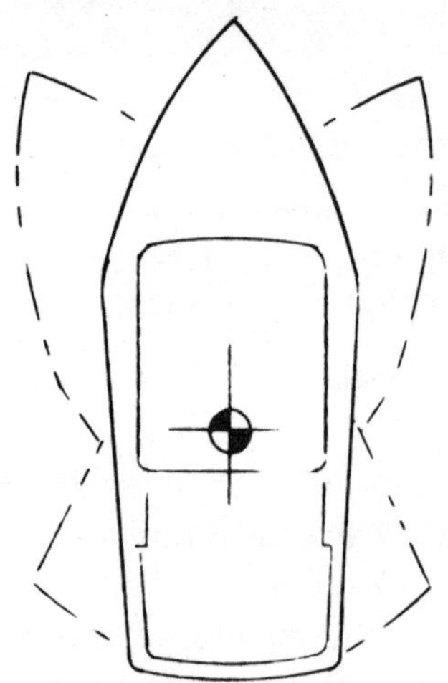

Center of Yaw

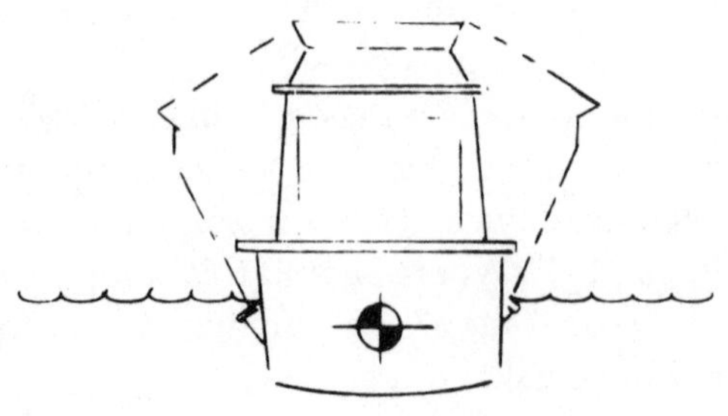

Center of Roll

Specialized autopilots are available for sailboats that are steered by tiller. These units are much more compact and usually house sensor, control, and motor in one housing, as shown by the representative device in Figure 21–6. Instead of the rotational output of the mechanical autopilot or the pressure output of the hydraulic one, the sailboat tiller model exerts its control by pushing and pulling to steer. The push/pull actuator connects to the tiller but in a manner that may be disconnected instantly when the skipper must take over.

The speed at which the boat is moving has a great deal to do with autopilot action. A small application of rudder at high speed may result in a sharp turn, while it may cause only a slight change in course when the vessel is moving slowly. (This can be understood quickly if it be compared to how an automobile steers.) Most autopilots have a front panel control with which to compensate for this steering phenomenon.

The larger autopilots have the potential for exerting considerably more force on the boat's steering system than a human can. This could become destructive, for instance if the rudder were to be forced against its stops by the steering motor. Limit switches are provided in some systems to guard against this happening. These switches disconnect the power just before the end of rudder travel.

The problems likely to be met when installing an autopilot can range from none to many. The simplest installation, because it really is no installation at all, is accomplished with a tiller steering sailboat device. The unit is merely placed in the cockpit between the port and starboard seats, hooked onto the tiller, and connected to the boat battery. A bit of checking and it is ready to go. (See Figure 21–7.)

The large powerboat autopilots with individual units are another thing entirely when it comes to installing them. Although manufacturers' instructions are explicit, most skippers will do well to turn to experienced technicians. Many checkout procedures are

Figure 21-6. The various units that make up one complete sailboat autopilot system are shown here. The specialized form of motor extends and retracts the arm that attaches to the quadrant. (Alpha Marine Systems)

Figure 21-7. This skipper is ready, autopilot dodger in hand, for any needed quick change in course. (Alpha Marine Systems)

complex and require special equipment for performing the step by step tests that verify correct operation.

A quick look at the steering systems normally installed in boats with which the autopilots must interface will prove helpful. The basic modus operandi of all hydraulic steering devices is explained by the simplified drawing in Figure 21–8. The steering wheel is locked to the same shaft as the pump. Turning the wheel causes the pump to force oil either to the left or the right, depending on the direction of rotation. The valve, with its combination of piston and spring-loaded balls, permits oil under pressure to reach the steering cylinder and push the piston that moves the rudder. At the same time, the valve permits oil from the other side of the piston to return to the pump and the reservoir. Another function of the valve is to block external forces on the rudder from affecting the steering wheel.

Of course, actual commercial hydraulic steering devices are a bit more complicated, but this does not necessarily make the adding of an autopilot any more difficult. In most cases the hydraulic autopilot is connected into the system with simple pipe tees so that its output goes directly to the steering cylinder and piston. Valves keep the manual system from interfering with the automatic—and vice versa. The piping and the tees are shown in Figure 21–9, while the installation on a boat of this unit is depicted in Figure 21–10.

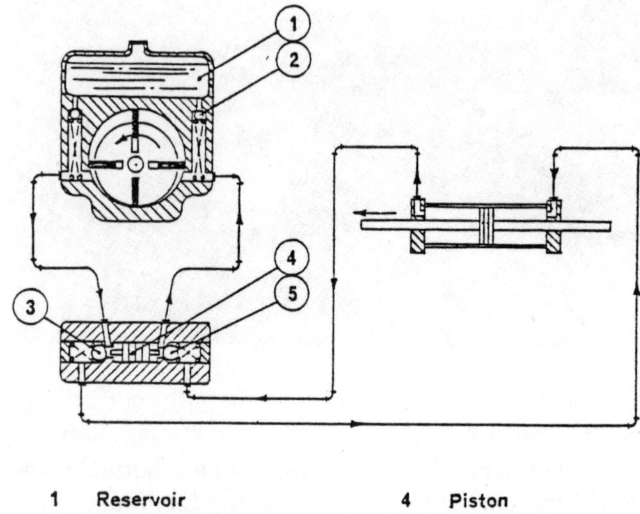

Figure 21-8. This schematic drawing explains the basic hydraulic steering system. (Wagner Engineering Co.)

The installation of a mechanical autopilot to a mechanical steering system is naturally entirely mechanical. The easiest is a drive to the shaft of the steering wheel—either by direct connection when the shafts are axially in line or by universal joints when they are not. This latter method is shown in Figure 21–11. An oft used practice places one sprocket on the steering wheel shaft and another, usually smaller, on the autopilot output shaft; a sprocket chain connects the two. An advantage of this method is the many variations in ratio of drive that may be made by suitably chosen sprocket sizes.

Most autopilots offer remote controls as an accessory. These, commonly called "dodgers" because of the quick course changes for which they are used, allow the skipper to be away from the helm and yet retain control of the steering. Some skippers use the dodger to achieve a form of "power steering" when the course-keeping ability of the autopilot is not being employed.

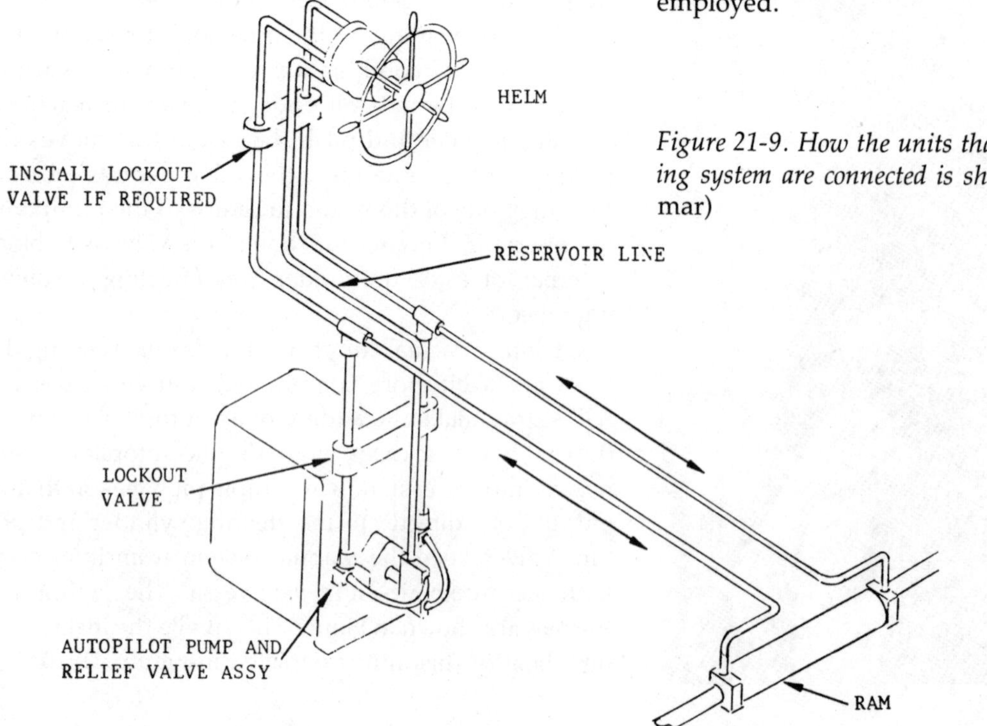

Figure 21-9. How the units that make up a hydraulic steering system are connected is shown in this drawing. (Benmar)

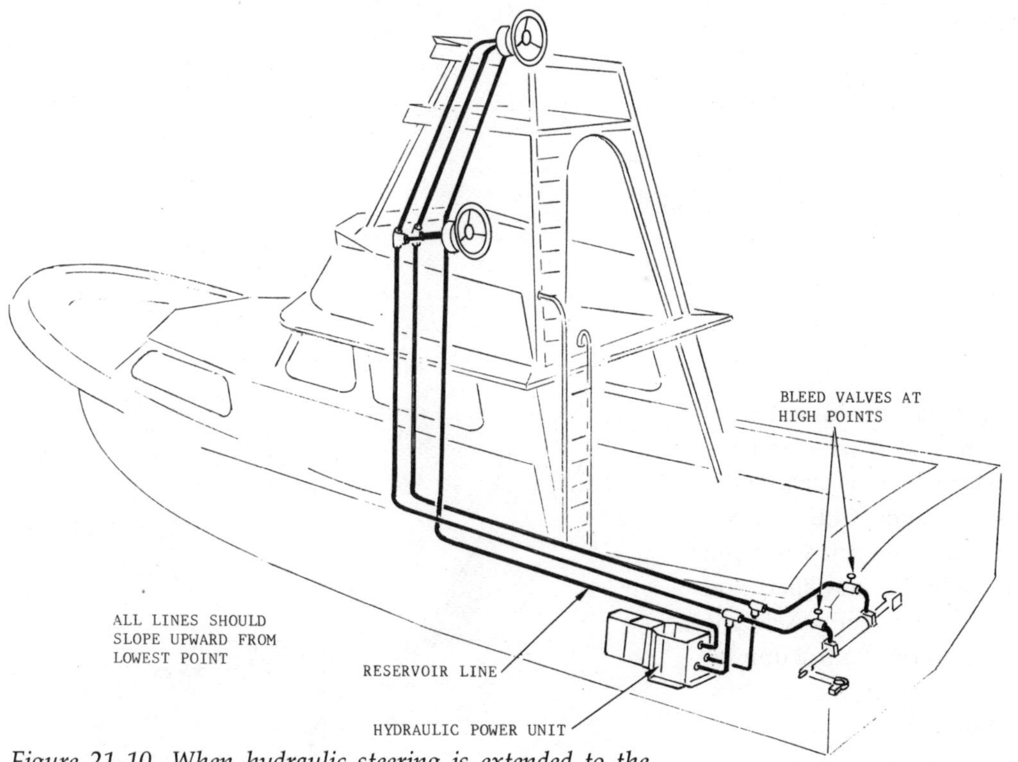

Figure 21-10. When hydraulic steering is extended to the flybridge, the piping connections are made as shown above. (Benmar)

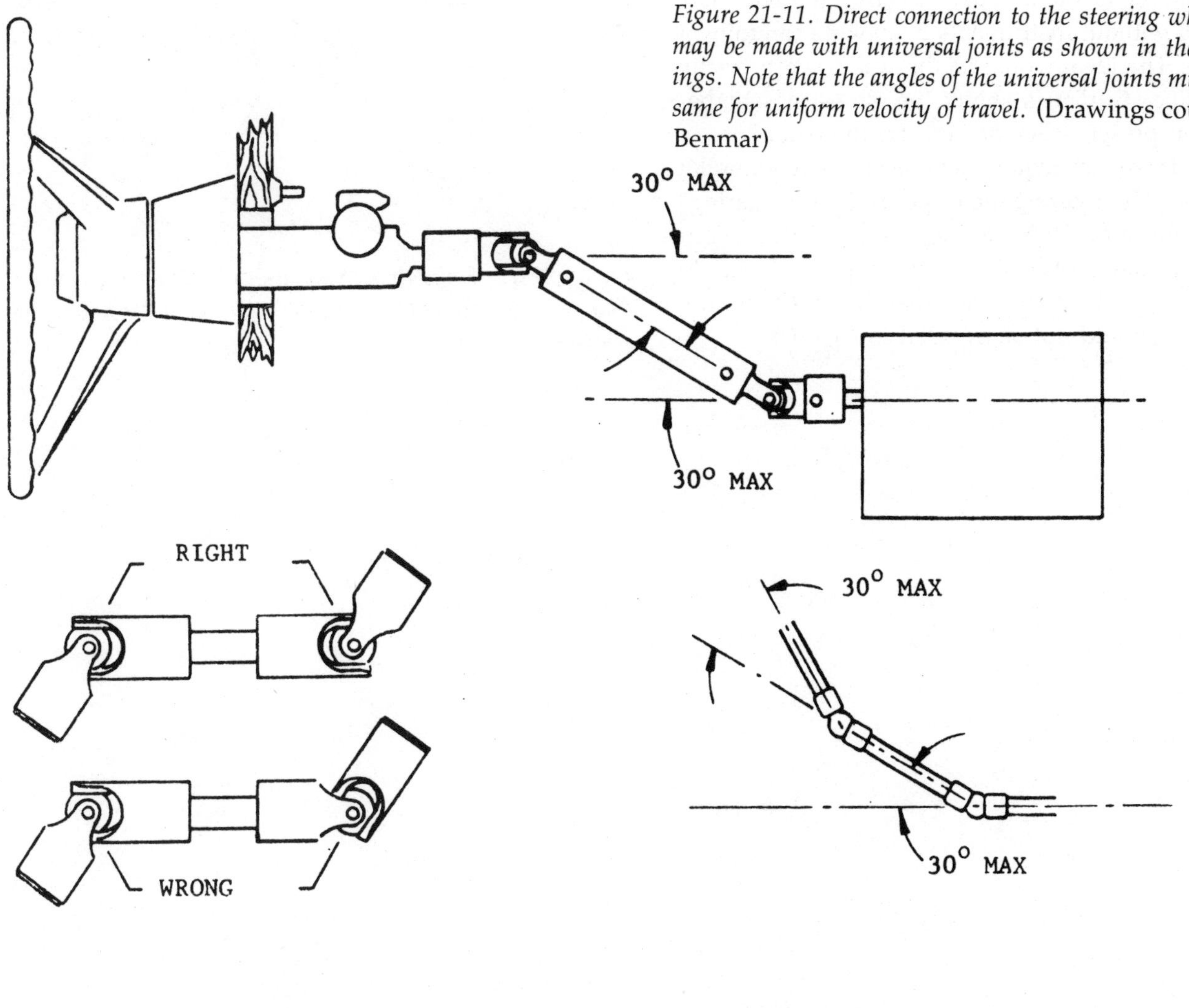

Figure 21-11. Direct connection to the steering wheel shaft may be made with universal joints as shown in these drawings. Note that the angles of the universal joints must be the same for uniform velocity of travel. (Drawings courtesy of Benmar)

30° MAX

30° MAX

RIGHT

WRONG

30° MAX

30° MAX

The sailboat skipper may even harness his autopilot to the wind instead of to a compass. A wind sensing vane like the one shown in Figure 21–12 is clamped to a convenient handrail. The vane aligns itself with the relative wind (the vector resultant of the true wind and the wind caused by the movement of the boat) and senses any deviation from the relative course set. The wind vane sends its error signals to the control unit for action, just as a compass does. As with the compass, the mast position is also avoided for the wind vane because ship roll and yaw are magnified there. A switch permits either the wind vane or the compass to do the directional sensing for the autopilot.

The efficacy of the autopilot steering the boat may be judged without instruments by looking carefully at the wake. The ideal would be the wake as a perfectly straight line. Curlicues should be almost absent because their presence is a demerit for the autopilot.

The maintenance requirement of a modern autopilot is minimal. The steering motor generally has bearings that are lubricated for life at the factory. The rest consists of eyeballing to see that everything is as it should be according to the owner's manual. Units that are not waterproof should not be exposed to weather.

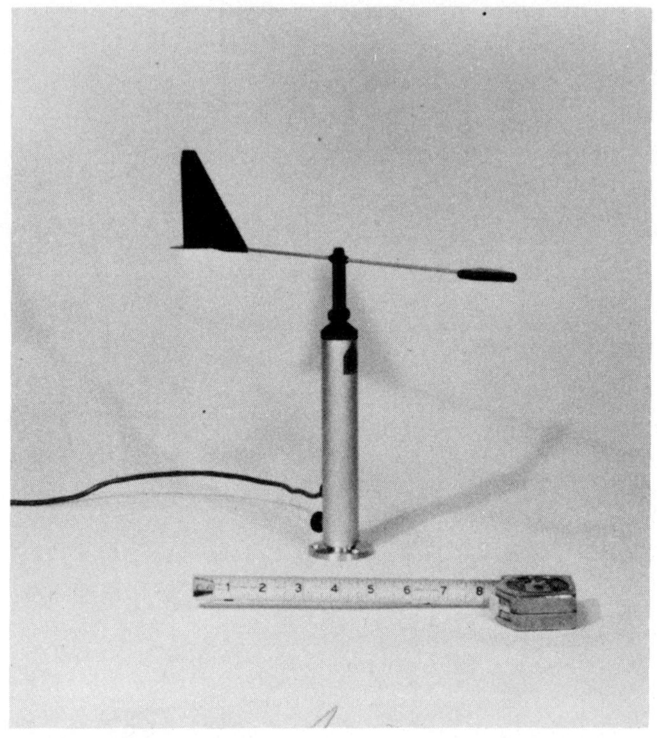

Figure 21-12. Sailboatmen may also use the apparent wind as a guide for the autopilot by adding a wind vane. (Alpha Marine Systems)

22. *Satellite Navigation*

The oceans of money poured into the space program have washed one bonus into the realm of the passagemaking skipper—satellite navigation. With the aid of extremely sophisticated electronics a special satellite passing overhead can provide a position fix accurate to a tiny fraction of one mile. And this magic demands little of the skipper other than plugging in a few local statistics and then reading latitude and longitude from an automatically presented printed tape.

The navigational satellites are a highly specialized form developed jointly by the U.S. Navy and Johns Hopkins University for this unique use. The program is officially called "Transit." Six Transit satellites are now in orbit, each encircling the earth once every 107 minutes at an altitude of approximately 600 miles, travelling at roughly 275 miles per minute. The satellites are all in polar orbit, meaning that each one crosses both the north and south poles on each complete pass. The spacing of the orbits is such that a ship anywhere on earth will experience at least six satellite passes per day from which to take bearings. This is true day and night, regardless of weather. Successive usable satellite passes may be as close together as one and one-half hours. See the drawing in Figure 22–1.

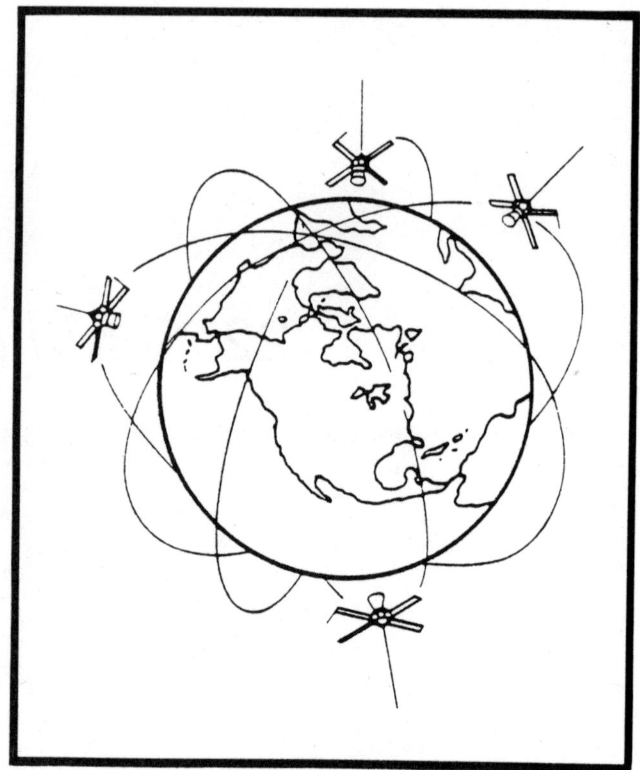

Figure 22-1. The navigational satellites follow the orbits pictured above and make positional information available anywhere on earth. (Navigation Communication Systems)

Despite the complexity of the equipment, the basic principle behind it all goes back more than 100 years to a scientist named Doppler. The story is that Herr Doppler, standing at the crossing waiting for a train to pass, noticed that the locomotive whistle's pitch was higher while approaching, then normal as it arrived directly in front of him, and lower as the train receded. This phenomenon is now called the "Doppler effect." Any frequency, whether audio, radio, or light, will seem higher to an observer when it is approaching him and lower when it is going away. The frequency shift will occur at the moment of passing.

Incidentally, the Doppler effect has become a basic tool in the modern measurement of speed, and many systems incorporate it. Radio is the most common medium whose frequency shift is measured in these devices in order to arrive at a speed reading, but sound and light also are employed. In fact, the change in the frequency of light is used by astronomers to determine how fast heavenly bodies are receding from the earth.

It should not be difficult to create a mental picture that will explain the simplicity of the Doppler effect. In the case of the locomotive whistle rushing to the observer, it is obvious that more cycles of sound will reach his ear per second. This is equivalent to saying that he hears a higher pitch than the whistle truly is. The reverse of this situation, when the whistle is rushing away, creates the illusion of a lower whistle pitch.

In the satellite navigation system, the Doppler effect

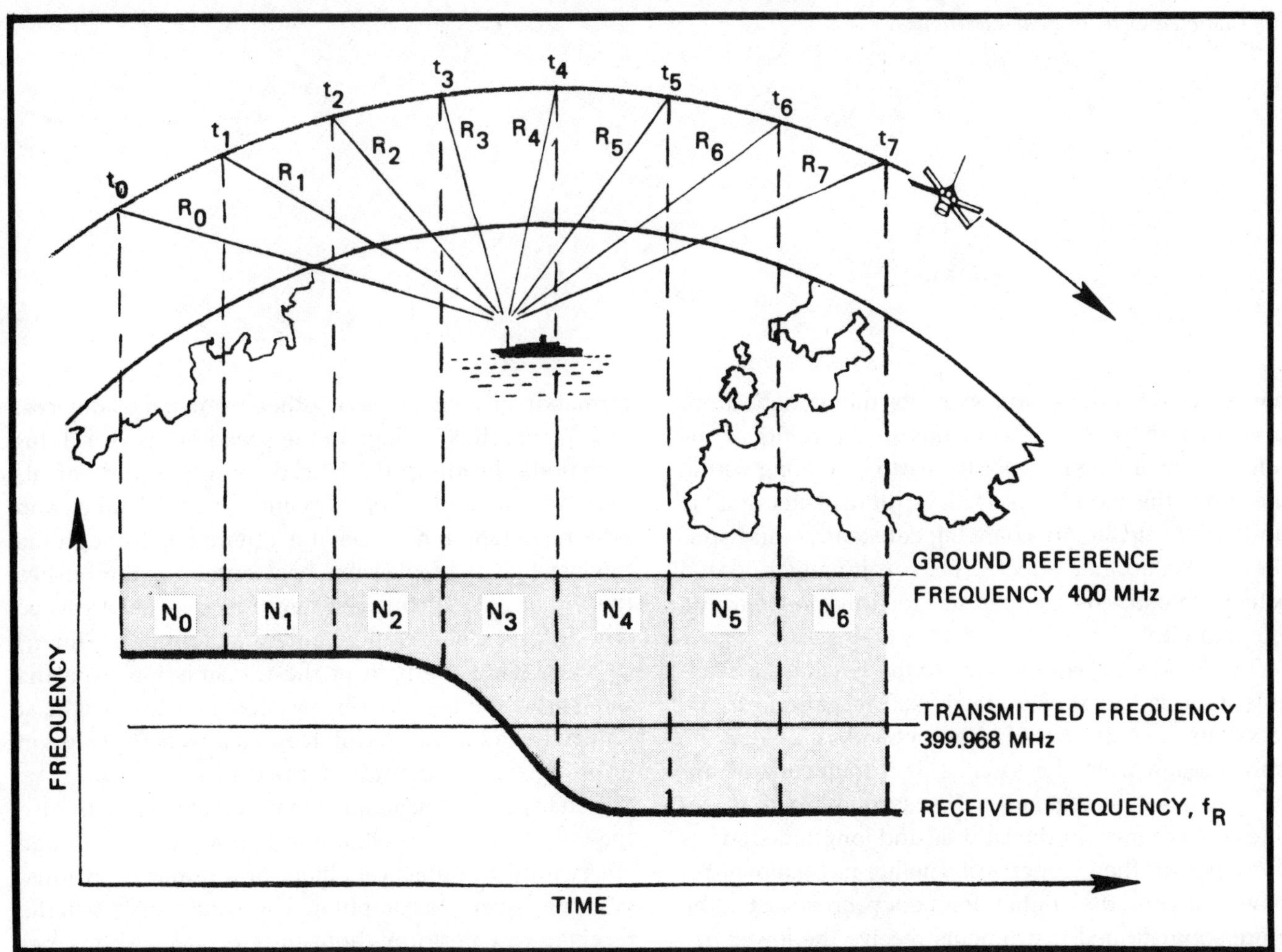

Figure 22-2. How the received signal changes in frequency as the satellite passes over is diagrammed above. (Navigation Communication Systems)

is the means for marking the exact instant when the satellite (and the ground point directly beneath it) are closest to the observing ship. Since the orbit of the satellite is known within very close limits, its position at that instant can become the basis for calculations that will emerge as the latitude and longitude of the inquiring ship. (The process is explained graphically in Figure 22–2.)

Each satellite transmits continuously on a frequency of 399.968 megahertz. This is an extremely high frequency and, since the Doppler effect may be handled more effectively on a lower one, a heterodyning process is introduced. A local oscillator in the receiver "beats" against the incoming transmission and creates a mirror image of it at a conveniently low frequency that accentuates the Doppler change. This is the frequency that rises and falls as the satellite comes and goes. (The heterodyne principle is in common use in super-heterodyne radio receivers and is even found in the front end of the satellite receiver. See Chapters 7 and 16.)

In order to be of practical position-finding value to a skipper, a satellite must meet three criteria: Its orbit must be known within a very small tolerance; its radio transmitter must maintain perfect frequency stability in order not to confuse the Doppler effect; and it must be governed by minutely accurate time. All three of these requirements are satisfied in the Transit navigational satellites. Quartz crystals and atomic standards hold the transmitted frequency at exactly 399.968 megahertz. A time pulse, for synchronization, is broadcast every two minutes plus or minus a few millionths of one second. The orbital data message sent out between time pulses is updated several times per day.

The need for constant revision of the orbital data describing the path is caused by gravity, or rather by the changing forces of gravitational pull. The earth is not a perfect sphere and even its minor deflections from that shape cause variations in the orbits of the satellites encircling it. Since the earth is rotating within the orbit, the mountains, valleys, and plains directly under the satellite are changing constantly—and so is their gravitational effect. The continually updated satellite broadcasts compensate for this problem and eliminate it.

Two modes of "connection" to the satellite are available. One is the approximately 400 megahertz transmissions just discussed. The second is a separate transmission from the satellite at a frequency of approximately 150 megahertz. This second broadcast refines the accuracy of the latitude and longitude figures presented to the skipper. All satellite navigational receivers accept the higher frequency broadcasts; the more sophisticated instruments receive the lower frequency transmissions as well. Price-wise, of course, the receivers with the double reception ability are much more expensive.

The satellite navigation system aboard ship is put into operation by supplying the receiver with local information. By switches or other front panel devices, the skipper must "plug" in the speed of his vessel, his directional heading, the latitude and longitude of his dead-reckoned position, Greenwich mean time, and also the height of his receiving antenna. (Although the modest height of pleasure boat antennas diminishes the importance of this last item for a yacht, it can be crucial for the skyscraping antenna of an ocean liner.) By visualizing the path of the transmission from the satellite to the boat, it can be seen that the height at which the radio wave is intercepted affects the solution of the imaginary triangles formed in space.

The involved calculations that must be made to turn the data from the satellite into latitude and longitude are beyond the practical ability of a human and must be turned over to a computer. The minicomputer in the receiver can perform thousands of calculations per minute and it arrives at the final answer by a repetitive processing scheme that reduces error step by step to a minimum. Mathematicians call this the "method of least squares." Figure 22–3 is a photo of a typical Transit satellite receiver.

Figure 22-3. A representative satellite receiver.
(Tracor Instruments)

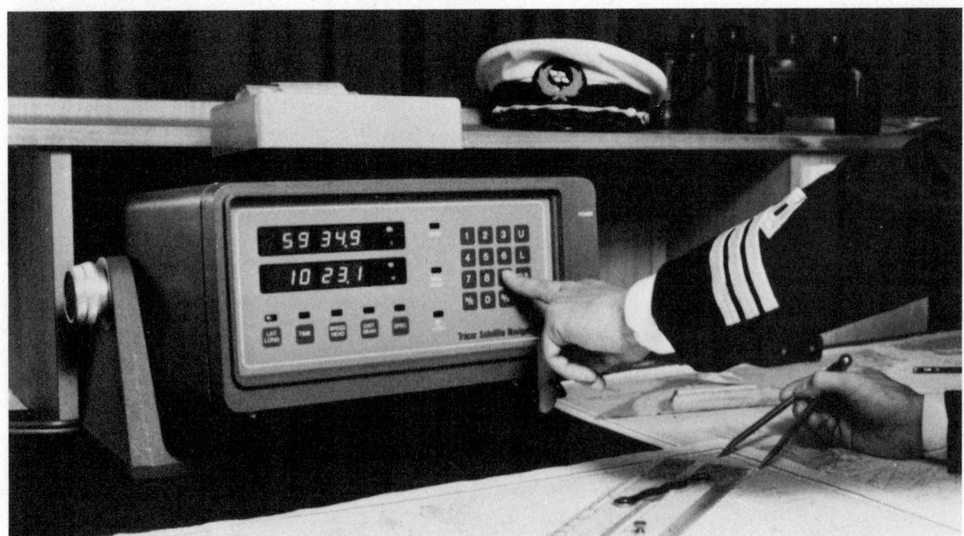

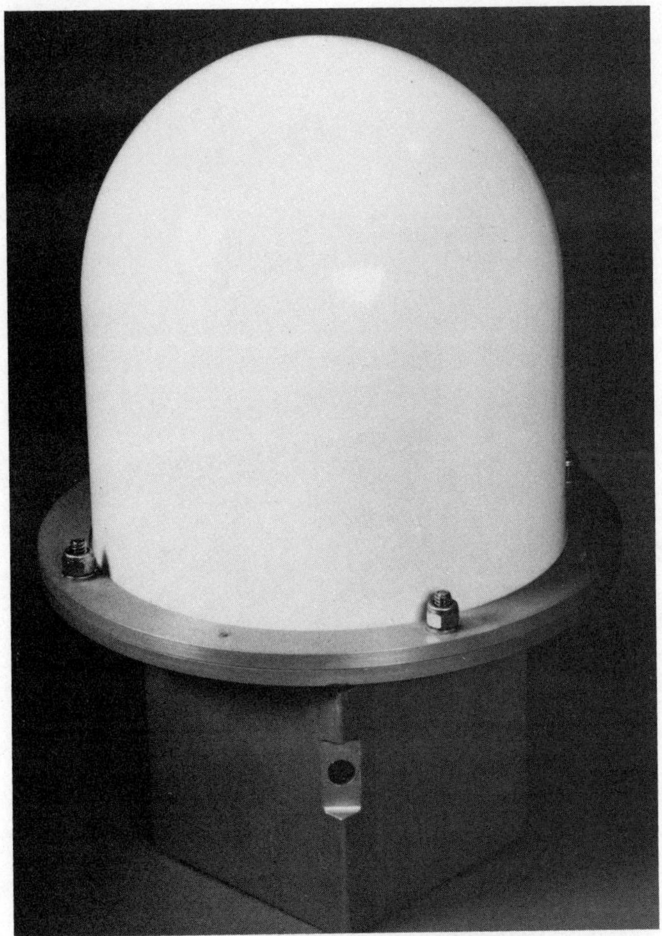

Figure 22-4. The antenna for the satellite receiver is housed in a plastic dome. (Satellite Communications Systems)

The satellite navigational system is an immense and successful government operation. Four tracking stations spaced around the globe keep the satellites from continuous observation and measurement. Computing centers transform the information thus obtained into orbital path parameters and so-called injection stations transmit this to the satellite memory for rebroadcast in the two minute intervals between time pulses. Subsequent injections change the memory as needed.

What actually happens as a result of all these computer machinations is the evolvement of a series of lines of position. However, although these LOPs are not plottable and are unlike the more familiar LOPs of simple plane sailing, they are usable grist for the computer mill. Within minutes the rushing electrons present the desired answer.

The antenna for the shipboard navigational satellite receiver is very small because of the extremely high frequency involved; the photo in Figure 22–4 shows a typical unit. The encasement in a foam-filled plastic dome hermetically seals the antenna and protects it from the ravages of salt sea air and primarily from the short-circuiting effects of salt encrustation. A preamplifier in the same enclosure boosts the received signal to compensate for the normal losses in the transmission line to the receiver. The two requirements for a good antenna installation are sufficient height and an unobstructed "view" around the full circle. Especially to be avoided are any nearby metal objects that could cause confusing rebroadcast or echoes.

Not all satellite passes are employable for precise navigation. Reception conditions may prove unsuitable and the elevation of the satellite, viewed from the ship, may be too high or too low (more than 75° or less than 15°.) This restriction is similar to the rejection in plane sailing of LOPs that cross at angles prone to larger errors.

Unlike Loran-C, Decca, and Omega (discussed in earlier chapters) satellite navigation does not provide the skipper with continuous fixes; information is received only during a satellite pass. However, modern satellite receivers keep on processing the data between passes and make fixes available on a dead-reckoning basis during intervals. The final results from all the systems are therefore much alike, with the bonus of greater accuracy from the satellites.

There is a slight margin for error in the satellite method and its origin lies in the local data that the skipper must plug in to initialize operation. For instance, if he mistates the speed of his vessel it will result in an ambiguity of approximately two-tenths of one mile in position for each knot over or under the exact figure. (Big ships feed speed information into the receiver automatically to avoid this problem.) Likewise, any course change of the ship during a satellite transmission will deteriorate the fix announced by the receiver.

Commercial satellite receivers give the skipper several options on the manner in which information is presented to him. Some units offer printed tape, others have illuminated digits and still others spell everything out on a cathode ray screen. By adding suitable extras all three forms of presentation may be had within a single system. All receivers have provisions for remote displays, for instance at the helm and also on the flying bridge or in the owner's cabin. The permanent record offered by printed tape may have legal and logbook value under certain conditions.

The versatility of the most sophisticated satellite receivers is almost beyond understanding. In addition to the basic positional data they present vital information obtainable in no other way. For example, they will identify the satellite by its code and predict its time of rise over the horizon, watch for a warn of arrival at preselected waypoints along the course, compute the set and drift of current or wind, monitor the functioning of its own internal circuits, and provide Greenwich mean time more accurately than most ship chronometers. And all this at the push of a button!

Solid state techniques of construction have made satellite receivers misers of power consumption. Liquid crystal diodes (LCD) and light emitting diodes (LED), both in the milliwatt range, are employed for readouts and pilots while transistors add their own frugal characteristics. The receivers accept direct current in a wide span of voltages or standard alternating current; the needed changes to working voltages are made internally. Some receivers contain a standby battery that maintains operation during power failure and thereby avoids the loss of data. The battery charges automatically under normal operating conditions.

Satellite navigation differs also from Loran-C and similar hyperbolic systems in the rapidity with which a fix is obtained. The Loran announcement of latitude and longitude is almost immediate once the receiver has tied itself in. By contrast, the satellite receiver may take several two-minute broadcasts before it is ready to declare its findings as exactly such and such a latitude and longitude. Figure 22–5 is a reproduction of an actual output tape from a receiver. Figure 22–6

Figure 22-5. The completeness of the information presented to the navigator by the satellite receiver is shown by the reproduction above of a receiver printout. (Navigation Communication Systems)

Figure 22-6. The screen of the remote cathode ray tube duplicates the information given on the printout. (Navigation Communication Systems)

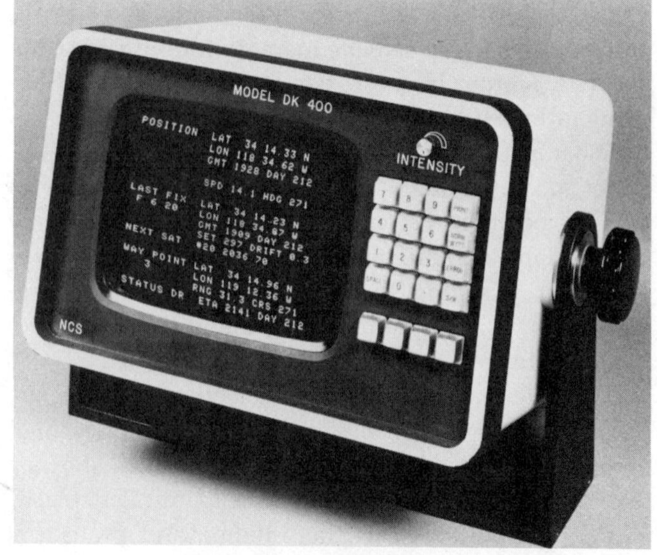

(a)

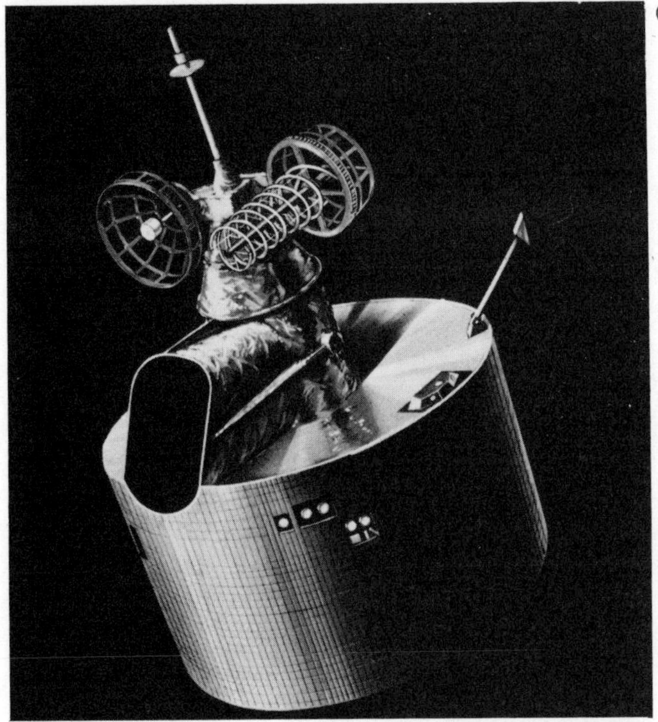

(b)

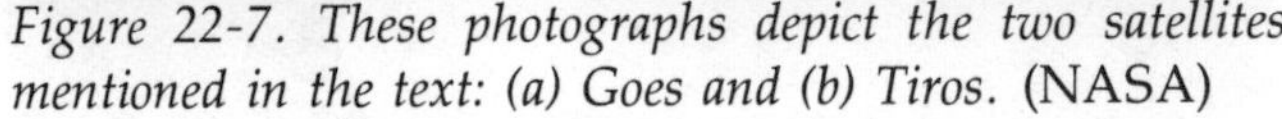

Figure 22-7. These photographs depict the two satellites mentioned in the text: (a) Goes and (b) Tiros. (NASA)

shows similar information presented on a cathode ray tube.

The satellite orbits form a concentric spherical cage within which the earth rotates. At the present time these orbit rings are approximately equally spaced angularly from each other, but there is no assurance or expectation that they will remain that way for great lengths of time. Natural forces eventually will swing these orbits as though they were pivoted on an axis through the earth's poles. These gradual changes will have no deteriorating effect on the service to the mariner because of the constant updating of the orbital information broadcasts.

The passage-making skipper will find that the deluxe satellite navigation receivers can relieve him of much tedious calculation and chart work. If he selects his destination, the instrument will promptly tell him the course to steer and the distance to run. It will give him these answers either for a great circle route or for a rhumb line, at his pleasure. As a final touch, the receiver will alert him upon arrival. An experienced human navigator could not do more.

Earlier, the need to plug in speed and heading was mentioned as a prerequisite to startup, and it was assumed that this would be done manually subject to normal human errors. Optional additional equipment is available from most manufacturers for doing this automatically. The common method of noting speed is to change it to a fixed number of circuit closures per knot and then to impress these pulses on the proper receiver circuit. Heading information is taken from a gyro compass in much the same manner.

All makers test their receivers for several days under actual and simulated conditions, "burn it in" as they call it, before shipment in order to achieve service reliability. The tests are carried out at various meaningful temperatures and under voltage conditions that resemble the not too well regulated power on a ship.

There is yet another manner in which satellite technology helps the cruising skipper who often finds himself far from land. These satellites, entirely separate and different from the Transit units, give him the latest weather in picture form, showing exact cloud cover. However, in contrast to the direct reception of navigational information, the weather data is best received indirectly from a processing station rather than directly from the satellite. It is simpler, better and less costly that way because of the complex procedure involved in preparing the data for general use.

Two satellites collect the pictorial weather information and broadcast it to earth. One is the "Goes" satellite; the other is the "Tiros" satellite. The Goes is stationary with respect to earth while the Tiros orbits the earth. (Figure 22–7 depicts the two satellites.)

Large parabolic dish antennas placed at strategic spots pick up the transmissions from Goes. Proper reception requires critical pointing and focusing of the dish; the margin of error is only a couple of degrees either way. Obviously, this is no job for a yacht, not even on glassy still water. Very large vessels that attempt this reception are equipped with ultra-efficient stabilizers.

The reception problem with Tiros is not much simpler. The same large dish antenna is needed, but with an additional complication—it must track the satellite closely. Again, even a large yacht may be dismissed out of hand as completely unsuitable.

What then is the way to go for a pleasure boat skipper who wishes to acquire this exceedingly valuable weather data? The answer is indirect reception from one of the many stations around the world that broadcast this material. The necessary receiver is a facsimile printer like the one shown in Figure 22–8.

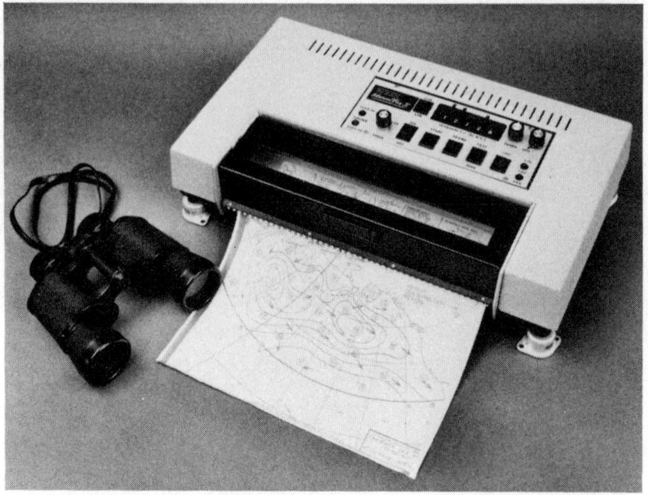

Figure 22-8. This receiver produces a complete weather chart in response to the signals it receives from the satellite. (Alden Electronic Recording Equipment Co.)

Indirect reception provides an almost invaluable advantage. The received picture comes with an overlaid grid that relates cloud cover and other details to identifiable land below. This grid is applied by experts prior to rebroadcast; few skippers would have the skill to grid correctly the raw picture from the satellite.

The rebroadcast picture is brought up to date every three hours. As an additional service, the transmitting stations also broadcast a wide range of surface weather information that includes wind and wave forecasts plus meteorological and oceanographic data that extend the usefulness of the cloud pictures.

The receivers require no special attention other than to be turned on and tuned in. Thereafter, the machine grinds out the pictures automatically. Electric current demands of these receivers are nominal, well within the supplying ability of any vessel large enough to justify installing this equipment.

The instrumentation may take either of two forms. In one, the facsimile printer that actually produces the picture is a separate unit that gets its input from a high frequency radio receiver tuned to the transmitting station. The second form is a complete unit that combines the radio receiver and the printer in one unit.

The radio imparts its information to the printer in the form of audio tone that varies between 1,500 herz and 2,300 herz. The 1,500 Hz transmission is the equivalent of black in the picture while 2,300 Hz is white; the intermediate frequencies cause varying shades of gray.

23. *Hailers, Alarms, and Accessories*

When a microphone, an amplifier, and a loudspeaker join together and go to sea, they metamorphose into a "hailer" with increased social status and a commensurate higher level of price. However, the ability to holler at people beyond the range of the natural voice, which a hailer makes possible, quickly erases memories of the cost.

An audio frequency transistorized amplifier whose output power may vary from a few to many watts, depending on purchase price, is the central unit of a hailer. In the hailing mode a microphone is at the input of this amplifier and a weatherproof external loudspeaker is at the output. In the listening mode the positions of these two units are interchanged and now the loudspeaker acts as a listening transducer and is placed at the amplifier's input. Early models used the microphone as an earpiece when listening, but present design replaces this with a small speaker at the console. The switching back and forth is actuated by pressing a microphone button.

With amplifier and speaker already available, all that needs to be added to turn the system into a signalling and fog horn is a feedback circuit. By feeding a portion of the output of the amplifier back into the input and choosing the right circuit components, a skipper may generate any desired signalling tone. (See Chapter 11.) A console pushbutton permits the usual port-starboard piloting blasts and a self-contained timer spaces fog signals legally. And that is not all! By connecting door and window intrusion switches the hailer becomes a burglar alarm with a loud warning. A seagoing chameleon. (A representative hailer is shown in Figure 23–1.)

Competition forces makers to add specific features to their hailers. One worthwhile addition is separate volume controls for talking and for listening; one monitors the strength of the blast emanating from the external speaker while the other governs the volume from the small speaker at the console. One advantage of this feature is the lessening of the feedback squeal that can become so annoying.

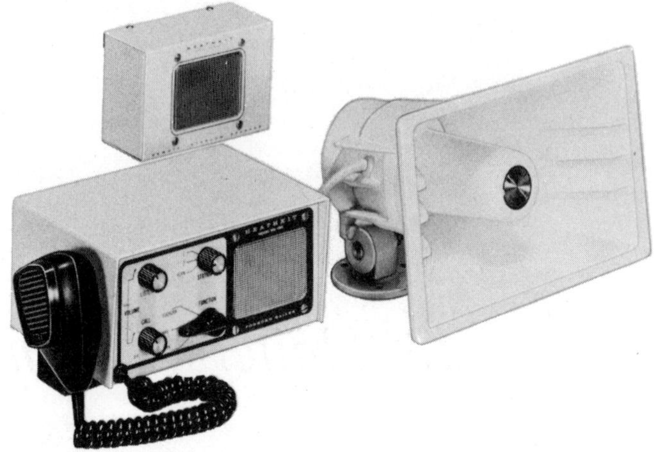

Figure 23-1. The units of a loud hailer that also act as a foghorn are shown above. (Heathkit)

The best speaker to use at the external position is the reflex horn type illustrated in Figure 23–2. The horn shape imparts a strong directive effect to the sound, thereby concentrating the audio power. These horns employ metal diaphragms in place of the paper cones of loudspeakers and are reasonably weatherproof—though certainly not waterproof.

Figure 23-2. The reflex horn achieves a long calibrated sound path within the confines of a compact unit.

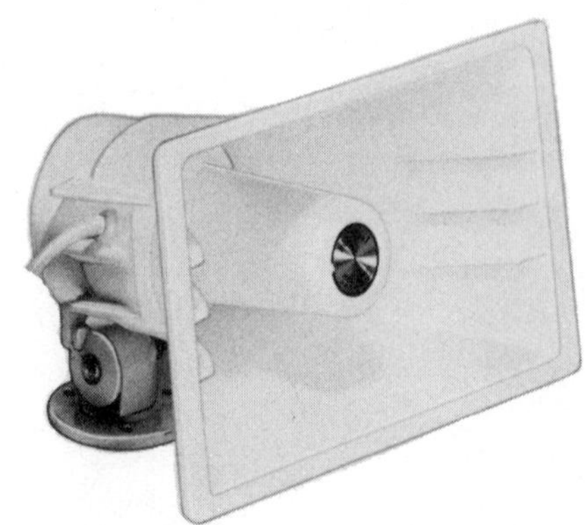

Horns and speakers are available with various impedance ratings; this parameter is an important item in selection. For best power transfer, the horn or speaker impedance should match the output impedance of the amplifier (this usually is found on the nameplate).

It may be a bit puzzling to realize that a loudspeaker also functions fairly well as a pickup microphone, but a look into how the critter works will make it clear. The center of the cone of a loudspeaker is attached to a light coil of wire that "rides" on the core of a strong permanent magnet. The varying current in the coil from the amplifier reacts with the magnetic field and causes the coil, and the cone to which it is attached, to move back and forth and create sound. Conversely, sound striking the cone moves the coil back and forth to generate a voltage that is amplified to recreate the sound. It's the old axiom that a conductor produces a voltage when it "cuts" a magnetic field—the underlying principle of all generators, alternators, and transformers.

Again, the hailer already incorporates all the individual units needed for a powerful intercom, except perhaps for an additional remote speaker, and so its added use for that service is obvious. As before, the speakers act alternately as inputs and outputs. One common use for an intercom aboard is to connect the flying bridge to the galley to hasten delivery of the indispensable cup of coffee for his majesty.

Installing a commercial hailer is easy and requires no skill. The main unit is connected to the battery lines with the usual fuse in series. The line to the external horn is best run away from other wires because it could pick up extraneous electrical noise in the listening mode. The current requirements of hailers are minimal on standby because of the transistorized circuits and run to several amperes when using full output power. The watts input (volts multiplied by amperes) needed for the hailer is approximately twice the rated watts output.

The national epidemic of sticky fingers has hit boating as a lucrative target and theft of valuable equipment is now commonplace. Lock and key alone do not seem to be sufficiently effective and many skippers are turning to burglar alarms to help secure their boats from intrusion. With a hailer already aboard, the adaptation of it to security duty is a simple matter involving only the addition of a few inexpensive components. A burglar then is greeted with loud yowling from the loudspeaker that is almost certain to scare him off whether or not help arrives.

Burglar alarm circuits are essentially simple doorbell hookups well within the ability of the average person with limited electrical skill. Of course, highly sophisticated alarm circuits that function with photoelectric cells, sound waves, radio patterns, and whatnot are offered commercially, but these are not the correct choice for the average pleasure boat. They are too complex for the variable marine conditions and they probably would not survive the buffetings of wave and weather.

The basic buidling blocks for simple burglar alarms are switches that respond to the opening of a window or door and the physical presence of a person coming aboard. These switches are available in two forms: open circuit and closed circuit. The open circuit device makes contact and closes the circuit when the event takes place, whereas the closed circuit device breaks the normally continuous circuit at that time. The closed circuit type is preferable because any attempt to circumvent the system by cutting the wires makes matters worse for the intruder by sounding the alarm instantly.

The magnetic switch, shown in its two parts in Figure 23–3, is the most widely used closed circuit device and the most adaptable to varying needs. The part with two screw terminals contains a reed element (also shown) that responds to the permanent magnet contained in the other part. When the two parts are close together, as they would be on a closed window or

Figure 23-3. One unit of this magnetic switch pair contains a magnetic reed switch. The other contains the magnet. (EICO Electronics)

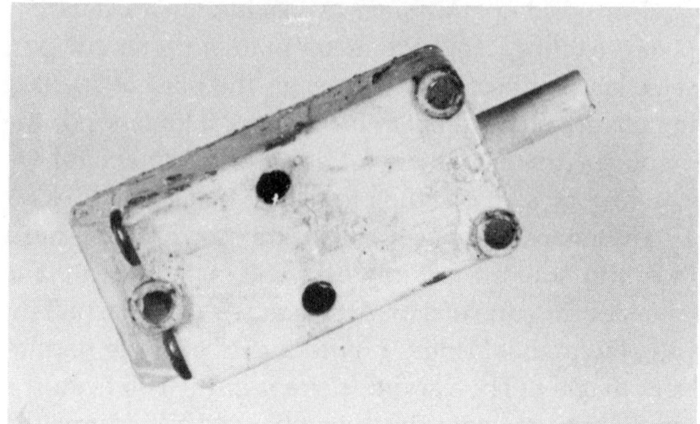

Figure 23-4. A spring-loaded plunger is held in by the closed door. Opening the door allows the plunger to project, thereby setting off the alarm. (EICO Electronics)

door, the magnetism causes the two reeds in the glass tube to touch each other and make electrical contact. Opening the door or window removes the magnetism and allows the reeds to resume their natural separated position, thereby breaking the circuit. The two-part magnetic reed switch may be adapted to protect almost anything aboard that a burglar would want to remove.

Another form of switch that is especially useful for guarding doors and hatch covers is shown in Figure 23–4. The spring loaded plunger must be held completely pushed-in for the circuit to remain continuous. If the hatch cover or door be opened, the plunger flies out and the internal contacts separate to break the circuit. This, too, is a closed circuit device.

Many ingenious intrusion sensors may be improvised. One such is comprised of a simple leaf spring contact and some black thread. The thread attached to the contact is stretched around the perimeter of the area to be protected in a manner that will cause an intruder to collide with it. Movement or breakage of the thread sounds the alarm. Another scheme runs a narrow strip of foil around windows so that smashing the glass will also rupture the foil and sound the alarm.

The skipper has a choice of three sources of power for his burglar alarm. He may run it from the boat battery, from a separate dry battery, or from the shore power line on the pier. This last is done through a suitable control box that transforms the shore line 120 volts to the required low voltage. Shore power alone is obviously vulnerable because the intruder could pull the plug and then go aboard undisturbed. Manufacturers have recognized this and the commercial control box provides space for a lantern battery that can take over and continue protection whenever shore power fails for any reason. (A control box that meets this description is shown in Figure 23–5.)

As stated earlier, burglar alarms are simple doorbell circuits and the components may be purchased separately as needed at any electronics store. However, some skippers may prefer the convenience of completely packaged systems that contain everything necessary for making an installation. Such kits are available at moderate price (a typical one is shown in Figure 26–6).

Looked at from the standpoint of battery drain, the open circuit system draws no current while it is quiescent. Consequently, the useful life of a dry cell in such a hookup is its shelf life. Only while the alarm is sounding must the battery supply whatever current the bell, buzzer, or other signal needs.

Figure 23-5. This control box permits the onboard burglar alarm system to be run from shore power, boat battery or self-contained dry battery. The switchover from one to another is automatic. (Eico Electronic Instruments)

By contrast, the closed circuit system draws current continuously because it is continuously in an "armed" condition. But the current draw is small, only enough to keep the relays in the system in the closed condition; dry batteries have no difficulty supplying this. In the combination shore current/battery current system the battery is on standby and is called upon for current only when the shore current is disconnected or goes dead. (For marine service all wiring should be done with plastic-coated wire; gauges from 18 B&S to 22 B&S are suitable for alarm circuits.)

Figure 23-6. This do-it-yourself burglar alarm kit contains the parts needed for a simple installation. (Eico Electronic Instruments Co.)

Speedometers are comparatively rare on boats, perhaps because they do not have the legal status accorded them on automobiles. Yet an accurate knowledge of the speed at which the vessel is traveling is a help in piloting, and a valuable parameter in gauging performance. A speedometer is the thing a novice skipper misses most in his transition from driving a car.

Commercially available speedometers for boats are of many types. Their method of operation ranges from hydraulic to mechanical to electronic. One sophisticated instrument employs the Doppler effect (described in Chapter 22 in connection with satellite navigation). Years ago, ocean liners used the taffrail log, and a common sight on board each noon was two seamen and a deck officer taking the daily speed reading at the stern rail.

The simplest speed-checking device is the nonmechanical hydraulic Pitot tube shown in Figure 23–7. In similar form the Pitot tube has a long history of use by aircraft. This instrument presents an open-ended tube to the stream and converts speed into a corresponding pressure that is read on a dial or a manometer calibrated in miles per hour or knots. The Pitot is certainly not an electronic device, but it is included here for reference.

Figure 23-7. The Pitot tube sensor for the speedometer works on the principle of differential pressure generated by its motion through the water. (Airguide Instruments Co.)

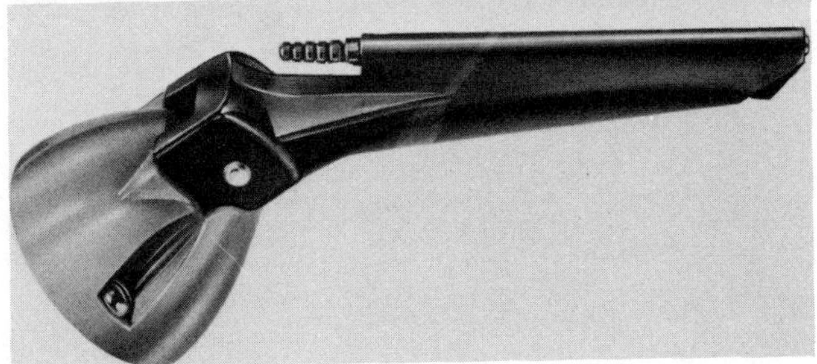

Modern marine speedometers offer their readings either in analog form, meaning a pointer over a calibrated dial, or else in digital numbers made possible by electronic technology. The pointer with its dial has something in its favor because a quick glance at its position tells the reading, as with the hands of a clock, whereas numbers must be individually recognized at short distance. Nevertheless, the trend is to digitalized equipment.

Measuring the speed of a boat presents a unique difficulty that is absent when making a comparable measurement in an automobile. The water may be under the influence of a current that is moving it either with or against the vessel, whereas the road on which the auto travels is a fixed reference. Consequently, when referring to the rate of travel of any ship, the qualifying "speed over the ground" or "speed through the water" must be added because the two may be different.

The most sophisticated (and perhaps the most accurate) marine speedometer employs the Doppler effect to accomplish its readings. The indicating dial and the control panel of such a device are shown in Figure 23–8, and Figure 23–9 is a diagram explaining the action.

A blister on the bottom near the keel at the bow houses a transducer that projects two one megahertz beams downward. One beam is angled forward, the other astern. In waters of nominal depth, these beams hit the bottom and measure actual speed over the ground. When water depth becomes too great for proper readings, the beams are considered to bounce off an imaginary layer of water closer below the keel. The instrument indicates which reference it is using so that the navigator can make necessary compensations.

This instrument also has a mile totalizer, akin to the odometer on a car, that keeps track of total distance traveled. As with all Doppler effect devices, the fre-

Figure 23-8. The dial indicator and the control panel for the doppler speed measuring unit are shown. (For doppler shift explanation see text.) (Simrad, Inc.)

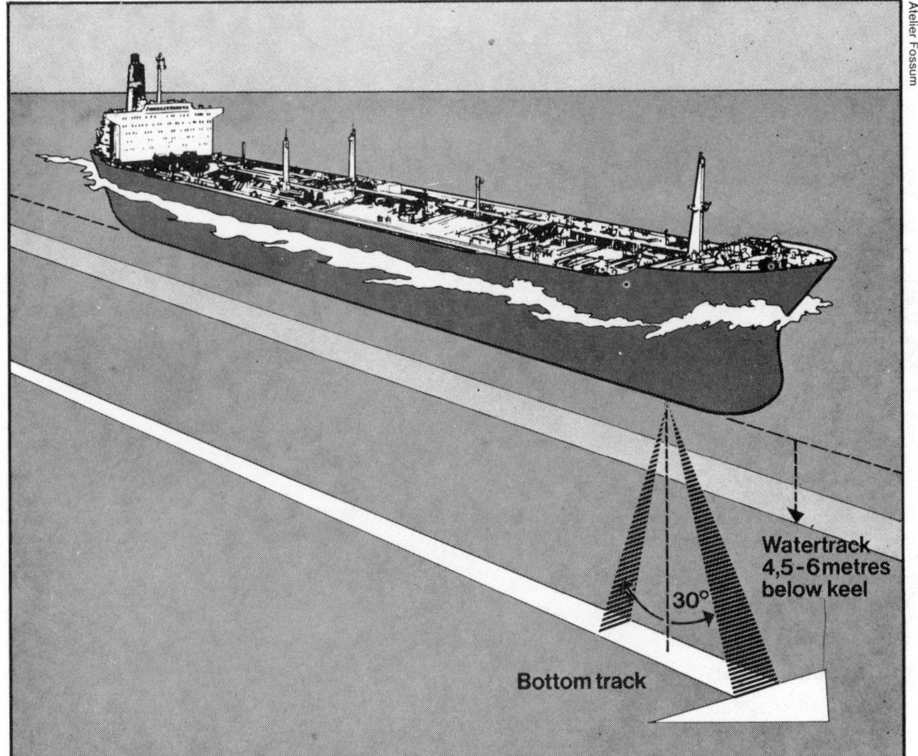

Figure 23-9. How the doppler method of measuring speed functions in relation to a vessel is explained in this diagram. The transmitted signal is echoed back at a different frequency depending upon the speed of the ship. (Simrad, Inc.)

quency of the returning beam, the echo, differs from that of the original beam sent down. The difference is caused by the speed of the ship and the amount of this difference is translated into knots or miles per hour.

A miniature paddle wheel is the sensing element for another type of boat speedometer. The paddles extend just enough below the hull to be engaged by the passing stream of water. The rotating paddle wheel generates pulses that are transmitted to an integrating circuit in the instrument and resolved into miles per hour. The reading is on a calibrated dial or, with more expensive devices, the reading is digital. (See Figure 23–10.)

The correct location of the speed sensor, whether it is the paddle wheel or other, has a direct bearing on the accuracy of the results. Several factors come into

Figure 23-10. One form of electronic boat speed sensor projects a small paddle wheel into the slipstream beneath the hull. A magnet in one paddle generates pulses in an adjacent coil and these are processed by the circuit into a reading of equivalent speed. (Datamarine International)

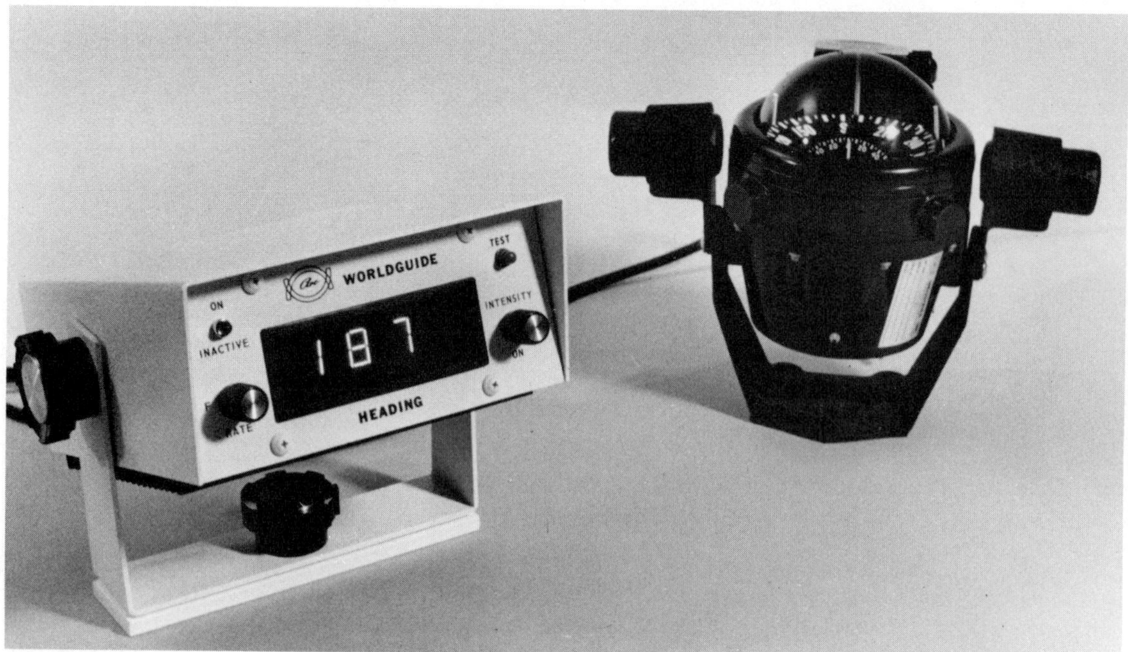

Figure 23-11. The indicator shown above repeats compass readings digitally. The compass (a) is equipped with quadrantal cylinders required when the unit is installed on a steel vessel. (b) This digital compass gets its information from a saturable reactor that has no moving parts and eliminates the magnetic compass sensor. (Wesmar)

play, and they are different for planing hulls and for displacement hulls. One problem is the boundary layer of water that attaches itself to the hull and moves with it, thus creating a false condition for the sensor. The thickness of this boundary layer is affected by hull shape as well as by hull fouling. The saving grace is that boundary layer conditions seem to be linear and therefore may be compensated by instrument calibration after installation.

The best spot for the speed sensor on a displacement hull is far forward where the water flow is still comparatively undisturbed and reasonably laminar. Obviously, the same location on a planing hull would be out of the water and useless at high speed.

Not even the compass has escaped the trend to digitalization. Modern models of this age-old direction indicator proclaim the headings in brightly lit numbers —and it must be admitted that this is less conducive to error in reading. Any helmsman who has strained his eyes to decipher the little degree marks floating past the lubbers line will bless a digital compass.

The trick of a digital compass is turned in either of two ways. In one the digital indicator is actually only a repeater of a remote standard magnetic compass; the pickup and transition are done electronically. The other method has no moving parts, no compass card. A saturable reactor (a wirewound iron core) senses the direction of earth's magnetic field, and thus of North, by the strength and polarity of the reaction. Complex circuits then convert this to the displayed digits. An instrument of each kind is shown in Figure 23–11.

Wind speed and wind direction are of supreme interest to sailboat skippers, but this knowledge is useful for powerboat skippers as well; it is good information to have displayed on the console. Again, a wide selection exists of commercial offerings of the necessary instruments. The masthead or the yardarm is the usual location for installation of the sensors.

Wind direction is taken off in the good, old-fashioned manner of the common weathervane. A small, carefully balanced low-friction vane turns with the wind and is the mechanical part of the sensor; attached to the vane shaft is the electrical portion. A small arm centrally fastened to the shaft has contact brushes at its outer ends that wipe the surface of a circular potentiometer. The position of the vane thus determines the resistance and therefore the current that will flow in each of the three segments into which the circuit is divided.

The indicator unit has three corresponding magnetic coils, positioned 120° apart, that control the motion of the pointer. It can be seen that the various combinations of magnetic strength of the coils can place the pointer at any position on the full circle that portrays the signals sent from the mast. The power for all this comes from the boat battery and the steady current is much less than one ampere. (A diagram that explains the action is shown in Figure 23–12.) The foregoing is only one form of modus operandi.

The reading of the indicator signifies the *true* wind only when the boat is standing still. At all other times, the information pertains to the *apparent* wind. The apparent wind is the vector combination of the direction of the true wind and the wind created by the direction of travel of the boat. The apparent wind is a basic operating parameter for the sailboater.

Wind velocity is measured by an anemometer with the familiar revolving halfcups. The hemispherical shape provides a double advantage—regardless of the direction from which the wind comes, the hollow side of the cup generates a push while the back creates a suction. The stronger the wind, the faster the speed of revolution. This unit usually is mounted in connection with the wind vane.

Figure 23-12. This schematic diagram depicts the wiring of a combined wind speed and direction indicator. The dashed box at the right shows what is at the masthead; the box at the left explains the indicator on the console. (Signet Scientific)

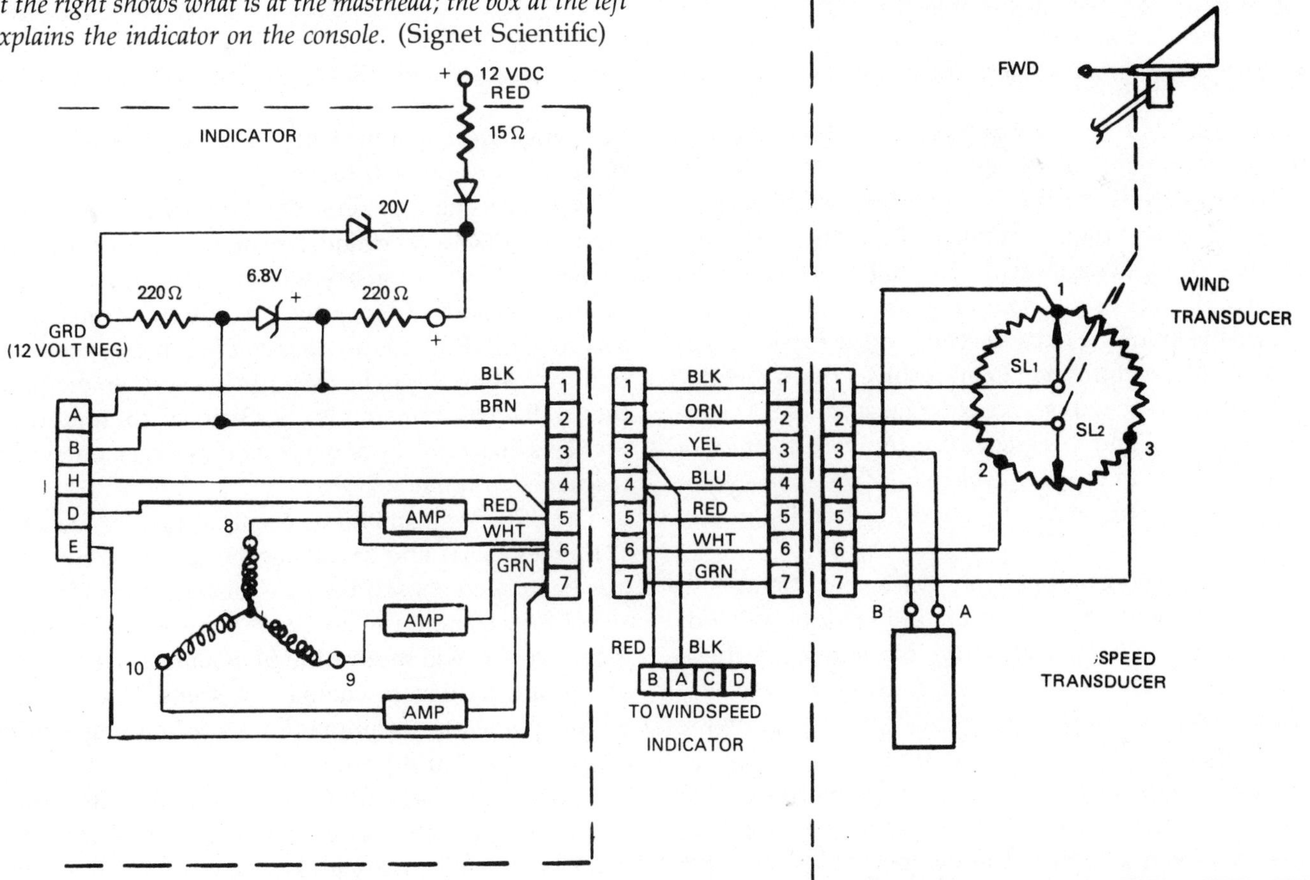

Figure 23-13. The skipper who wins sailboat races is tuned into even the slightest nuances of the wind and often he has instruments to help sharpen his senses. A group of wind detecting instruments is shown above. (Signet Scientific)

The speed of the revolving cups may be sent down to the instrument console either in the form of a voltage or of a series of pulses to be transformed into a dial reading or a digital output. The simplest combination is a tiny generator on the cup shaft and a voltmeter calibrated in miles per hour or knots. Another system allows the spinning cups to send a pulse at each revolution while a counting circuit resolves the frequency of pulses into speed or velocity. The accuracy of readout is only slightly degraded by the vector effect of direction of ship travel versus wind direction—except when these two are directly opposite. (See Figure 23–13.)

Racing sailboat skippers are extremely attentive to very slight changes in direction and velocity of wind, and for them a special series of meters of these parameters has been developed. In essence these auxiliary meters act as sensitive verniers that magnify the small changes that are almost undistinguishable on standard indicators. The extreme sensitivity also shows up instantaneous variations that are meaningless, however, and these are damped out. Often this damping effect is adjustable to satisfy the many requirements of sailboat racing.

Most cruising skippers who venture far from land carry an EPIRB as additional insurance against the possibility that an emergency will find their regular communication radio inoperative or otherwise unable to secure help. (EPIRB is an acronym for *E*mergency *P*osition *I*ndicating *R*adio *B*eacon.) In such an emergency, the EPIRB will transmit an SOS signal continuously, for days if need be, and provide a target on which rescuers can home.

The EPIRB transmissions occur on two frequencies: 121.5 megahertz and 243.0 megahertz. These frequencies have been chosen because they are monitored by all aircraft, commercial and governmental, and assure that someone will receive the SOS almost within minutes and will direct rescuers to the scene. Coast Guard rescue planes are equipped with a homing device tuned to the EPIRB transmissions.

A representative EPIRB unit is shown in Figure 23–14. The device operates from a sealed battery pack that has a shelf life in excess of one year, and the entire

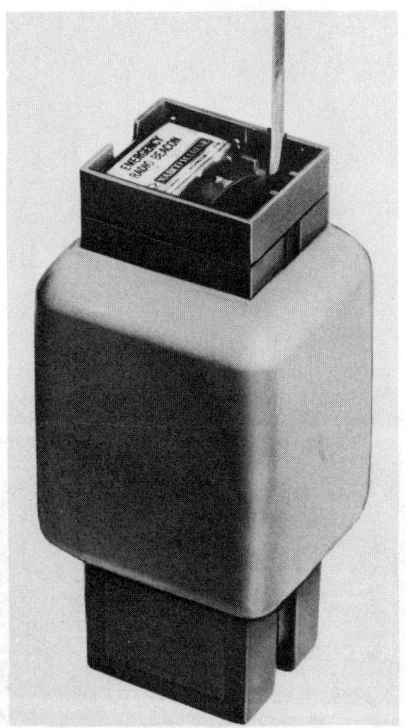

Figure 23-14. The EPIRB (emergency position indicating radio beacon) is a valuable adjunct to safety afloat for long distance passagemakers. This floating transmitter automatically sends emergency messages on channels monitored by aircraft and others. (Narco Marine)

unit floats. In the yacht model, transmission is initiated with a switch; the commercial model is automatically released from a sinking ship and becomes operative when it hits the water. Having an EPIRB aboard is mandatory only for commercial vessels carrying more than six passengers.

A common source of unpleasant surprises for a helmsman leaving a pier is his lack of knowledge of exact rudder position. Quite possibly the rudder could be at an angle antagonistic to the maneuver he intends to make—with embarrassing results. A rudder angle indicator on the console saves him from this.

The rudder indicator unit consists of a center-zero meter calibrated in degrees to port and to starboard, a sensor at the rudder post, and the connecting cable. The sensor is a potentiometer whose contact arm is mechanically linked to the rudder post and swings with it to either side of center. A bridge circuit keeps the meter at zero potential when the arm is centered; the varying resistances established when the arm swings move the meter needle to left and to right to indicate port and starboard. (The basic circuit of such an instrument is shown in Figure 23–15.)

It is said (and with some truth) that the best fume detector is the skipper's nose. Yet the skipper's nose cannot constantly be in the bilge to warn of danger, and so an electronic vapor detector becomes a good investment in safety. On a vessel whose tanks store gasoline or whose cooking fuel is gas, a vapor detector installation is almost mandatory.

Present vapor detectors function on either of two principles. In one, a tiny sample of the vapor is actually combusted harmlessly on a heated platinum filament;

Figure 23-15. (a) The position of the rudder is indicated on this dial at all times. (Wagner) *(b) One form of circuit for accomplishing rudder position indications is shown here for 12 volts.*

(a)

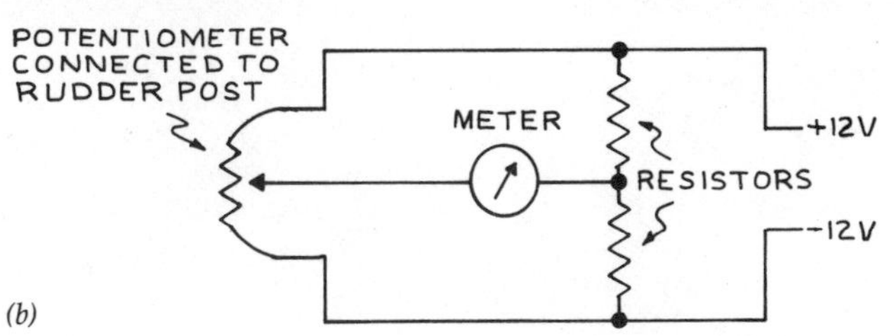

(b)

the consequent change in resistance of the filament causes the meter reading or the alarm. In the other type no similar combustion takes place; a sensor whose resistance changes when it adsorbs flammable vapors initiates the alarm. Manufacturers of both types recommend periodic testing of the installations and they give instructions for doing this safely. (See Figure 23-16.)

A warning system that should be on every powerboat rings a bell if the engine lubrication pressure fails. This is accomplished very simply with a pressure switch similar to those that actuate the stop lights on automobile brakes. An audible warning of excessively high engine temperature is equally desirable.

(a)

(b)

Figure 23-16. All skippers of pleasure boats should be constantly on the alert for accidental fire, but those on gasoline-powered craft must be especially careful. A vapor detector that continuously monitors the bilge for flammable vapors is excellent safety insurance. A remote unit can take the vapor alarm to the skipper's quarters. (Aqua Meter Instrument Corp.)

PART FIVE

INSTALLATION AND MAINTENANCE

24. *Antennas, Grounds, and Transmission Lines*

Two identical wire hoops, perhaps one foot in diameter, each split to create a tiny gap in the circumference, were the world's first wireless antennas. (It was "wireless" then because the word "radio" had not yet come into use.) One hoop was connected to a conglomeration of electrical equipment consisting of batteries, spark coil, Leyden jars, and helixes on a table. The other hoop was entirely unconnected and in the hands of Heinrich Hertz a few feet away.

Closing the switch on the table brought a stream of sparks across the gap in the hoop there and, almost unbelievably, sympathetic sparks jumped the gap of the distant hoop held by Mr. Hertz. The world's first radio "message" had been sent—and received.

Since then, radio antennas have assumed many shapes and sizes: The long wire high up between the masts of a large ship; the metal backstay of a sailboat; the vertical metal whip so common on pleasure boats and automobiles; the loop of a radio direction finder; the parabolic dish of a tracking station; and the rotating element of a radar. All serve the specific intended needs; there is no one perfect antenna that is equally proficient for all services.

Historically, all antennas may be classified into one of two groups, each named after one of radio's pioneers. The determining factor is whether or not an antenna employs ground (the earth) as part of its radiating function. Those that use ground are Marconi antennas; those that do not are Hertz antennas. Each type displays technical features that make it more adaptable than the other for certain installations. (See Figure 24-1.)

Figure 24-1.

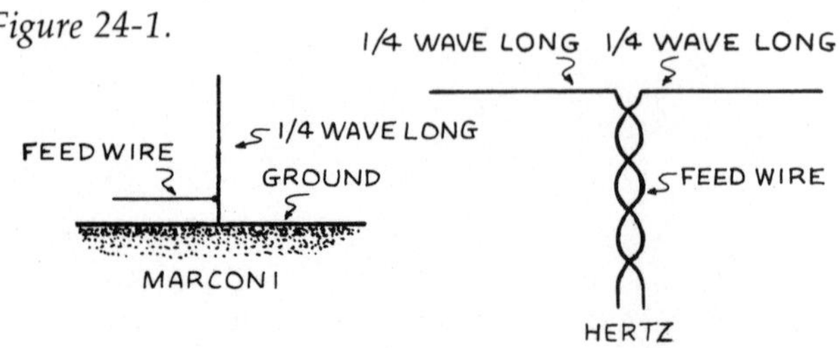

The first parameter that applies to all antennas is size or length because this is dependent upon the frequency for which the antenna is to be resonant, that is, most sensitive. The frequency of a radio wave and its length in space are interdependent and derive from the velocity at which radio waves travel, which, in free space, is approximately 300,000,000 meters per second, the same as light. This relationship between frequency and length is quite simple and is solved by dividing the velocity by the frequency in megahertz. Thus the wave propagated by a transmitter operating at a frequency of 150 megahertz is 2 meters long (300,000,000 ÷ 150,000,000 = 2).

Length is taken into account because an antenna must be specific fractions or multiples of one wavelength long in order to function efficiently. There are half-wave antennas, quarter-wave antennas, multiple-wave antennas—in each case the designation of length applies to the physical dimension of the antenna's radiating portion. However, the simple division exampled above gives the theoretical dimension the antenna would have if it functioned in true free space—and such a condition cannot be achieved in the actual world. A practical antenna is affected by its surroundings and as a result its elements are slightly shorter than the calculated theoretical length.

An antenna transforms the electrical energy sent to it by the transmitter into combined electrostatic and electromagnetic fields that we know as radio waves. This transformation is reversible: Radio waves striking the antenna generate electrical energy that is sent on to the receiver where the intelligence of the message is retrieved.

The energy sent out by a transmitting antenna and the energy captured by a receiving antenna may differ by factors of millions. The energy put into the antenna by a powerful commercial radio station may amount to millions of watts, whereas the average energy picked up by a receiving antenna may be only a few millionths of one watt.

Antennas for land use are usually designed to be directive, to concentrate their effectiveness into only a small angle of the circumference. The familiar multi-element television antenna (Yagi by name) is an example; it is aimed at the sending station. Antennas for marine use obviously must be omnidirectional, because the message may come from any point of the compass. The imagined appearance of such omnidirectional effectiveness may be visualized as a large horizontal doughnut without a hole and with a vertical whip antenna at its center. (This is shown in Figure 24–2.) That venerable comparison of a stone dropped into a pond still applies: Radio waves radiate in all directions from the omnidirectional antenna just as the ripples do from the stone.

But why do radio waves spread out from the antenna when the transmit button is pressed? Surely no radiation would occur if alternating house current were applied to the same antenna. The answer is "frequency." The transmitter supplies energy at extremely high frequency; the house current alternates at a mere sixty cycles per second. There is a minimum threshold frequency that must be crossed before radiation takes place.

House current would also create electrostatic and electromagnetic fields about the antenna, but they would have plenty of time to collapse back into the wire before the next alternation of current took place.

Figure 24-2. The radiation from a vertical whip antenna is omni-directional. If the radiation could be seen when looking down at the antenna from above, it would appear as a huge doughnut with the antenna at the center.

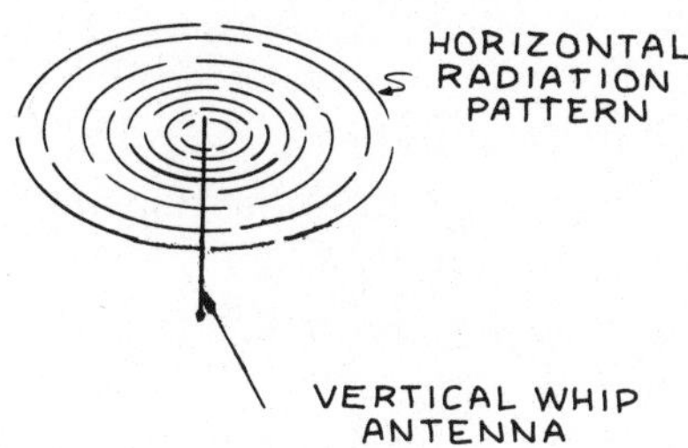

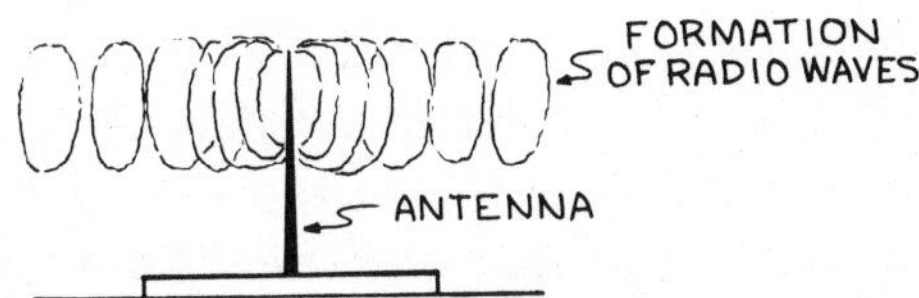

Figure 24-3. At the high frequencies of radio transmissions, the field created about the antenna does not have time to collapse back but are "snapped off" and projected into space.

Not so with high frequency current from the transmitter. There is not time for one field to collapse back before the next field is formed. Each field is cut off from returning and is forced to sail out into space as a radio wave. (See Figure 24–3.)

A finite amount of energy is put into the radio wave by the transmitter. As the wave advances, it can be seen that this energy must be suffused through greater and greater spacial volume and that consequently the strength of the wave decreases rapidly. This is the reason for the microscopically small energy intercepted by the average receiving antenna—and this explains the need for high sensitivity in radio receivers.

The present universally used VHF-FM radio communication band offers a fortuitous condition as far as antennas are concerned, because the wavelength, as already explained, is short and the antennas therefore are of manageable size. This does not hold true down around two megahertz, a single-sideband area. Here a half wavelength is 75 meters long and beyond the realm of practical construction for a whip-type boat antenna. The solution lies with "loading coils," small inductances at the bottom, middle, or top of antennas much shorter than the desired fraction of the operating wavelength. The loading coils make the antenna appear electrically longer than it physically is.

Antenna orientation affects the mode of travel of the radiated radio wave; this is known as "polarization." A vertical antenna emits vertically polarized waves while the emissions from a horizontal antenna are horizontally polarized. In general, vertically polarized waves are received best by vertical antennas and horizontally polarized waves by horizontal antennas. However, this is not always strictly true, because the polarization of a wave may be changed on its travel through space by intervening reflections. The likelihood of change in polarization is negligible at low radio frequencies and increases quickly as the frequency of transmission rises into the very high and ultra ranges.

The earth is surrounded by a shell of ionized particles called the "ionosphere." Radio waves either pierce this shell and are lost in space or bounce back and forth between it and the earth to achieve long distance communication. Whether they do one or the other depends chiefly on the angle at which the antenna radiates them, on their frequency, and on the condition and height of the reflecting ionosphere layer.

Three layers of the ionosphere have most effect on radio transmissions, and tables exist that predict the heights of these layers above the earth and their potency. Commercial interests use these tables to determine the best time of day and the best frequency for achieving successful contacts. The drawings in Figure 24–4 explain how ionospheric reflection enables a radio sky wave to be received at a distant point. The radio ground wave is of course effective for the initial short distance, but falls short of the point at which the reflected sky wave touches down.

The "skip distance" is the terrain between the transmitter and the area on earth where the sky wave comes down and may be received. The area between there and the end of the ground wave is a silent zone in which that radio transmission cannot be received. As may be seen, the skip distance may vary for various reasons. The sky wave also may bounce back and forth more than once. "Skip" may have phenomenal results; low power citizen's band transceivers often are heard hundreds of miles away.

Figure 24-4. The earth exists within a shell of charged particles called the ionosphere that has the ability to reflect radio waves as shown. However, at a critical angle (a) the transmissions pierce the shell and are lost in space.

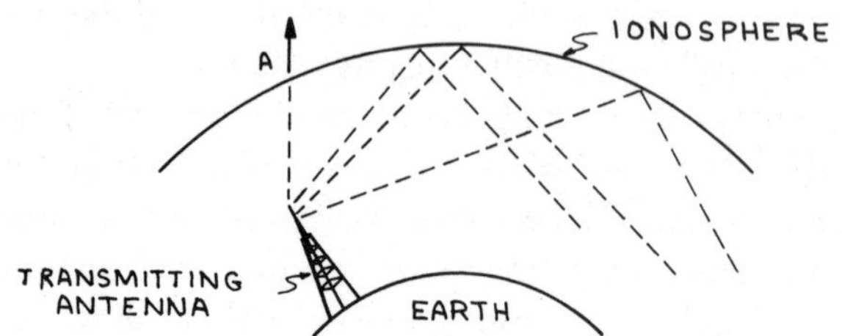

High power marine radio transmitters, especially those in the single sideband group, are designed for multi-frequency operation as a means of overcoming the vagaries of skip. By selecting frequencies that are most effective in combating the conditions caused by ionosphere height, time of day, sunspot activity, and other natural factors, reliable communication is maintained. A so-called "critical frequency" is the determinant of whether a wave is reflected or is allowed to pass through into space, and unfortunately, this too is variable.

Thus the choice of antennas for pleasure boat communication springs from the types of radio equipment carried aboard. In many cases the vessel's installation will include citizen's band, single sideband, and VHF-FM, thereby necessitating three separate antennas because of the wide disparity in working frequencies. Since whips most likely will be chosen, these antennas will look very much alike except for their varying lengths, although internally they will differ.

If a vertical metal rod one-fourth wavelength long were connected to ground at its lower end, the result would be a Marconi antenna for the frequency represented by that wavelength. The ground connection is vital because the ground acts as a "mirror" to create the lower half of the response pattern; this is shown in Figure 24–5. The energy from the transmitter is fed to such an antenna a short distance up from its lower end in order to achieve the desirable "impedance match" discussed later.

A Hertz antenna for the same frequency, devoid of a

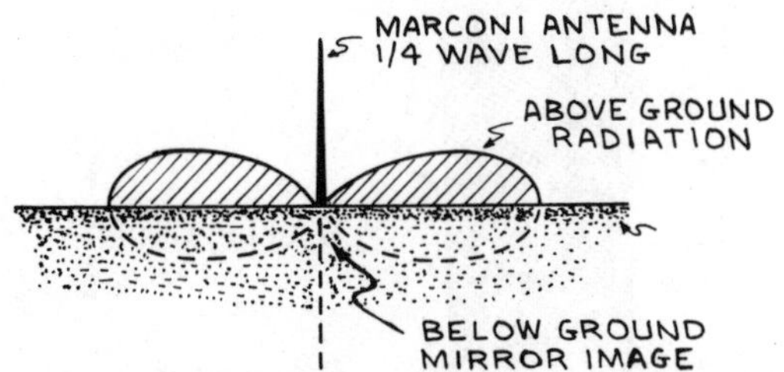

Figure 24-5. The ground acts like a mirror for the radiation from a marconi antenna one-quarter wave long and makes the pattern appear as though from an antenna one-half wave long.

ground connection, would require two rods, each one-fourth wavelength long and arranged colinearly as depicted in Figure 24–6. For obvious reasons it is called a "dipole." The transmitter's energy is fed to each rod at its central end and the total overall length of the antenna is one-half wavelength. Current flow in this antenna is first in one direction then in the other as the alternations take place at radio frequency.

It can be seen from the foregoing that both the Marconi antenna and the Hertz antenna require radiating elements that are, actually or in effect, one-half wavelength long. However, in the basic Marconi, only the one-quarter wavelength dimension is actual; the second quarter is the mirror reflection provided by ground. In the Hertz the full one-half wavelength dimension is physically there.

Both antennas project radio waves into space because, as already stated, the electrostatic and electromagnetic fields are formed in such rapid succession that the only course of action open to them is complete separation and then radiation.

Figure 24-6. The radiation from a Hertz dipole antenna is at right angles to the antenna length with virtually zero radiation from the ends.

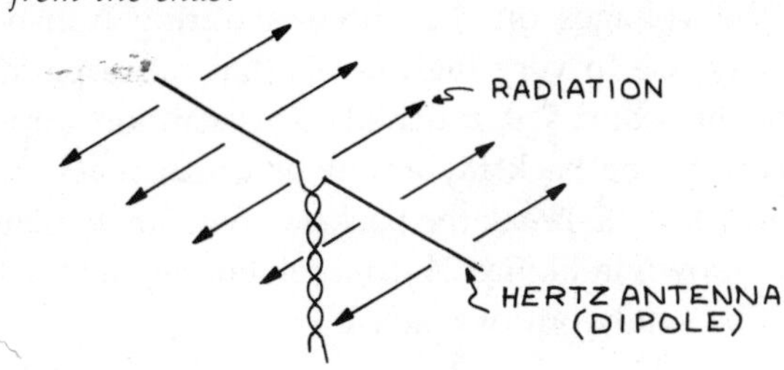

The point on its radiating elements where an antenna is fed with energy from the transmitter is critical. This point differs for the Marconi and the Hertz. The reason is the varying voltages and currents along the antenna and the consequent changes in impedance along its length. (This is shown in Figure 24–7.)

When power is fed at resonance to the half-wave antenna, each end is at maximum voltage, with zero volts at the center. Conversely, the current at the center is maximum with zero current at both ends. This is another way of saying that the impedance is very low at the center and very high at the ends. In fact, the base

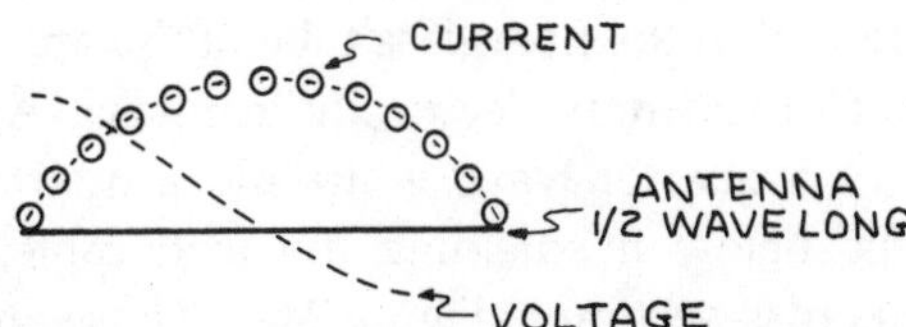

Figure 24-7. How the voltage and the current vary along the length of a half-wave antenna is shown in this diagram.

of the quarter-wave Marconi antenna (which is equivalent to the center of a half-wave dipole or Hertz) may be connected directly to ground because, as already stated, its impedance there is practically zero.

Manufacturers of marine whip antennas rate the effectiveness of their products in decibels (dBs) "gain." Here the magic number is three, because power is doubled every three dBs. A three dB antenna effectively doubles the radiation of power from the transmitter; a six dB would double that, and a nine dB unit would double it once more.

The statement that an antenna can increase the radiation of power could be misunderstood and needs clarification. An antenna is a passive device; it cannot actually add to the energy sent to it by the transmitter. The fact that a three dB antenna doubles the radiated

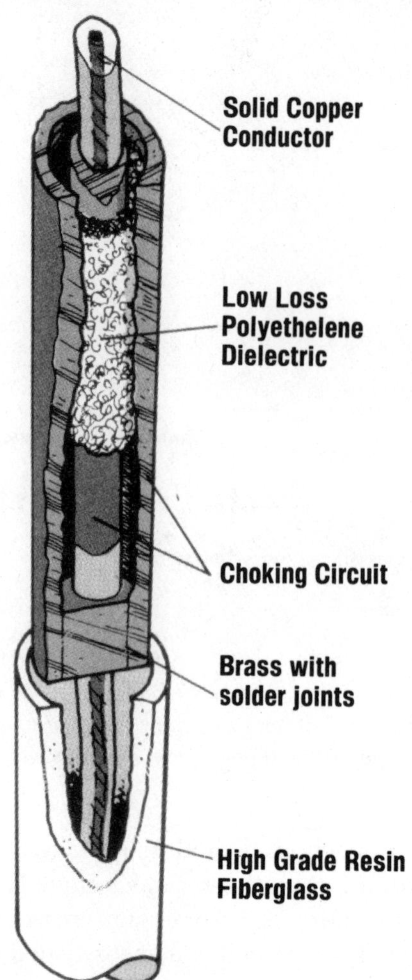

Figure 24-8. This cross-sectional drawing shows what takes place inside one type of whip antenna. (Shakespeare Co.)

power simply means that it lops off radiation in undesired directions and adds this recovered portion to the desired direction. As the dB rating of the antenna goes up, the pattern of the radiated energy becomes narrower. A searchlight does somewhat the same thing: The light from a small bulb, which ordinarily would be distributed over an entire sphere, is concentrated into a narrow beam of tremendous candlepower—but the total amount of light available has not been changed.

Compressing the radiation pattern of an antenna, and thereby increasing its gain, may be achieved by several methods. One is by the addition of a number of reflective and directive elements, as is done in the Yagi type so popular for television reception, but such an antenna is not suitable for pleasure boat marine service. The more apt method is to place several dipoles colinearly within the fiberglass tube of the antenna, all connected to function electrically in phase. A variant consists of a quarter-wave whip and a quarter-wave brass tube below it shielding the feed cable; this is shown in cross-section in Figure 24–8. These are representative of the common fiberglass antennas commercially available to the boatman.

The adverse effects of a high dB VHF-FM antenna should also be considered. The sharply flattened radiation pattern of a nine dB antenna may alternately shoot the signal into the sky and into the water when the vessel rolls—to the detriment of communication. (See Figure 24–9.) The wider pattern of the six dB antenna has made it a standard choice for powerboats. For sailboats, the three dB unit mounted at the masthead has become standard; the high location makes up in part for the lower dB.

Antenna design for VHF-FM service presents few problems because the range of frequencies from one end of the band to the other is so small. The lowest and highest frequencies are not far enough off center resonance to cause any great loss in efficiency by using the fixed antenna system. The opposite is true when the mode of operation is single sideband (HF-SSB). Here the frequency spread across the various channels is approximately fourteen megahertz and means must be provided for tuning the antenna circuit to resonance for each band. The most effective means is an antenna coupler that can tune the antenna sharply for each frequency. Traps and loading coils without a coupler are useful over only a small spread of frequencies.

The ideal location for every radio antenna is at a point that affords it unencumbered "view" in all directions. This is seldom achieved in practice and the next best thing must do. The undesirable conditions to be avoided are nearby metal objects or stays that absorb and waste some of the energy radiated by the antenna. Danger to personnel also should be borne in mind because the voltages on the antenna during transmissions may rise to very high levels that cause painful, slow-healing burns if touched. This caution applies especially to the backstay antennas of sailboats. (The solution here is to break the backstay with an insulator, such as shown in Figure 24–10, high enough above the deck to be safely out of reach.)

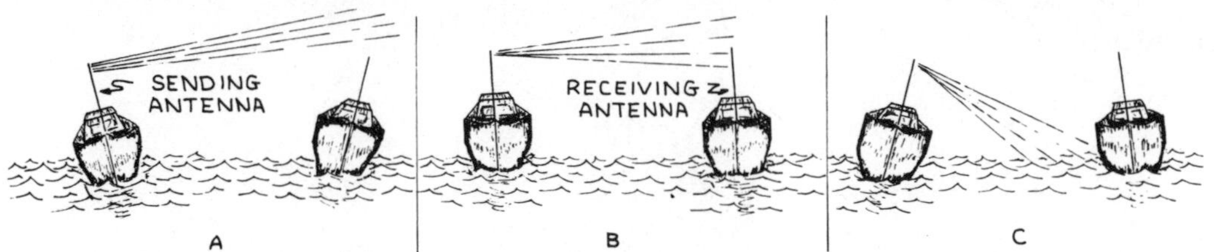

Figure 24-9. The problem that occurs when a boat with a narrow beam antenna rolls in a heavy sea is shown in these drawings. At (a) the transmission is hitting the sky above the receiving antenna; at (b) the receiver is getting the message while at (c) the transmission is falling short. Interrupted communication is the result.

Figure 24-10. The metal backstay (preferably stainless steel) of a sailboat makes an excellent antenna when modified as shown. A standard whip is a good idea also to have aboard in case the backstay is lost. (Motorola)

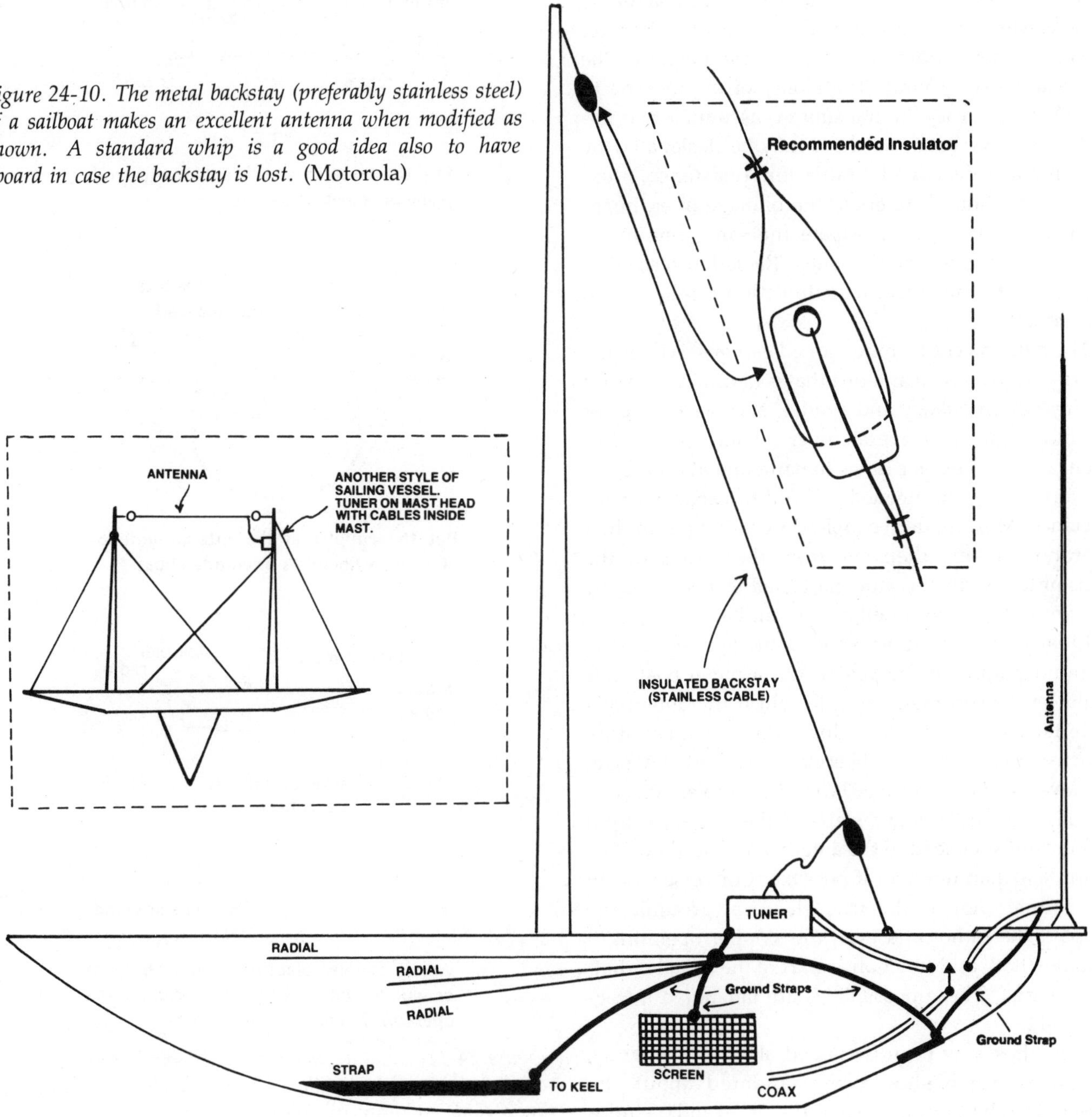

Theoretically, an ideal antenna radiates all the power fed to it and wastes none within itself; actually, some waste occurs because of the effects of resistance, reactance, surrounding objects, and the height of the antenna above ground. The ability of an antenna to absorb power from the transmitter, assuming that impedances have been matched, may be designated by an equivalency called "radiation resistance," expressed in ohms. It refers to the ohmage of an imagined load that would dissipate the same amount of power that the antenna radiates. Radiation resistance varies with the antenna's length in relation to wavelength.

The antenna coupler mentioned earlier serves as the interface between an antenna that is naturally resonant at only one frequency and a radio, such as an SSB for instance, that must operate over a wide span of frequencies. The coupler tunes the antenna at each band by introducing the needed amount of capacity or inductance. Most modern couplers are automatic in their operation, taking their cue from the tuning of the transmitter. With the antenna tuned to resonance by the coupler, maximum efficiency can be achieved.

Ideally, the coupler should be located at the "ground plane" of the vessel. On a metal boat, this could be the top metal deck to which the antenna is fastened. Ground plane location becomes more abstruse on a wooden or fiberglass boat and may be at the level of the waterline. With the coupler correctly placed, the true antenna starts at the coupler output and includes the lead to the antenna; this lead must be short, less than one foot if possible. Conversely, with the coupler not at the true electrical ground, the coupler, the radio, and even the skipper operating it become "hot" and function as part of the actual radiating antenna. These possibilities are illustrated in Figure 24-11.

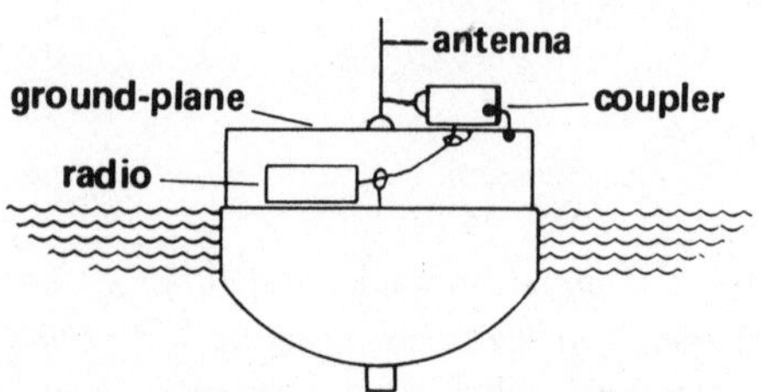

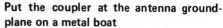

Put the coupler at the antenna ground-plane on a metal boat

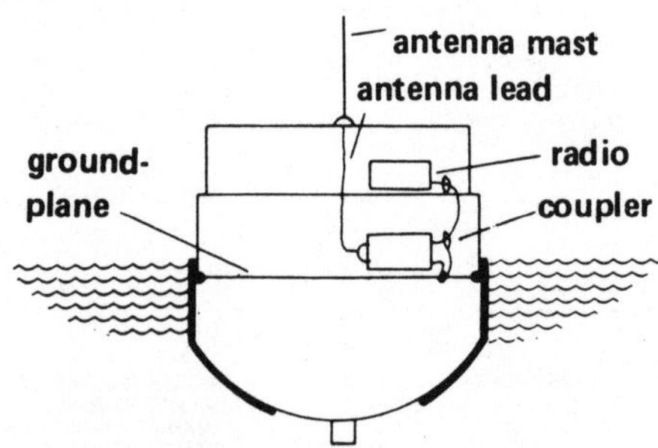

Put the coupler at the antenna ground-plane on a fiberglass or wooden boat

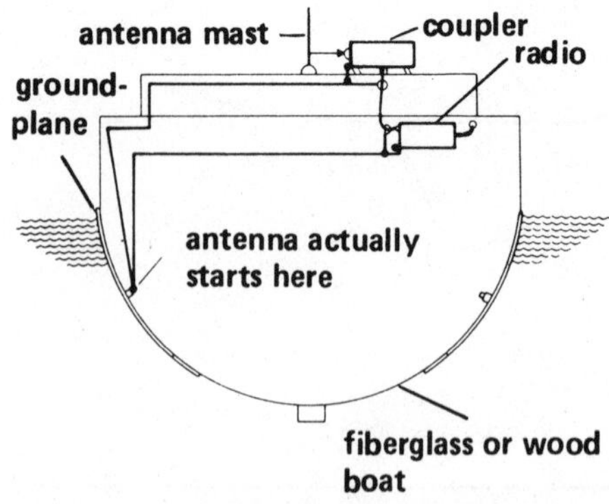

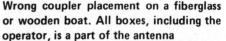

Wrong coupler placement on a fiberglass or wooden boat. All boxes, including the operator, is a part of the antenna

Figure 24-11. Locating the antenna coupler of an SSB receiver is a critical part of the installation. The sketches above are a guide. (Communication Associates, Inc.)

Much has now been discussed about the antenna; almost as much should be mandated about the "ground." The performance of the finest radio equip-

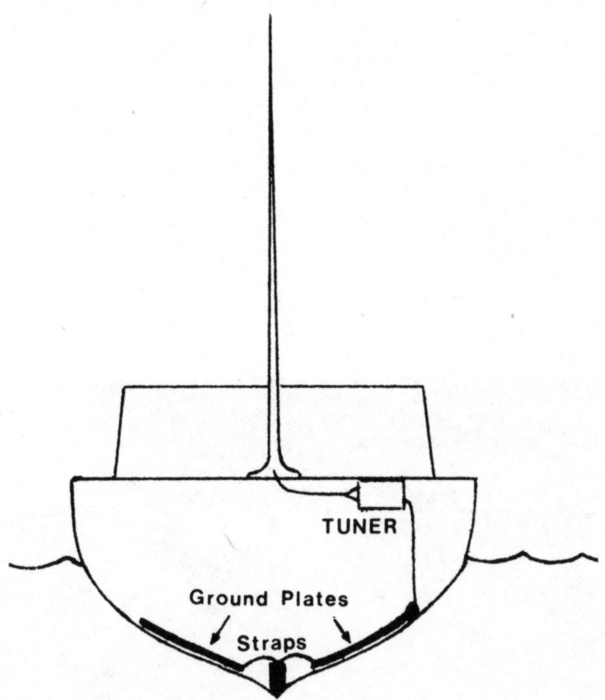

Figure 24-12. Ground plates may be located inside the hull, as shown above, when an outside immersed location is not convenient. The hull acts as the dielectric of the capacitor formed by the plates and the water. (Motorola)

ment can be cut down to mediocrity by a poor antenna or an insufficient ground. The exception, but only as far as a ground is concerned, is the VHF-FM radio; the very high frequency at which it operates makes a ground, other than the negative battery lead, unnecessary. A proper ground is also able to give protection from lightning when it is connected to a spike at the masthead.

A boat is fortunate in that the water in which it floats is an electrical extension of the ground, especially so if the water be salt. A sailboat with a metal keel has a ready-made electrical connection with the water and thus with the ground. Even if the metal keel be enclosed in fiberglass, the connection is capacitive and valid (the keel acts as one plate of a capacitor, the fiberglass is the dielectric, and the water is the other plate), but such an arrangement will not provide lightning protection. Powerboats have their connection with ground through their shafts and propellers.

However, despite the usefulness of the foregoing methods of grounding, an actual ground plate fastened to the outside of the hull below the water line is still the best. A usual form is a copper sheet with an area of at least ten square feet. Much smaller plates have been on the market with the erroneous claim that their porosity makes up for the smaller size. This porosity may lower the direct current resistance, but it has no value at the high frequencies of radio communication. As with the encapsulated keel of the sailboat, the copper ground sheet also may be placed inside the hull and used capacitatively. (See Figure 24-12.)

The ground so necessary for forming the mirror image described earlier as essential to the functioning of a Marconi antenna may be produced artificially with what is known as a "counterpoise." The counterpoise is a metallic structure of large area placed below the antenna and insulated from ground. It "connects" with actual ground capacitatively, just as described for the encapsulated keel; it is treated in the circuit as the actual ground would be. Often the counterpoise consists of radial metal rods at the base of the antenna.

The effectiveness of a whip antenna radiating at the comparatively low frequencies used in single sideband may be improved by the addition of conductive areas such as metal screening to the ground. (This is shown in Figure 24–13.) Note that all metal masses aboard are

Figure 24-13. Ground screens are a help in getting the grounding system as efficient as possible. (Motorola)

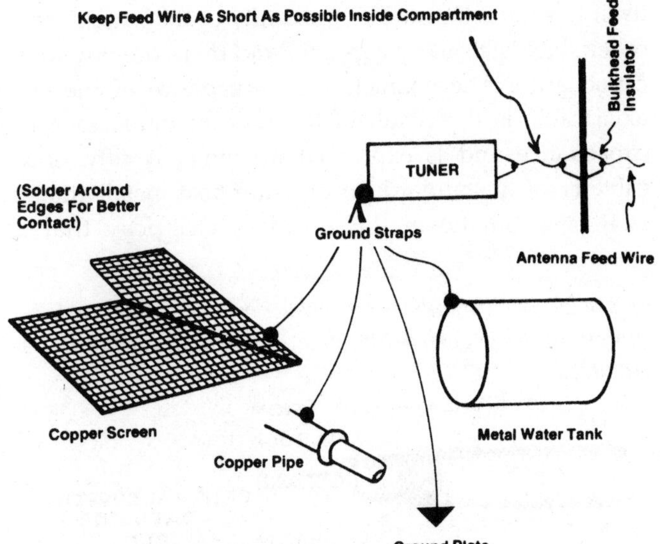

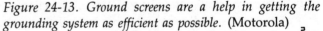

also connected to the ground terminal by individual wide copper straps. The screen area should extend in all directions for at least the distance equivalent to antenna length. In some fiberglass boats the screen is encapsulated by the builder; in others it may be placed under the carpet.

The energy put out by the transmitter must be conducted to the antenna with as little loss as possible. On pleasure boats this task generally falls to coaxial cable, shown in cross-section in Figure 24–14. A central copper conductor is held concentrically within an outer cylindrical copper braid by insulating spacers or even by a filling of foam plastic; an outer cover of vinyl protects the cable. Coaxial cables are identified by their "characteristic impedance" and this, in turn, is determined by the outer diameter of the inner wire, the inner diameter of the braid, and the type of insulation. The characteristic impedance remains the same, regardless of the length of the cable.

Impedance occurs in alternating current circuits, while direct current systems exhibit only resistance. Impedance is a combination of resistance, capacitance, and inductance—and a coaxial cable has all three although the resistance is negligible. The central wire itself has inductance. The proximity of this wire to the outer shield acts like a capacitor and therefore provides capacitance. The characteristic impedance of the coaxial cable is the result of the ratio of inductance to capacitance and is expressed in ohms. A fifty ohm cable is fairly standard for pleasure boat installations.

It is axiomatic in the world of electricity that a generator or transmitter transfers its energy to a connected line most efficiently when both are at the same impedance. Transmitter output has been standardized at fifty ohms, and so a fifty ohm coaxial cable is the proper choice.

Figure 24-14. The internal construction of a coaxial cable is shown in this peeled-off view. The shield usually is grounded.

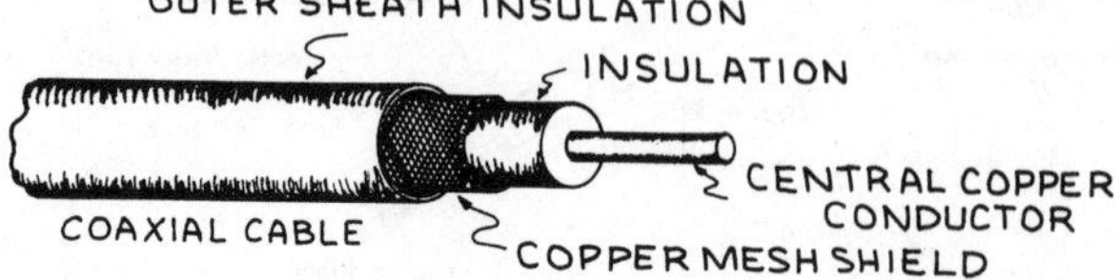

Figure 24-15. Marine antennas for television take various forms. Shown above is a folded dipole that contains its own preamplifier.

The coaxial cable runs into a different problem at its antenna connection unless only one small span of frequencies is used, as in VHF-FM, where the antenna is pretuned. For the many frequencies of single-sideband operation, the antenna varies widely in its impedance, and the oft-mentioned coupler must be added to retain efficiency. This coupler serves a double function: It tunes the antenna to resonance for the frequency being used and then provides the necessary impedance match for the coaxial cable.

The modern pleasure boat carries several radio transceivers; several antennas jutting up topside is a common sight. The caution here is that the antennas should be separated from each other by as much distance as is practicable. It is quite possible for a nearby antenna to leach energy from the transmitting antenna and also to spoil its radiation pattern in the bargain.

Antennas may attract lightning, but their light construction makes them unsuitable for carrying the heavy currents that a lightning strike engenders. The best protection is had with a metal spike at the masthead connected by copper strap to the *immersed* ground plate. Note again that the capacitive inside ground plate has no value for lightning protection.

What has just been said refers to commercial whip antennas. A properly installed long-wire antenna, such as, for instance, a sailboat might use, could be adequate for lightning protection. A good idea is to ground this antenna with a switch when it is not in use for communication.

Another common sight aboard boats today is a television antenna. One form is a dipole made with heavy gauge, wide metal strips, illustrated in Figure 24–15. This antenna contains a radio frequency amplifier to boost the incoming signal before it gets to the TV receiver. The power for this amplifier is sent up via the coaxial lead-in cable.

25. *Lightning and Static*

Viewed from a distance, lightning is a beautiful example of nature's wonders; experienced close by, lightning is cataclysmic and fraught with terror and destruction. Lightning is of special interest to the boatman because his vessel, as the only upward protuberance on a flat expanse of water, becomes a possible target in a storm.

A stroke of lightning encompasses so much raw electric power that a comparison with man-made generators only emphasizes man's impunity. One estimate attributes several trillion kilowatts to a stroke. Yet in most circumstances this destructive giant may be dissipated harmlessly into the earth if correct preparations have been made to receive him.

Everyone is familiar with Benjamin Franklin, his kite, the wet conducting string, and the key at the end of it from which he drew an electric spark during a lightning storm. The only remarkable facet of this experiment is the fact that Mr. Franklin could walk away from it instead of having been burned to a cinder. The act was foolhardy in the extreme. The best advice about lightning is to stay away from it and, if this be impossible, to follow approved technical procedures.

Nature generates the electricity of lightning and stores the charge on clouds that act as huge capacitors whose dielectric is air. The electricity may be the product of friction, of rapid condensation, of inherent water vapor, of falling rain drops, of induction, or of other natural phenomena. When the charge grows beyond the ability of air to insulate it, the air "breaks down" and the familiar lightning stroke occurs. The discharge continues with rapid multi-strokes until the cloud and its surroundings are back in equilibrium.

Observation tends to confirm that the first manifestation from a highly charged cloud is an almost invisible "leader stroke" to a protruding earth object. The ionization resulting from this provides a conducting electrical path for a violent return stroke that is the intensely bright lightning flash. Several additional strokes may follow the same path instantly until full discharge takes place. The tremendous electrical stresses excite the atoms of the air to the point of radiation, and this is the source of the light.

Sharp objects extending upward beyond their surroundings will sometimes show a luminous brush discharge during a thunderstorm. This was thought to be a religious omen by medieval sailors who named it St. Elmo's fire. Actually, St. Elmo's fire is a warning that the atmosphere is highly charged and that a leader stroke may be forming as a prelude to a lightning strike.

Thunder, although almost as frightening as the lightning stroke itself, is a harmless sound wave. The heat of the long electric spark raises the temperature of the air so violently that a far-reaching compression wave of sound is formed. The sound of thunder extends for miles and is a means for determining the distance to the actual lightning. The time in seconds between the flash and the sound is divided by five to give the approximate distance in nautical miles.

We owe the lightning rod to the fact that Mr. Franklin came out of his experiment alive—he invented it. The sharp-pointed rod above a building or at the top of a mast discharges the surrounding electrostatic field and sends the current harmlessly to ground over the connecting heavy wire. On the debit side, there is also the suspicion that the rod may attract a lightning strike—but if it does, the expectation is that the current will be dissipated without incurring damage.

The old adage that electricity always follows the easiest path explains why high sailboat masts are a favorite lightning target. The mast, especially if it be a metal one, offers less resistance than the equivalent air path to ground. The otherwise recommended practice of putting the VHF antenna at the top of the mast has cost sailboatmen many antennas; in that position the antenna also becomes a lightning rod, and most antennas cannot take that.

The vagaries of electric conduction by ionization

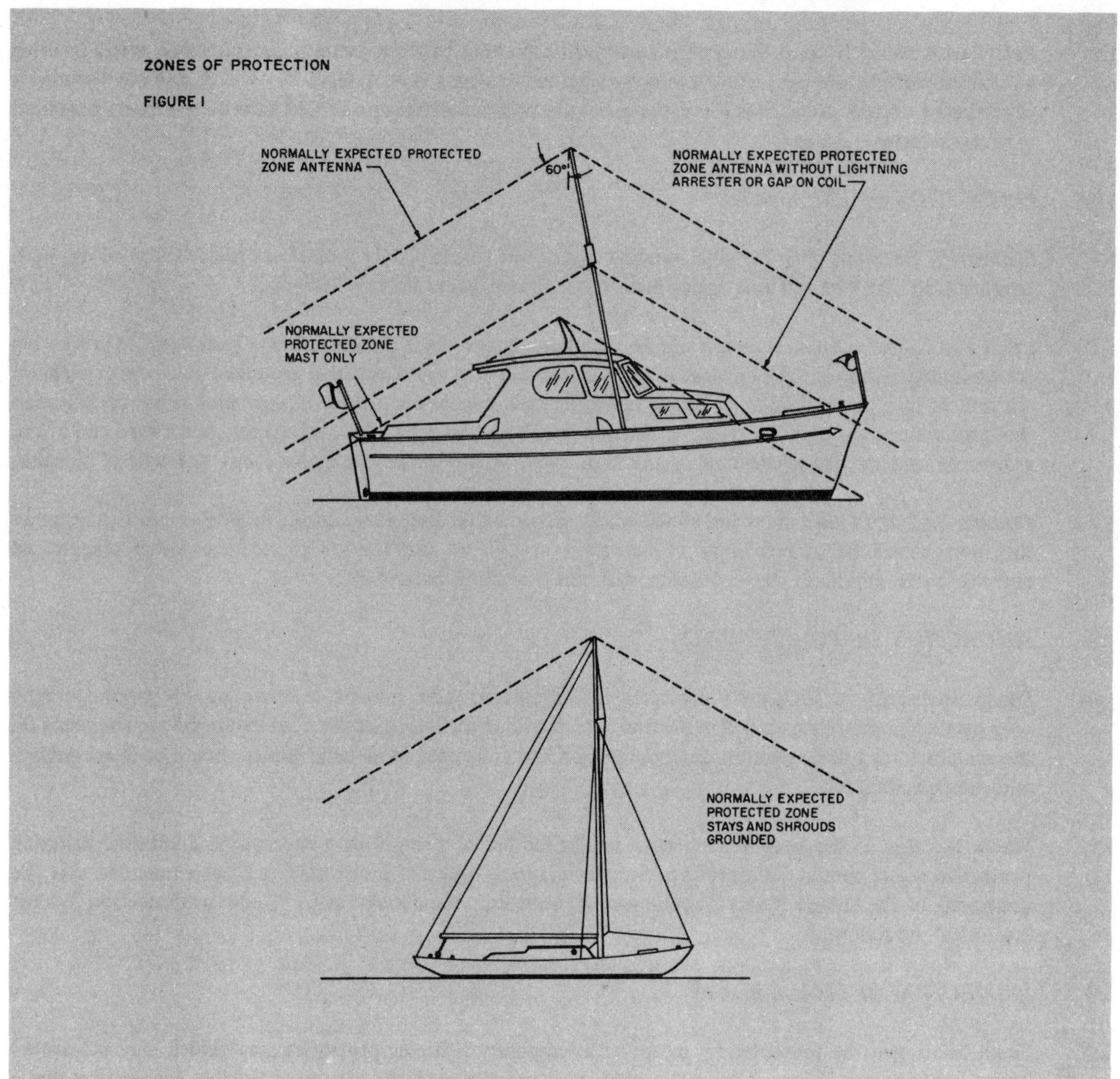

Figure 25-1. A properly grounded lightning spike at the mast top provides a 60 degree "cone of protection" as shown. It is evident that the higher the mast, the greater the protected area at the base of the cone. (American Boat and Yacht Council)

also have a good side: They permit a "cone of protection" from lightning for any vessel with a properly installed lightning rod. (This is illustrated in Figure 25–1.) Everything within the 60° cone whose apex is the lightning rod is reasonably safe from being struck.

As already stated, to assure protection, such a lightning rod must connect directly with the ground, in this case the water. The ground plate must be external and submerged. An inside ground plate, perfectly satisfactory for radio purposes, will not do, and in fact presents another hazard. Holes have been punctured in hulls as the inside ground discharges to the outside water.

Years ago when long wire antennas were the vogue,

a lightning ground switch was a standard part of the installation. This was a heavy single pole, double-throw switch; in one position in connected the antenna to the radio, and in the other position the antenna was connected directly to ground. When heavy static in the radio receiver indicated the coming of a storm, the switch was thrown to the grounding position. Some modern whip antennas are still being installed with such a switch.

Today's radio receivers with their sensitive diodes and transistors are vulnerable to atmospheric electricity. Even without a direct strike, the intense electrostatic field surrounding an antenna in the vicinity of a storm may induce a high enough voltage into the antenna system to burn out semiconductors. Under severe conditions it is wise to disconnect the radio and connect the antenna to ground. Of course, this should be done early on, before there is actual danger.

In days gone by, it was customary to install a lightning arrestor in the downlead from the antenna. This device is a small, knife-edge gap mounted on an insulator; the spacing of the gap is only a few thousandths of one inch. One plate of the gap is connected to the antenna lead, the other to ground. The theory held that any excess voltage coming from the antenna would be jumped to ground before doing harm to the receiver. With today's ultra-sensitive semiconductors, it is possible that even the voltages that got by such gaps could be destructive to receivers. (Of course, the gap had to be great enough to prevent the transmitter from jumping it.)

Getting down to the nitty gritty of lightning protection installation: The connection between the mast spike and the ground plate should be at least No. 8 AWG copper wire or else one-inch-wide copper strip at least one-thirty-second-inch thick. It should be run in as straight a line as possible, avoiding all sharp bends. There is an ambivalent answer as to whether or not this wire should be insulated, and in most installations it is not.

Excessive potential differences could exist between the grounding wire and nearby large metal objects; this could result in destructive side flashes. The preventive is to connect these objects to the ground wire so that the potential difference is reduced to zero and flashes cannot take place. Projecting stacks, davits, handrails, and searchlights that extend above the cabin top—all should be tied to the grounding wire with No. 8 or equivalent strip. It is assumed that engines and tanks are already grounded in the normal manner.

Sailboats have an easier time becoming safe from lightning. Grounding the metal stays and shrouds forms an internal area that is safe. Separate metal railings and metal pulpits should be tied in. So also should metal mast tracks and metal boom tracks.

Earlier discussion of antennas expressed some doubt as to the value of commercial whips for lightning protection. In the case of spirally wound whip antennas, this doubt becomes a complete negation; such antennas have no value as lightning protectors. Antennas with loading coils that do not contain lightning arrestors are effective only for the length below the coil.

Personnel behavior has a great deal to do with the effectiveness of lightning protective methods in preventing injury. During a lightning storm all people aboard should remain inside. They should refrain from contact with the large metal objects that are tied into the grounding system. Operating a searchlight that extends up through the cabin top could be especially dangerous. As stated before, the radio is best disconnected if the vessel finds itself in an actual storm area with its highly charged atmosphere.

Even though a lightning strike has been dissipated safely by the grounding system, it is possible that unseen damage has been done to the compass and other sensitive instruments. The intense magnetic fields created by the heavy lightning current may have thrown the compass completely awry, and it is wise to have it checked.

Figure 25-2. This photo shows how lightning discharges bridge the gap between clouds and earth.

When amplitude modulation and telegraphic code were the only forms of radio communication, the cracks and bangs of atmospheric static were a warning to the skipper that a storm was near. The immunity to static of modern frequency modulation receivers has deprived the boatman of this warning, but the need for caution remains.

26. *Power for Electronics*

By far, the most common source of electric power for the electronics aboard a pleasure boat is the vessel's storage battery, in other words, the 12 volt direct current supply. The alternating current from the pier and that from an onboard generating plant come into the picture peripherally because they are the main means of keeping the battery charged—with some help, of course, from the alternators on the propulsion engines.

Luckily, the power requirements of modern solid state electronic devices are small and easily supplied. The first radio receivers were the so-called "crystal sets" that needed no externally supplied current. Then came the multi-tubed transceivers that gulped power—even demanded current to keep vacuum tube filaments alive when on "standby" and not actually in use. Now electronics aboard have reverted to an economy level because transistors sip power only in small fractions of watts. Sailboats with enforced long intervals between battery chargings may now carry a wide selection of electronic navigating equipment. Even radar is not proscribed if it be low current solid state.

Two types of storage batteries are available on the commercial market: the lead-acid and the nickel-cadmium; only the former is in universal use on pleasure boats. The lead-acid battery consists of alternate plates of lead dioxide (positive) and red lead (negative) immersed in an aqueous solution of sulphuric acid; the nominal terminal voltage of one cell is 2 and the usual package of six cells is rated at 12 volts. The nickel-cadmium cell has positive plates of nickel hydroxide and negative plates of cadmium immersed in an aqueous solution of potassium hydroxide; the nominal terminal voltage of a cell is 1.2.

These batteries are called "storage" because they have the ability to be discharged and then to be recharged for another subsequent discharge. However, the name is a misnomer because the cells do not "store" electricity. They undergo reversible chemical changes that in one direction are caused by the charging current and, in the other direction, generate the current that the cell supplies.

Lead-acid storage batteries present two hazards that must be guarded against. One is from the acid itself that is highly corrosive to man and to material. The second is from the explosive hydrogen gas evolved during charging. Both hazards of the batteries are minimized by simple common sense in handling them.

The batteries are heavy and obviously require solid support able to withstand the rigors of rough seas. The location of the batteries in the boat should make it easy to maintain and test them at frequent intervals and should provide adequate ventilation. An insulating cover over the boat batteries is well worthwhile and prevents accidental short circuits from dropped tools. (The illustration in Figure 26–1 shows a typical lead-acid storage battery in cross-section.)

The most reliable method for testing the state of charge of a lead-acid storage battery is with a hydrometer. The specific gravity of the electrolyte solution changes with the electrical condition of the battery, and this is read easily with a hydrometer. The relationship between charge and specific gravity is given in the table in Figure 26-2. Special voltmeters are available for testing the state of charge, but these are not so reliable as the hydrometer.

Storage battery maintenance is minimal and consists of keeping the electrolyte up to the mark and checking the state of charge. The level is maintained by adding water; theoretically, this should be distilled water, but manufacturers are fairly well in agreement that water suitable for drinking is adequate. Acid never is added unless there has been loss of electrolyte by actual spillage. Handling full strength sulphuric acid is dangerous; if acid must be added, it is safer to use the dilute, ready-mixed electrolyte.

Polarity does not exist in alternating current circuits, but is indispensable in direct current circuits such as the 12 volt supply from the storage battery. To facilitate identification, the positive post of the battery usually is

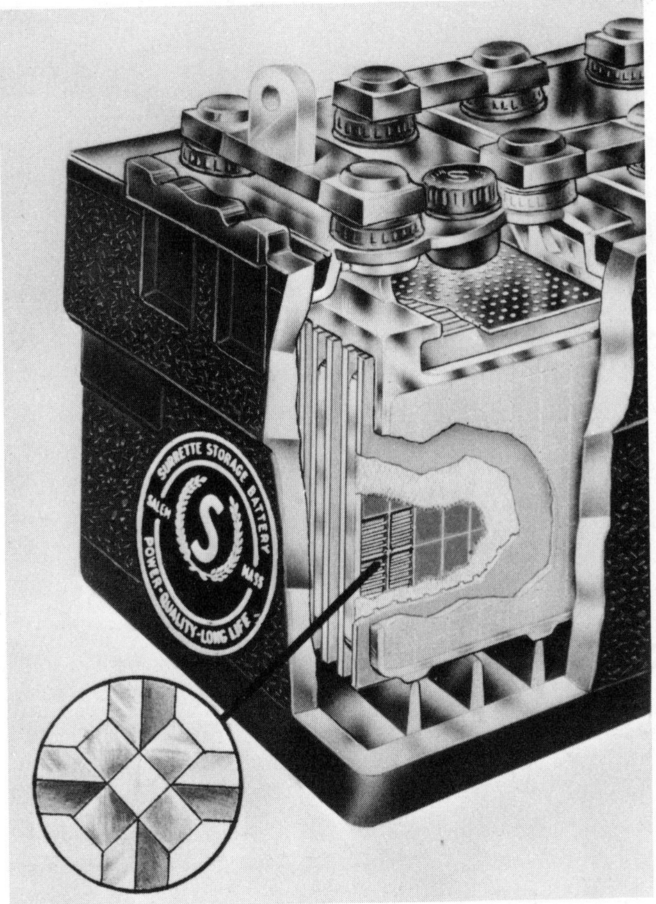

Figure 26-1. The internal construction of a standard storage battery is shown clearly in the above cutaway view. (Surrette Storage Battery Co.)

Figure 26-2. The readings in the table are for the battery at the industry standard temperature of 80°F. For extremes of temperature, add .004 for every 10 degrees above 80° and subtract .004 for every 10 degrees below 80°F.

Initial Standard Full Charge Reading at 80°F		*State of Charge*
1.260	1.280	
1.260	1.280	Fully Charged
1.230	1.250	75% Charged
1.200	1.220	50% Charged
1.170	1.190	25% Charged
1.140	1.160	10% Charged
1.110	1.130	Discharged

marked with an indented plus sign. Sometimes this is hard to discern on old batteries and then the different diameters of the two battery terminal posts become the tell-tale sign. (Figure 26–3 shows the dimensions.)

The approved method of caring for storage batteries has changed over the years. At one time "trickle charging" was the common procedure—a small charging current was maintained continuously. Then it was "constant current" charging in which the maximum recommended amperes were fed into the battery for a carefully measured period of time. Today on pleasure boats the best practice is to install a charger that shuts itself off automatically when the battery is fully charged.

These modern, so-called automatic chargers can handle several batteries simultaneously. Diodes (see Chapter 3) isolate the batteries so that one cannot discharge into another. (A representative automatic battery charger circuit is shown in Figure 26–4.) The charger is installed as a permanent unit at a location with good ventilation and is left connected to the batteries on one side and to the alternating current supply at the other.

The internal circuits of the charger sense the battery voltage at all times and compare it with a reference voltage that represents full charge. When a different voltage exists, as would be the case with a partially discharged battery, electronic switches turn the charging current on. When the battery comes up to its charged condition and the difference voltage disappears, the switches turn the current off.

The system is excellent and practically infallible when only a single battery is connected to the charger. When several batteries are connected, a slight discrepancy appears. The sensor responds only to the highest of the battery voltages or to an average of them. But the difficulty is minor and easily overcome by more careful and more frequent battery inspection and testing.

Pleasure boats spend most of their lives at a pier, and so the alternating current available thereon becomes the vessel's primary source of power. This shore current is used indirectly and also directly. Indirectly the alternating current, by powering the battery charger, enables bilge pumps and other battery-operated devices to function. Direct use is for major

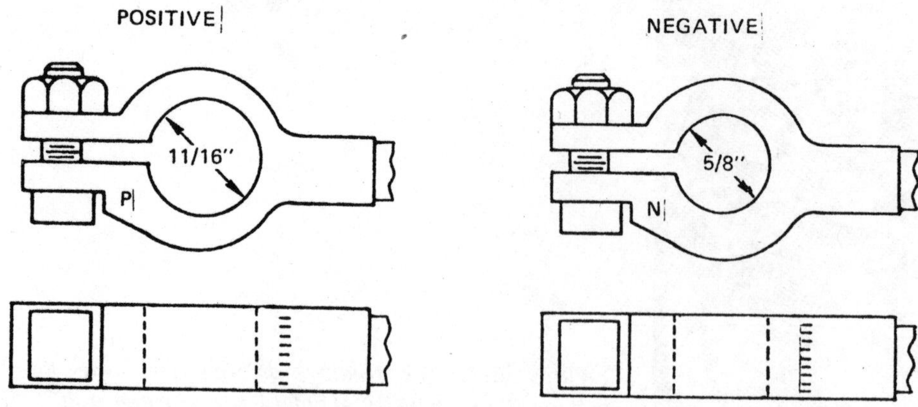

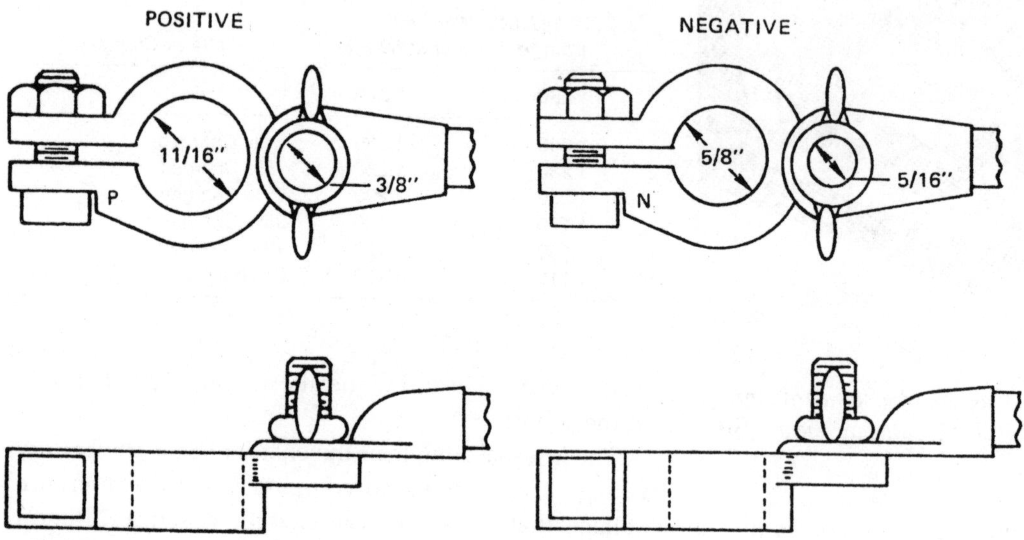

Figure 26-3. All storage battery terminals have been standardized to the dimensions shown above. Note that the positive terminal is larger than the negative and this provides an easy method of polarity identification.

appliances, such as air conditioners, that require too much "juice" to be practical for battery lines. The familiar umbilical cable, like the one shown in Figure 26–5, is on multiple view in every marina. These shore cables are fitted with male and female plugs that have coded arrangements of contacts that designate the permitted amperage.

Onboard generating plants are being installed on smaller and smaller boats as the weight of this machinery gets less and the design more compact. The smallest of these units runs on gasoline and may be rope-started like small outboard motors. (See Figure 26–6.) From these midgets the output capacity climbs and 15 kilowatt (or kilo-volt-ampere) units are common. The larger generating plants have diesel engines.

Onboard generating plants permit the use underway of the appliances that formerly were available at dockside only. They also allow battery charging when the propulsion engines are not running and thereby remove restrictions on the use of 12 volt electronics. Many of these plants start automatically when any demand for current is made. Alternating current produced aboard does not have the frequency stability of shore power. This does not affect small appliances, but television receivers may experience "flop-over" and electric clocks especially will lose their accuracy.

Generating plants make noise, and now a silent, comparatively new substitute is trying hard to come aboard to replace them. The interloper is the "inverter." Of course, the designation "substitute" is not truly earned, because the plants generate power while the inverters make use of what already is there.

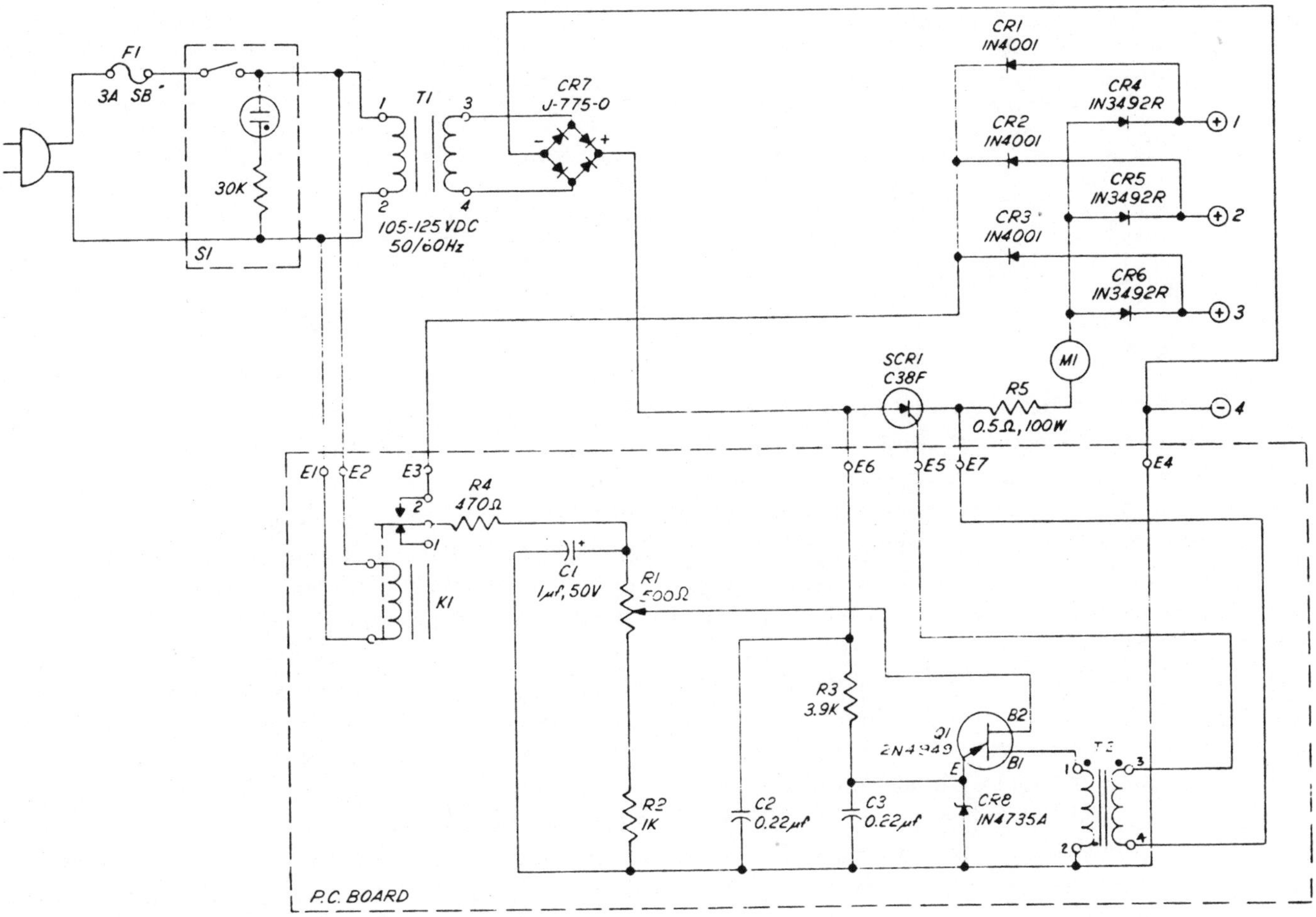

Figure 26-4. The circuit above is representative for automatic battery chargers. Provision is made for up to three batteries. Note the circuit within the dotted rectangle that monitors battery condition. (Ray Jefferson)

Figure 26-5. The cable that feeds shore power to a boat tied to the pier is a familiar sight in all marinas. The connectors at cable end have different configurations for various amperage ratings. (Harvey Hubbell, Inc.)

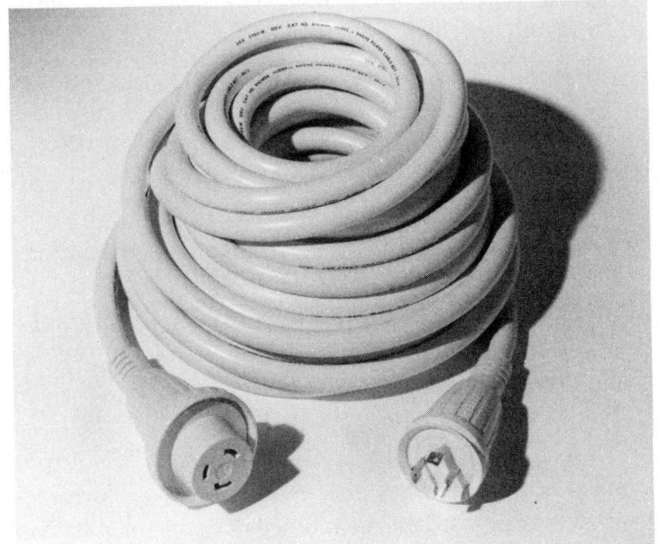

Figure 26-6. This compact portable generator could be a life saver when everything goes dead aboard. A yank on the starting cord brings 120 volts alternating current and 12 volt direct current. (Honda)

An inverter takes direct current and transforms or inverts it to alternating current. Modern solid state inverters do this by switching the direct current on and off to simulate an alternating current that is acceptable to a transformer. (Transformers function only on alternating current.) The transformer raises the voltage to the required 120 volts or 240 volts for which most alternating current devices are designed. The actual switching is done solid state by transistors and silicon controlled diodes (see Chapter 3) without the use of any mechanical moving parts.

Early inverters did not pay much attention to frequency stability and the switching was done at random. Modern units, such as the one shown in Figure 26–7, hold the desired 60 hertz frequency to plus or minus one hertz! This accuracy is attained through the use of quartz-controlled oscillators, tuned to the standard household 60 hertz, that mastermind the switching. (See Chapter 7.)

As stated before, inverters do not generate power and thus must have access to a source of energy. For the so-called "static" inverter, this source is the vessel's storage battery; the "dynamic" inverter uses the alternator on the propulsion engine as its source.

Figure 26-7. This unit converts the output of the alternator on the propulsion engine into standard voltage alternating current. Other units are available that invert battery current without engine running. (Dynamote Corp.)

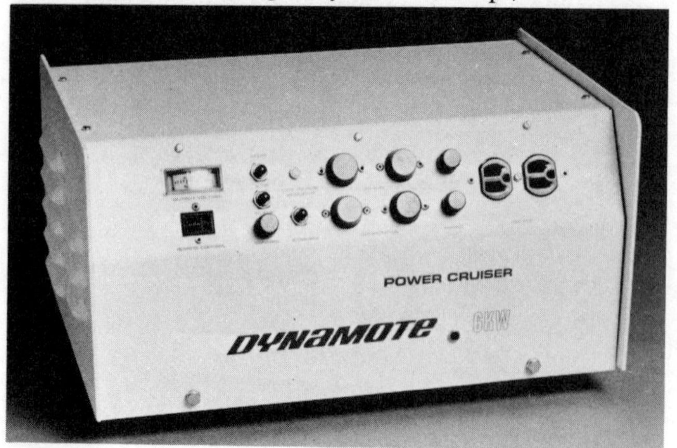

Bear in mind that every transformation of energy results in loss that appears as heat or some other undesired byproduct. Thus, if a certain number of watts is to be delivered by the interter, a much greater number of watts must be drawn from the source. In the case of the static inverter, the capacity of the storage battery therefore sets the level of possible activity. The alternator on the propulsion engine is the limiting factor for the dynamic inverter; the question is how much more power may be drawn from it than the manufacturer originally intended.

Inverter makers urge that a separate battery be hooked exclusively to the inverter and that the engine battery not be used. This is good advice and precludes the possibility of a skipper finding himself with a battery too discharged to start the engine. Obviously, from what has just been written, the inverter battery must be large enough to take care of expected alternating current loads.

The idea behind the large outputs of the dynamic inverters is to employ the engine alternator far above its normal operating condition. Whether or not this is good and permissible for any given machine must be decided by its manufacturer, and an inquiry is suggested. The speed-up is accomplished by a change in pulley ratio, the added output by a change in the method of regulation. The usual regulator is bypassed and a regulator within the dynamic inverter increases the field current of the alternator. Again, the extra output is not "free"; it puts an extra load on the engine and is paid for with extra fuel. A six kilowatt (or kilovolt-ampere) drain on the inverter may take perhaps ten horsepower of the engine to fulfill. As usual, "nothing for nothing" is nature's rule.

The output from a standard shore power line, seen on an oscilloscope, is a smooth sine curve. The output from most inverters is a sharp, square wave. For practical, everyday use, this difference in wave form has no significance. The pure sine wave form may be restored when required by means of filters available as options.

The spendthrift concept of power is normal for powerboaters, but annoys the skippers of sailboats who usually must get along on very little. Attempts have been made to solve this situation by draining some power from the wind and from the sun, but none has been particularly successful. Solar cells and windmills are the means employed.

Proponents of solar power cite the trillions of watts the sun pours down upon the earth every day, but they seem to forget to mention the huge areas involved. For any area that is small enough to be practical for solar cell coverage, the total power available is nothing to get excited about. Solar cells presently available on the market are expensive; their power output is minimal. Such small power sources may be sufficient to trickle charge a storage battery or to run a small transistor radio, but are not very meaningful aboard. The sun does not shine all the time and so solar cells must be mated to storage batteries.

The wind is more able to produce usable power, but it too is intermittent and must work with a battery as a team. Small windmills that may be set atop the mast or on a taffrail are available at moderate cost. The generators or alternators that constitute their electricity producers are similar to standard automotive units. A good wind can become the source of several hundred watts. With the sailboat at anchor, there is no problem; underway these windmills will result in a drag that most sailors will not countenance.

One recent addition to the sources of power may be called a "watermill", in contrast to a windmill, because it is immersed and depends upon the motion of the boat underway relative to the water. Of course it, too, represents a drag. This is just another instance of nothing for nothing.

The only feature common to all the sources of electric power that have been mentioned is their incompatibility to one another. The alternating current from shore and that from the onboard generating plant cannot be mixed. The easiest way to keep them separate is with a standard "ship/shore switch" that permits the connection to the boat system of only one at a time. Alternating current from any source cannot mingle with direct current. Separate wiring is the answer here.

Solid state transceivers have fostered great simplification of the power supplies needed for electronic circuits. Whereas vacuum tube equipment needs direct current at voltages running into the hundreds, transistors are satisfied with small fractions of that voltage. Current requirements also are much smaller, making filtering out noise easier.

27. *Safe Wiring Practices*

Safe electrical wiring may be attained on a pleasure boat from three standpoints. First, the wire must be of sufficiently heavy gauge and must have a large enough cross-sectional area to carry the proposed currents without heating. Second, the insulation between conductors of opposite polarity, or greatly differing voltages, must have high enough dielectric strength to forestall breakdown. Third, the routing and fastening of the wiring must prevent exposure to bilge and seawater and must make it impossible for personnel to come in contact with harmful voltages during normal shipboard routine.

The heating that occurs when wire is of insufficient gauge for the current being carried must be explained. All wire has *some* resistance to the passage of electricity. The resistance of a heavy wire (low gauge number) is minimal; that of a light wire (high gauge number) is appreciable. When current passes through a resistance, heat is generated. Technicians call this effect the "I^2R loss" (pronounced eye-square-are), and the condition may be cumulative until the temperature rises high enough to melt the wire. Note that the I, designating amperes of current, is squared; this means that doubling the current quadruples the heat.

Figure 27–1 shows the correlation between the various parameters of copper electric wire. The most important of these, for the skipper doing some new or replacement wiring, is the current carrying capacity. A "mil" is one-thousandth of one inch and is a measure of diameter. A "circular mil" is a measure of cross-section and is arrived at by multiplying the diameter in mils by itself. The gauge number designations are in accordance with the American wire gauge (AWG) that is standard in the industry.

Insulation on the wire used in pleasure boat circuits has a function in addition to its original purpose of keeping electricity where it belongs—it identifies circuits. This color coding follows the standards listed in the table of Figure 27–2. When the search to pin down trouble leads to a cable of wires, it is a great help to know, by its color, which wire does what. New boats invariably will follow the code; older craft may not.

The transition from wire to binding post where the connection is made to an electronic device may become a hidden source of trouble. Corrosion is the most likely cause. A thin film of oxide resulting from environmental conditions forms a high resistance barrier through which current cannot pass or, at best, passes only partially. Alternately tightening and loosening the terminal screw often cuts through the oxide and restores the low resistance path. However, the best recourse is to dismantle the connection and burnish wire, screw, and contact surface down to shining metal.

Corrosion is not the only culprit. Vibration on a powerboat also has a hand in the problem. Wires soldered and crimped into lugs often break away just enough to cause that bugaboo of the serviceman, the intermittent contact, while to casual observation everything is normal. Where this is suspected, the lug should be removed and resoldered or recrimped with the proper tool after thorough cleaning.

Soldering has lost its general acceptance as the best manner in which to make a permanent electrical connection. So-called "wire-wrap" has usurped its place. In this method, as shown in Figure 27–3, a mechanical tool wraps copper wire so tightly around a post with sharp edges that abrasion and intimate contact result. It seems difficult to consider this superior to good soldering, but the phone company with its millions of connections is in the forefront of those who think it is.

The bonding of all onboard large metal objects to a common ground was mentioned in Chapter 25 as a factor in lightning protection, but this practice is more than that; it is also a means of making marine wiring safe for personnel. The metal cases of electrical and electronic equipment should be included in the grounding for maximum safety. One possible harmful element exists in this all-inclusive bonding: It may tie the various boats along a pier together electrically with resultant galvanic corrosion problems.

Figure 27-1. The American Wire Gauge table lists the specifications for copper wire.

Gage No.	Diameter in mils at 20 deg. cent.	Pounds per 1,000 ft.	Feet per pound	Feet per Ohm* 20 deg. cent. (= 68 deg. fahr.)	50 deg. cent. (= 122 deg. fahr.)
0000	460.0	640.5	1.561	20,400.0	18,250.0
000	409.6	507.9	1.968	16,180.0	14,470.0
00	364.8	402.8	2.482	12,830.0	11,480.0
0	324.9	319.5	3.130	10,180.0	9,103.0
1	289.3	253.3	3.947	8,070.0	7,219.0
2	257.6	200.9	4.977	6,400.0	5,725.0
3	229.4	159.3	6.276	5,075.0	4,540.0
4	204.3	126.4	7.914	4,025.0	3,600.0
5	181.9	100.2	9.980	3,192.0	2,855.0
6	162.0	79.46	12.58	2,531.0	2,264.0
7	144.3	63.02	15.87	2,007.0	1,796.0
8	128.5	49.98	20.01	1,592.0	1,424.0
9	114.4	39.63	25.23	1,262.0	1,129.0
10	101.9	31.43	31.82	1,001.0	895.6
11	90.74	24.92	40.12	794.0	710.2
12	80.81	19.77	50.59	629.6	563.2
13	71.96	15.68	63.80	499.3	446.7
14	64.08	12.43	80.44	396.0	354.2
15	57.07	9.858	101.4	314.0	280.9
16	50.82	7.818	127.9	249.0	222.8
17	45.26	6.200	161.3	197.5	176.7
18	40.30	4.917	203.4	156.6	140.1
19	35.89	3.899	256.5	124.2	111.1
20	31.96	3.092	323.4	98.50	88.11
21	28.46	2.452	407.8	78.11	69.87
22	25.35	1.945	514.2	61.95	55.41
23	22.57	1.542	648.4	49.13	43.94
24	20.10	1.223	817.7	38.96	34.85
25	17.90	0.9699	1,031.0	30.90	27.64
26	15.94	0.7692	1,300.0	24.50	21.92
27	14.20	0.6100	1,639.0	19.43	17.38
28	12.64	0.4837	2,067.0	15.41	13.78
29	11.26	0.3836	2,607.0	12.22	10.93
30	10.03	0.3042	3,287.0	9.691	8.669
31	8.928	0.2413	4,145.0	7.685	6.875
32	7.950	0.1913	5,227.0	6.095	5.452
33	7.080	0.1517	6,591.0	4.833	4.323
34	6.305	0.1203	8,310.0	3.833	3.429
35	5.615	0.09542	10,480.0	3.040	2.719
36	5.000	0.07568	13,210.0	2.411	2.156
37	4.453	0.06001	16,660.0	1.912	1.710
38	3.965	0.04759	21,010.0	1.516	1.356
39	3.531	0.03774	26,500.0	1.202	1.075
40	3.145	0.02993	33.410.0	0.9534	0.8529

*Length at 20 deg. cent. of a wire whose resistance is 1 ohm at the stated temperatures.

Figure 27-2. Strict adherance to the suggested color code for boat electrical wiring pays dividends if trouble-shooting becomes necessary at a later date. A typical engine wiring system also is shown. (American Boat and Yacht Council)

COLOR	ITEM	USE
Yellow w/Red Stripe (YR)	Starting Circuit	Starting Switch to Solenoid
Yellow (Y)	Generator or Alternator Field	Generator or Alternator Field to RegulatorField Terminal
	Bilge Blowers	Fuse or Switch to Blowers
Dark Gray (Gy)	Navigation Lights	Fuse or Switch to Lights
	Tachometer	Tachometer Sender to Gauge
Brown (Br)	Generator Armature	Generator Armature to Regulator
	Alternator Charge Light	Generator Terminal/Alternator Auxiliary Terminal to Light Regulator
	Pumps	Fuse or Switch to Pumps
Orange (O)	Accessory Feed	Ammeter to Alternator or Generator Output and Accessory Fuses or Switches
	Accessory Common Feed	Distribution Panel to Accessory Switch
Purple (Pu)	Ignition	Ignition Switch to Coil and Electrical Instruments
	Instrument Feed	Distribution Panel to Electric Instruments
Dark Blue	Cabin and Instrument Lights	Fuse or Switch to Lights
Light Blue (Lt Bl)	Oil Pressure	Oil Pressure Sender to Gauge
Tan	Water Temperature	Water Temperature Sender to Gauge
Pink (Pk)	Fuel Gauge	Fuel Gauge Sender to Gauge

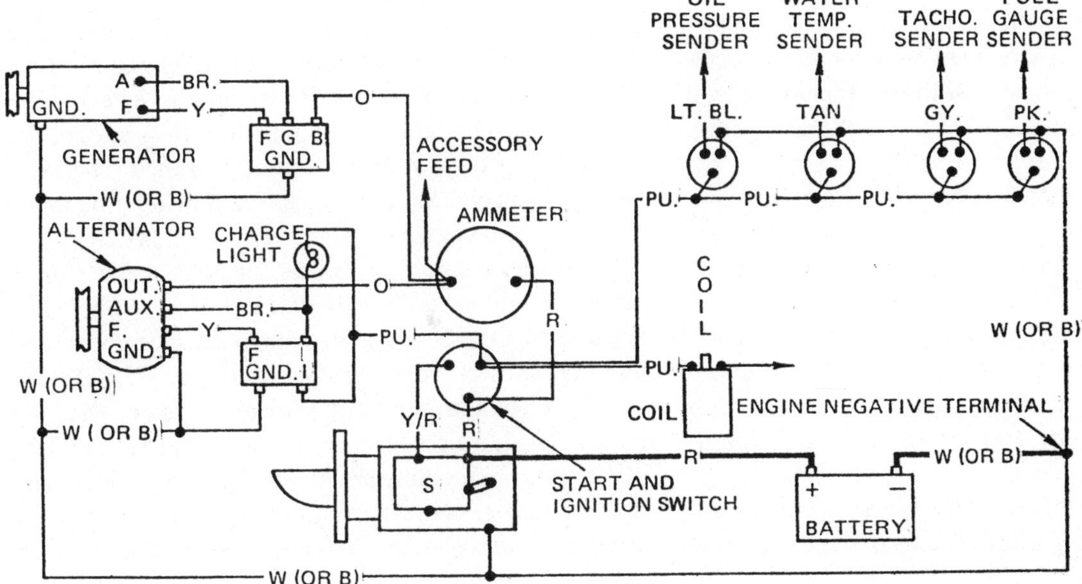

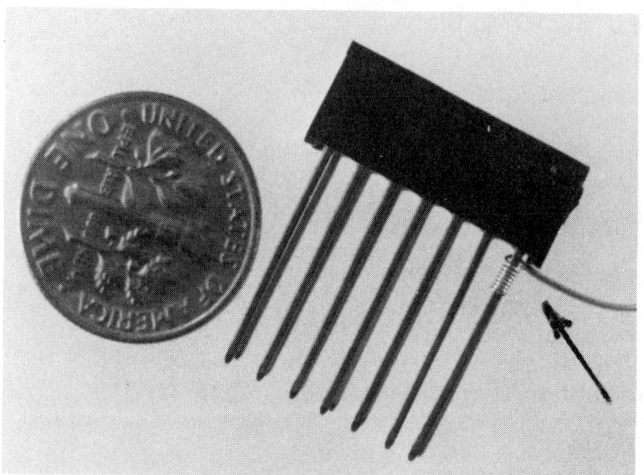

Figure 27-3. In some services the wire wrap system of connecting to terminals has won preference over time-honored soldering.

Figure 27-4. The Isolator insulates the vessel from the pier circuit as far as destructive galvanic currents are concerned, but does not interfere with safety grounding. (Mercury Marine)

The solution to these problems is a recently developed device called an "isolator." It does what its name implies—isolates the vessel from others along the pier and thus breaks the circuit needed for galvanism. The isolator does not remove the protection rendered by the ground wire in the shore cable; it permits emergency currents to pass without hindrance, thus maintaining the purpose of the arrangement. An isolator is shown in Figure 27–4. The complete bonding system is diagrammed in Figure 27–5.

Of course the best way of all to insure freedom from the dangers of the shore connection is to install an isolation transformer. Only the primary of this unit is connected to shore. The secondary, that supplies the ship with power, is completely separate. The only thing common to primary and secondary is the magnetic field that links them. But isolation transformers are heavy, bulky, and costly and therefore seldom found on pleasure boats.

It is helpful here to touch briefly on the electrochemical process known as galvanic corrosion, because this wasting away of metals may confront the skipper anywhere about the boat. When two dissimilar metals are connected together and immersed in any fluid that conducts electricity, a galvanic battery is formed, and one of the metals will waste away. One metal becomes an anode and the other a cathode. The anode is the active one that disappears. The table in Figure 27–6 is called a "galvanic series" and it identifies anodic and cathodic partners in any combination. The uppermost metal in the table always is the anode. The further apart the two metals are in the listing, the stronger the reaction and the quicker the disappearance. The table should be a constant guide, especially when one metal is used as a fastener for another.

Pleasure boat wiring, on all but the smallest open boats without cabins, is divided into two distinct classifications. One is the low voltage direct current system that springs from the storage battery. The other is the high voltage alternating current system that derives from shore power or from an onboard generating plant. In generalizing about the two, it may be said that the former is incapable of electric shock, while the latter may impose shocks that can be lethal under certain circumstances. Obviously, each requires different precautions for the maintenance of safety.

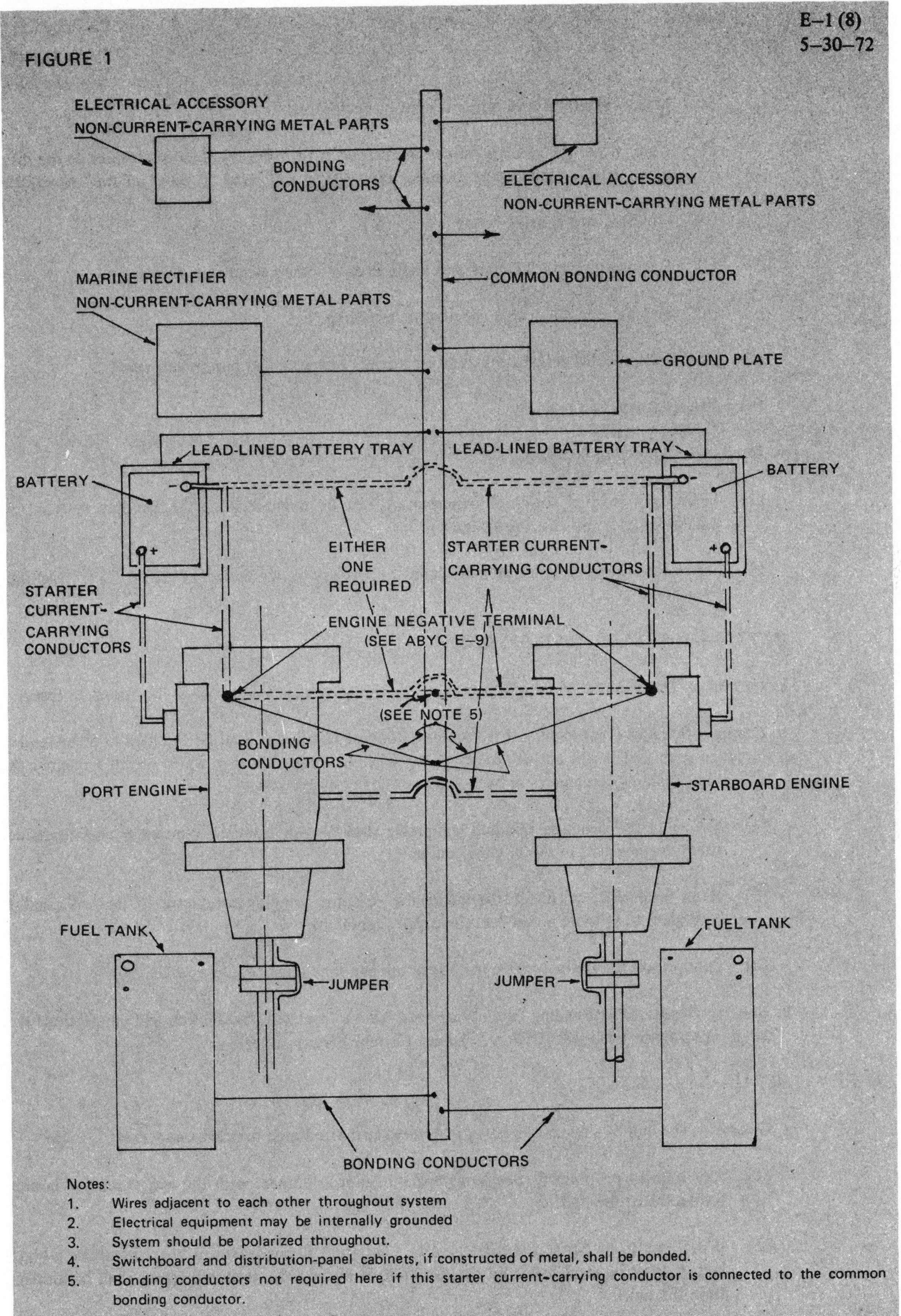
E–1 (8)
5–30–72
FIGURE 1
ELECTRICAL ACCESSORY
NON-CURRENT-CARRYING METAL PARTS
BONDING
CONDUCTORS
ELECTRICAL ACCESSORY
NON-CURRENT-CARRYING METAL PARTS
COMMON BONDING CONDUCTOR
MARINE RECTIFIER
NON-CURRENT-CARRYING METAL PARTS
GROUND PLATE
LEAD-LINED BATTERY TRAY
LEAD-LINED BATTERY TRAY
BATTERY
BATTERY
EITHER
ONE
REQUIRED
STARTER CURRENT-
CARRYING CONDUCTORS
STARTER
CURRENT-
CARRYING
CONDUCTORS
ENGINE NEGATIVE TERMINAL
(SEE ABYC E–9)
(SEE NOTE 5)
BONDING
CONDUCTORS
PORT ENGINE
STARBOARD ENGINE
FUEL TANK
FUEL TANK
JUMPER
JUMPER
BONDING CONDUCTORS
Notes:
1. Wires adjacent to each other throughout system
2. Electrical equipment may be internally grounded
3. System should be polarized throughout.
4. Switchboard and distribution-panel cabinets, if constructed of metal, shall be bonded.
5. Bonding conductors not required here if this starter current-carrying conductor is connected to the common bonding conductor.

Figure 27-5. The bonding system recommended by the American Boat & Yacht Council is shown in detail above.

Figure 27-6. This table predicts the corrosive action on two dissimilar metals connected together and immersed in sea water. The further apart the metals are in the table, the greater the degeneration. (American Boat & Yacht Council)

Galvanic Series of Metals in Sea Water

ANODIC OR LEAST NOBLE—ACTIVE

Magnesium and magnesium alloys
CB75 aluminum anode alloy
Zinc
B605 aluminum anode alloy
Galvanized steel or galvanized wrought iron
Aluminum 7072 (cladding alloy)
Aluminum 5456
Aluminum 5086
Aluminum 5052
Aluminum 3003, 1100, 6061, 356
Cadmium
2117 aluminum rivet alloy
Mild steel
Wrought iron
Cast Iron
Ni-Resist
13% chromium stainless steel, type 410 (active)
50-50 lead tin solder
18-8 stainless steel, type 304 (active)
18-8 3% NO stainless steel, type 316 (active)
Lead
Tin
Muntz metal
Manganese bronze
Naval brass (60% copper—39% zinc)
Nickel (active)
78% Ni.-13.5% Cr.-6% Fe. (Inconel) (Active)
Yellow brass (65% copper—35% zinc)
Admiralty brass
Aluminum bronze
Red brass (85% copper—15% zinc)
Copper
Silicon bronze
5% Zn.—20% Ni—75% Cu.
90% Cu.—10% Ni.
70% Cu.—30% Ni.
88% Cu.— 2% Zn.—10% Sn. (Composition G-bronze)
88% Cu.— 3% Zn.—6.5% Sn.—1.5% Pb (composition M-bronze)
Nickel (passive)
78% Ni.—13.5% Cr.—6% Fe. (Inconel) (Passive)
70% Ni.—30% Cu.
18-8 stainless steel type 304 (passive)
18-8 3% Mo. stainless steel, type 316 (passive)
Hastelloy C
Titanium
Platinum

CATHODIC OR MOST NOBLE—PASSIVE

The direct-current system, almost universally at 12 volts, is characterized by its ability to supply extremely heavy currents, such as the several hundred amperes needed for engine starting. Such amperage can melt wires and start fires, but it will not give an electric shock to personnel handling it. The voltage of the alternating current lines is 120 volts and often also 240 volts; such potentials are a hazard to anyone coming in contact with them and require an experienced approach.

The primary danger in all systems is a short circuit that provides an alternate low resistance path and entails an abnormally heavy flow of electricity. The protective measure is to break the circuit before wires melt, arcs occur, and fires start. This protection is offered by fuses and circuit breakers that function without human aid.

Fuses and circuit breakers are purposely provided as "weak links in the chain." The low melting point link in a fuse melts because of the heating effect of overcurrent. The action is rapid enough (when a proper fuse is being used) to prevent the wire in the rest of the circuit from becoming warm. Fuses are one-shot devices and must be replaced when blown.

Circuit breakers are employed because they are not destroyed like fuses when overcurrent occurs. Circuit breakers open the line when too heavy currents accidentally are drawn, and are "reset" easily by moving a lever or pushing a button when the trouble is corrected. The activating principle of circuit breakers is either thermal or magnetic. The overcurrent either generates heat that triggers the switch or else forms a magnetic field strong enough to cause the switch to open. Both fuses and circuit breakers are calibrated for the number of amperes that makes them act. (See Figure 27–7.) Some circuit breakers cannot be reset until the trouble is corrected.

A further protective device for the alternating current system of a pleasure boat is a "ground fault circuit interrupter," shown in Figure 27–8. This unit cuts off the flow of electricity to the boat when the existence of current in the normally currentless ground wire is an indication that something has gone awry and may prove a hazard to personnel. The ground fault circuit interrupter is more sensitive than the associated fuses and circuit breakers and goes into action before they do.

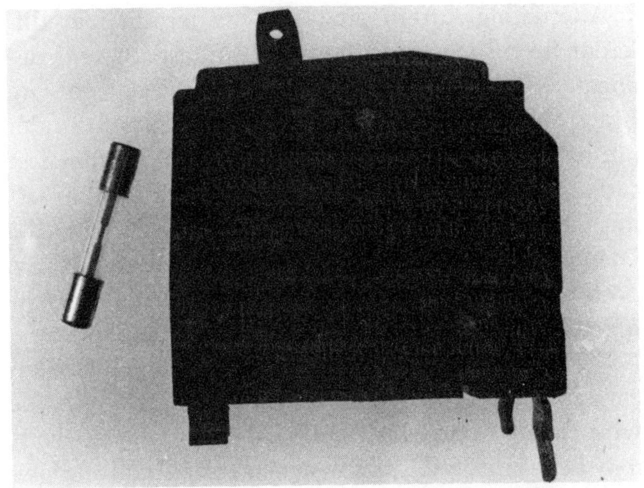

Figure 27-7. The fuse is a one-time device that must be replaced after it is "blown." The circuit breaker may be reset without damage to itself.

Figure 27-8. The ground fault interrupter monitors the circuit continuously. It breaks the circuit automatically if it discovers a small unintended current to ground. (Harvey Hubbell, Inc.)

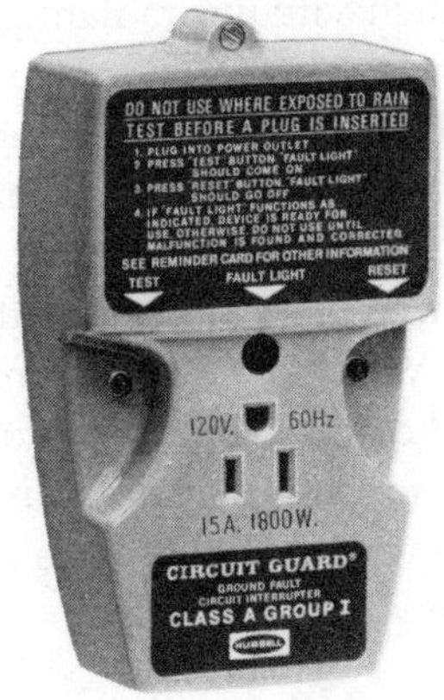

Alternating current circuits have a terminology different from their direct current counterparts and delineate wires uniquely in the color code. A green wire is a grounding wire and does not carry current under normal conditions. However, a ground wire that *does* carry current normally always is a part of an alternating current system, and it is white; this wire is the "neutral" of electricians' slang. The reason for the presence of the neutral is that all commercial alternating current supplies connect one wire to ground.

The neutral may be one of a two wire circuit feeding current at 120 volts. It may be the central one of a three wire circuit from which both 120 volts and 240 volts may be taken. In this arrangement the neutral and one outside wire yield 120 volts while 240 volts may be had by connecting to the two outside wires without the neutral. The green grounding wire is the fourth conductor in this three wire scheme.

It can be seen that the ground from the shore must be maintained as the ground on the boat and that a reversed connection could have a potential for danger. In most cases, polarized plugs and receptacles, such as those shown in Figure 27–9, assure that correct connections are maintained. Polarity indicating devices are an additional safeguard and are found on many pleasure boats; these warn, by light, sound, or both, when a reversed connection is made.

Most pleasure boats with sizable enclosed cabin space have branch circuits radiating from the main panel that receives the alternating current from shore or from an onboard generating plant. Each branch must have overload protection, either fuses or circuit breakers, in its "hot" wire where it connects to the main supply. These branches power the lights and the receptacles into which appliances are plugged. As a further guide to keeping polarity uniform, the connecting screws of each receptacle are marked by being either silver color or gold color. (The neutral wire goes to the silver screw.)

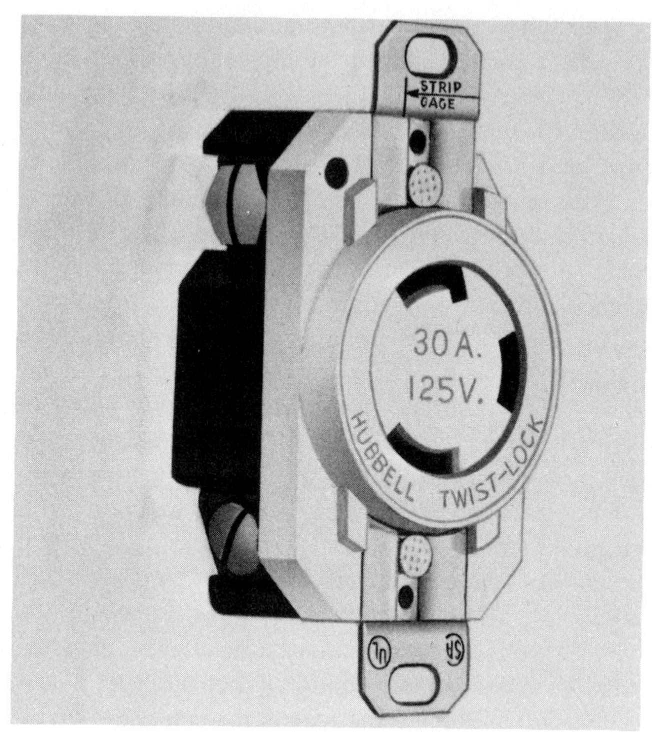

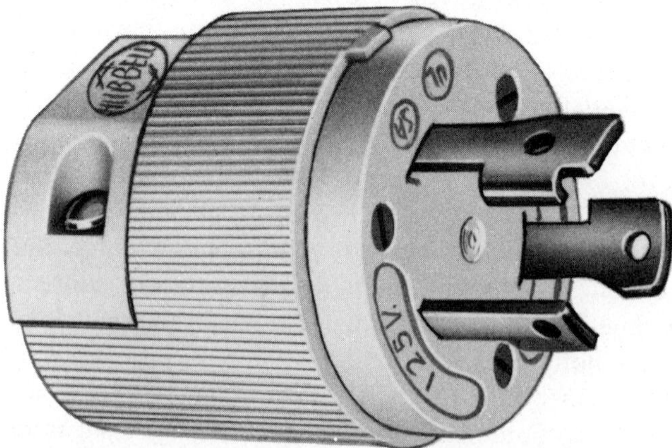

Figure 27-9. The geometry and size of the male and female contacts of the plugs and receptacles of the shore cable determine the allowable amperage to be drawn from the pier. (Harvey Hubbell, Inc.)

TYPE	DESCRIPTION	INSULATION TEMPERATURE RATING	APPLICATION
SO	Hard Service Cord —	60°C (140°F)	General Use except for Machinery Spaces
	Oil-Resistant Compound	75°C (167°F) & higher	General Use
ST	Hard Service Cord —	60°C (140°F)	General Use except for Machinery Spaces
	Thermoplastic	75°C (167°F) & higher	General Use
STO	Hard Service Cord —	60°C (140°F)	General Use except for Machinery Spaces
	Oil-Resistant Thermoplastic	75°C (167°F) & higher	General Use
*SJO	Junior Hard Serv. Cord —	60°C (140°F)	General Use except for Machinery Spaces
	Oil-Resistant Compound	75°C (167°F) & higher	General Use
*SJT	Junior Hard Serv. Cord —	60°C (140°F)	General Use except for Machinery Spaces
	Thermoplastic	75°C (167°F) & higher	General Use
*SJTO	Junior Hard Serv. Cord —	60°C (140°F)	General Use except for Machinery Spaces
	Oil-Resistant Thermoplastic	75°C (167°F) & higher	General Use

*NOTE: *Junior Hard Service Cords are acceptable under 33 CFR 183 and are rated at 300 volts instead of 600 volts.*

AMPACITY OF INSULATED COPPER CONDUCTORS
See Note (2) (4)

CONDUCTOR SIZE AWG	NOMINAL CM AREA See Note(1)	30°C (86°F) AMBIENT 3 CONDUCTORS	30°C (86°F) AMBIENT 2 CONDUCTORS	40°C (104°F) AMBIENT 3 CONDUCTORS	40°C (104°F) AMBIENT 2 CONDUCTORS
16	2,580	10	13	8	11
14	4,110	15	18	12	15
12	6,530	20	25	16	20
10	10,380	25	30	20	25
8	16,510	35	40	30	35
6	26,240	45	55	35	45
4	41,740	60	70	50	60
2	66,360	80	95	65	80

NOTES:

(1) *To recognize stranded conductors made of AWG size elements, the actual nominal cm area may differ from the specified nominal cm area, but by no more than 7 per cent.*

(2) *Current ratings are for not more than 2 or 3 current-carrying conductors in a flexible cord as indicated. Reduce the current rating to 80 per cent of values shown for 4 to 6 current-carrying conductors.*

(3) *A conductor used for equipment grounding and a neutral conductor which carries only the unbalanced current from other conductors, as in the case of normally balanced circuits of three or more conductors, are not considered to be current-carrying conductors.*

(4) *The ampacity of shore cables shall be based on 30°C (86°F) ambient.*

Figure 27-10. Flexible electric cords and cables are subject to severe service aboard because of the constant movement and the hostile environment. Correct choice is essential. The recommendations of the American Boat & Yacht Council are given above. "Ampacity" means current carrying capacity in amperes.

It has been said that most problems with portable electric appliances ashore originate in the flexible connecting cords. This statement is even truer afloat where the marine environment speeds deterioration. (A guide to the selection of flexible connecting wires is given in Figure 27–10. Two grades are listed for each type; the junior variety is derated in voltage from 600 to 300.)

Perhaps the rule most violated by skippers doing their own alternating current wiring is the requirement that all current-carrying connections be made within junction boxes or similar enclosures. In other words, free-standing twisted connections are taboo if compliance with the safety standards is to be maintained. If the junction boxes are not weatherproof, they must be protected from the elements. Wires are to be connected to terminals with lugs either of the ring or the captive spade type, and the lugs are to be attached to the wire by use of the manufacturer-specified crimping tool.

Error and consequent danger may be possible where alternating current branches and battery current branches terminate in nearby receptacles unless clear demarcation exists. To achieve this isolation, receptacle slots and plug prongs must be so different for each system that plugs for one cannot be entered into receptacles for the other. Without this precaution, plugging a 12 volt lamp into a 120 volt outlet could generate a miniature explosion.

Two locations of possible danger in the use of 120 volt electrical appliances aboard are the galley and the lavatory. The receptacle for an electric stove must be so placed that the unit may be plugged in without causing the connecting cord to cross either the stove, the sink, or a work area. Because a person in the lavatory easily grounds himself by touching commode, basin, or shower, the receptacle for this area should be located so as to minimize possibility of shock, and the circuit should have extra protection.

Shore cables that bring power from the pier to the boat undergo considerable fatigue and strain as the vessel rocks with ripples and rises and lowers with the tide. This cable has a male plug for the pier and a female plug for the power inlet on the side of the boat. The preferred design, such as that shown in Figure 27–11, locks the cable to the boat with a threaded retainer that is weatherproof. Cables are graded for the maximum amperes they may carry and adherence to this loading is automatically enforced by the nature of the coded plugs and receptacles that are illustrated in Figure 27–9.

All approved cables contain a green grounding wire and this, as discussed earlier, has its bad side as well as its good. Because of this wire, all vessels along the pier plugged into the same power circuit are electrically connected and open to the caprices of galvanic corrosion. Woe to the skipper with an aluminum propeller if all his neighbors have bronze wheels! The solution is the isolator already mentioned and pictured in Figure 27–4.

It is the general practice on pleasure boats to have the direct current (storage battery) system supply power for the running and anchor lights, the bilge pumps, and usually also for the electronic navigational equipment. Separate branch lines for each of these units radiate from a central panel that the battery feeds. Although engine starting motors also operate from the battery, they are on totally separate circuits because of the extraordinarily heavy currents they draw when activated.

Figure 27-11. The ubiquitous shore cable is the umbilical cord between pier and boat. (Harvey Hubbell, Inc.;

Each of the circuits leaving the panel does so from a protective device that is either circuit breaker or fuse, although fuses of the glass tube type are preponderant. Fuse ratings are determined by the load on the line, the amperes of which are stated on the manufacturer's labels. "Blown" fuses are easily detected because the lead strip within the glass tube no longer is continuous. A good central panel is shown in Figure 27–12; all circuits are marked.

The common starting point of the direct current system is considered to be the point at which the negative cable from the storage battery is bonded to the engine block. This also is the point to which the bonding system and the radio ground plate are connected in order to eliminate any differences of potential that could cause galvanically destructive currents to flow.

Figure 27-12. This main electric distribution panel is the origin of all electric wiring aboard. Each circuit breaker is marked for identification. Meters keep tabs on amperages and voltages. The entire panel and its wires are located for easy access. (Yacht *Yes Dear*, owner Rennie Miller.)

There is an important difference between automobile 12 volt wiring and pleasure boat 12 volt wiring. In the auto, only one wire, the positive "hot" one, supplies each circuit; the return or negative conductor is the metal chassis. On the boat, each circuit has its own two wires, the positive and the negative; even a metal hull is not used as a return. This sometimes puzzles automobile mechanics newly gone afloat, but the two wire method is less vulnerable to the hazards of the marine environment.

A magnetic field exists around all wires carrying direct current; this could cause compass deviation when these circuits are routed too close to that instrument. Twisting the nearby positive and negative wires together cancels the magnetic field and protects the compass from this influence. A simple check is made by watching the compass closely while the currents are switched on and off; if this action causes a needle movement, there is trouble that should be hunted down and eliminated by twisting and rerouting. This caution does not apply to wires carrying alternating current because their magnetic fields are constantly reversing.

A voltage drop occurs whenever an electric current passes through a conductor, and therefore the output voltage always is less than the input. The difference between input and output may be negligible or it may be great, depending upon the resistance of the conductor, and thus the secret of keeping it low is to use wires of adequate cross-section. A 2 volt drop in a 120 volt line is only a loss of less than 2 percent, but that same 2 volt drop in a 12 volt line represents a loss of more than 16 percent. The moral is that wire size becomes ultra-important with low-voltage circuits carrying appreciable currents.

The wiring found inside electronic instruments is a different world from the boat's power distribution circuits just discussed. In most cases the power levels will be miniscule. Actual point-to-point discrete wiring from one component to another (the "hard" wiring described in earlier text) seldom will be found in modern equipment. Major connections are affected by printed circuits. (See Chapter 13.) Often a large number of components are combined into one miniature housing called an integrated circuit. (See Chapter 14.)

The road map that simplifies all this complexity is the schematic wiring diagram. To save space and promote clarity, the industry has standardized on a set of symbols such as those reproduced in Figure 27–13. Many of these symbols are almost self-explanatory; others are easily memorized. These "shorthand" representations identify the components included in the circuit of any given piece of electronic equipment.

Many owner's manuals contain a schematic diagram of the electronic instrument described. The ability to recognize the various components of the diagram goes a long way toward understanding the functioning of the circuit.

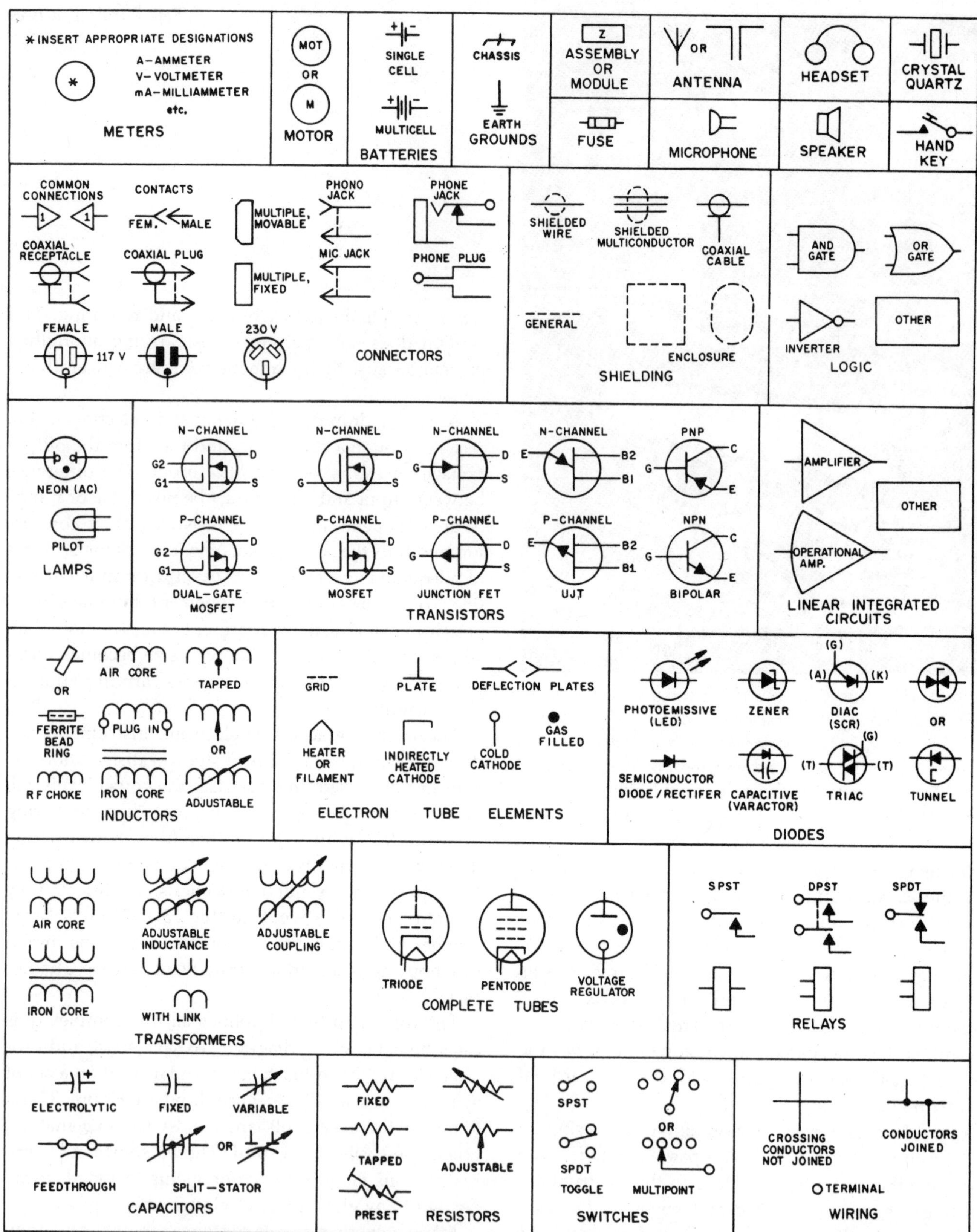

Figure 27-13. Electronic components are represented on wiring diagrams by "shorthand" symbols such as those pictured above. (ARRL Radio Amateur's Handbook)

28. *Electronic Maintenance*

Alas, everything on a boat, even the vessel itself, requires maintenance and then more maintenance, and electronic equipment is no exception. But at least maintaining electronics is neat, clean, and light work in contrast to the grease-spattered clothes and grimy hands that result from a maintenance tour of the engine room. Moreover, modern electronic devices are compact and so the entire operation is performed within an easily observable small area.

Electronics, be it the heart of a radio transceiver, a Loran set, or whatever, performs one or more of the following functions: It amplifies, it rectifies, it switches, and it generates oscillations—all in addition to the ancillary duties of filtering, buffering, and tailoring electric waves to a desired shape. Thus a prime requisite in maintenance is to determine the exact purpose of the circuit under consideration and then to understand the consecutive steps by which that purpose is achieved.

All electronic circuits require direct current as a basic input. It is logical, therefore, to consider this power source as a prime subject of maintenance attention. Is the voltage correct? Is it able to supply the current needed? Is the electricity delivered "clean," that is, free from undesired spikes and noises picked up extraneously?

The direct current source often is in two parts, the voltage from the battery and the secondary voltage consistent with the needs of the transistors and other sensitive elements. The change from one to the other may be complicated by the presence of oscillators, inverters, rectifiers, and filters that have been discussed earlier. Simple maintenance for all of these under normal conditions entails little more than removing dust and dirt and keeping connections tight and clean. Corrosion here is anathema.

Heat is generated wherever electric current is dissipated, and this heat must be removed before it builds up and raises the temperature to levels that are disastrous to semiconductor devices. It is customary to transfer the heat to the ambient atmosphere by means of heat sinks, venting louvers, and in some applications even electric fans. All of these methods depend upon a free flow of air. Conscientious maintenance sees to it that air flow is not impeded by dirt or by proximity to objects that form a barrier. Normal air flow without a fan is by a "chimney effect": Air enters at the bottom, is heated, rises to the top, and exits. (See Figure 28–1.)

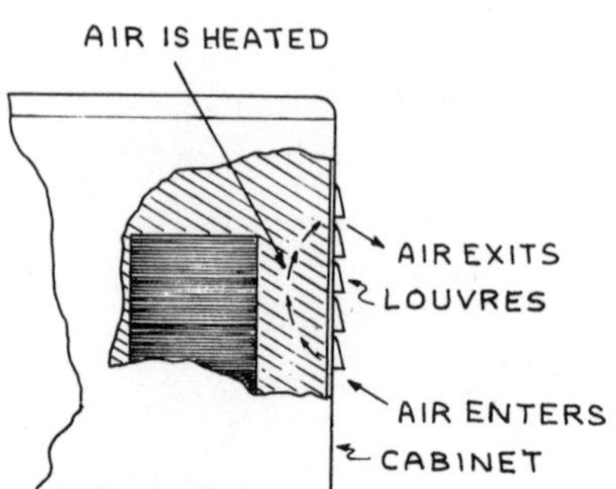

Figure 28-1. Air cooling of louvered electronic cabinets depends upon the "chimney effect." Air enters at the bottom, is heated, and exits at the top, taking instrument heat with it.

The metal rod portion of the antenna, exposed as it is to the weather, becomes the most endangered link in the chain of devices that form the radio communication system. Corrosion and pitting soon take over unless periodic maintenance includes thorough cleaning followed by a wipe with an oily or waxy rag. Surface corrosion assumes greater importance with wires that carry radio frequency currents than it does with ordinary wiring—although of course it is not desired there either. Radio frequency currents travel only on the surface of conductors and do not traverse the entire cross-section of a wire as does the more common form of electricity used for power and light.

Switches, with their vulnerable contacts, frequently become a source of trouble that often is blamed elsewhere. The surfaces of the contacts become coated with an almost invisible oxide that is an effective insulator. This oxide is the product of a simple chemical reaction between the metal and the moist, salt-laden atmosphere. Where the contacts are accessible, polishing them with extra fine sandpaper is a positive cure. Hidden switch contacts can usually be brought back to electrical efficiency by opening and closing the switch rapidly a number of times in succession.

Radar maintenance requires a bit of mechanical knowhow in addition to electronic knowledge because of the motors, gears, and belts that turn the antenna and swing the electron beam in the cathode ray tube. The hazard of lethally high voltages also exists. Furthermore, anything that affects the integrity of the radar transmission is off limits to the skipper and must be performed by a licensed radar technician.

First of all, power should be off during maintenance work. Dust and dirt are removed by vacuum or air blast and high voltage terminals are cleaned with carbon tetrachloride or with alcohol to prevent future arcing. The static field around high voltage terminals attracts dust that eventually becomes the path for corona discharge and actual arcs, both ruinous to insulation. Humidity inside the radar housing may be a problem; many units contain an incandescent bulb, continuously kept lit, to supply the heat to evaporate the moisture.

Figure 28-2. This technician is servicing the antenna unit of a radar designed in the World War II era for use on Navy ships. This radar is still found today on many large yachts and is prized by navigators for its large screen and good resolution.

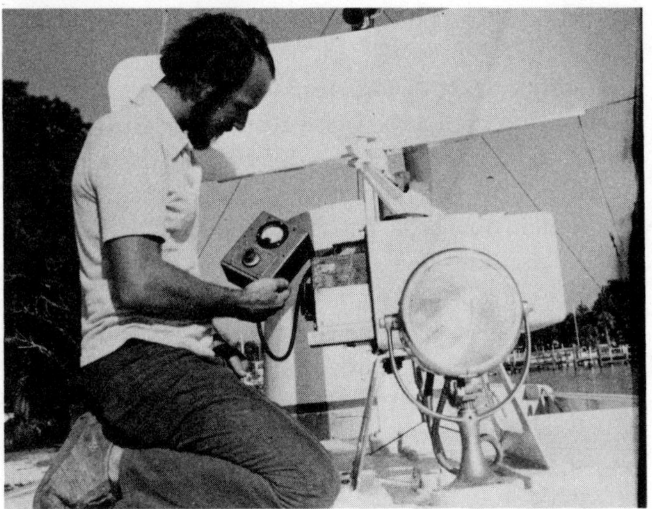

Most motors built into radars are ball bearing types, lubricated for life at the factory, and need no maintenance beyond being kept clean because they are brushless. Slip rings, where used, should be cleaned of any carbon tracking. Screws, perhaps loosened by vibration, should be checked for tightness. (See Figure 28–2.)

Many small craft radars are housed in plastic radomes. The cautions that are expressed for the maintenance of all plastics apply here too. The surface should be wiped with detergents only and not with any abrasive cleaners. Radar antenna housings should never be painted, because most paints degrade the radar transmissions. Finally, the manufacturer's instructions should always take precedence on specific maintenance details.

Extreme caution is required if it becomes necessary to handle the cathode ray (picture) tube. The glass of this tube is under great stress because of the high vacuum within and even a slight shock could set off a shattering implosion. Protective eyeglasses and gloves are recommended. The high voltage terminal of the tube should be grounded before anything is done in order to empty the capacitors of what could be a lethal remaining charge.

Electrical noise generated aboard can seriously deteriorate the performance of depth sounders, radio direction finders, and communication receivers, although the VHF-FM transceiver is less affected than other types of marine radios on other bands. The noise suppression devices originally installed may have lost their effectiveness because of corrosion, loosened connections, or other causes, and so a checkup should be included as part of the maintenance procedure. Often, incipient trouble can be spotted with a careful eyeball inspection.

Electrical noise is generated by countless things happening aboard. The prime offender, of course, is the gasoline engine. The spark plugs, the distributor, and the voltage regulator are all busy little transmitters of radio waves that travel through the wiring and through the air into receptive electronic devices. The diesel engine, acquitted on most counts of noisemaking because it uses no sparks, nevertheless has alternators, generators, and voltage regulators that can be noisy. Even some electronic units, depth sounders for instance, cause noise interference to other electronics.

At least, the noise sources mentioned above can be pinpointed and checked, each at its origin. Some cracklings and sputterings are much harder to find. They may originate at a corroded terminal or at a wire not sufficiently tightened under a screw and allowed a tiny movement when under vibration. Often these are "needle in a haystack" situations that require patience and point-to-point checking.

Suppression in the high tension ignition circuit is commonly accomplished by adding resistance that acts as a damper on the radio waves. The resistance may be in the plug, in a separate plug-in module, or in the cable in the form of a carbonized string that replaces the usual wire. Since resistance also dampens the desired hot spark, a more modern form of ignition cable contains a spiral winding over a ferrite core that acts as a radio frequency choke and barrier.

Voltage regulators that depend upon vibrating contacts are loud noisemakers. The cure resides in capacitors as shown in Figure 28–3. The capacitors are preferably of the coaxial type in which the main conductor feeds directly through. All manufacturers' instructions warn against connecting any capacitor to the field terminal of the regulator.

Any electrical device aboard that makes and breaks its circuit, and that includes commutators as well as contacts, becomes a source of radio interference. This

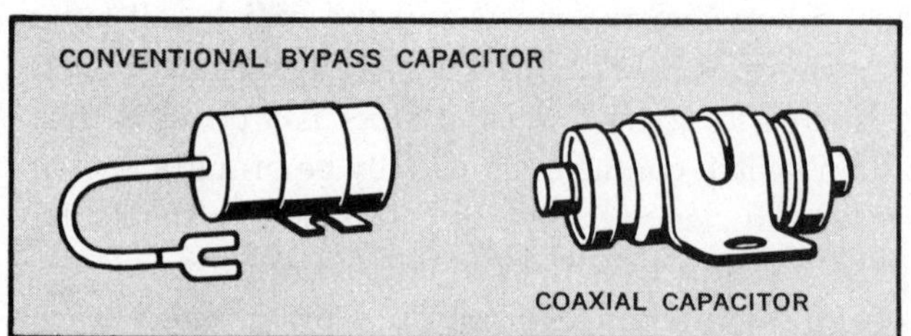

Figure 28-3. The secret of eliminating interference to the radio from onboard electrical equipment is to shunt the interfering waves to ground before they can be radiated. This is done with capacitors, preferably coaxial capacitors. The difference in appearance between conventional and coaxial capacitors is shown above.

interference is not sharply defined, as would be a regular radio transmission, but spreads broadly over the spectrum. As a consequence, these undesired waves enter electronic equipment despite tuned circuits.

The property of capacitors that allows alternating current to pass while blocking direct current comes to the rescue here. Especially felicitous is the fact that a capacitor offers less and less resistance to this passage as the frequency increases. (See Chapter 4.) Thus the relatively high frequency of the noise easily is shunted to ground and made harmless without impeding the needed direct current. How this is accomplished in practice is shown in detail in the drawings of Figure 28–4.

A simple search coil that may be used to ferret out elusive sources of radio noise is shown in Figure 28–5. This is a simple coil of wire placed on the end of a broomstick to facilitate poking into difficult spaces that contain suspects. The coil is attached to the radio receiver as shown.

Boat wiring also has a bearing on the electrical noise that affects electronic instruments. Wires from noisy devices such as tachometers and depth sounders should not be closely cabled with other conductors. Two conductors close together effectively form a capacitor that happily transfers the noise from one to the other.

Figure 28-4. How capacitors are placed on various pieces of equipment in order to shunt radio interference to ground is shown in these drawings. (a) on a generator, (b) on an alternator, (c) on a voltage regulator, (d) on console instruments, (e) on the ignition coil. At (f) a capacitor with clips is handy for trial connections. (Champion Spark Plug Co.)

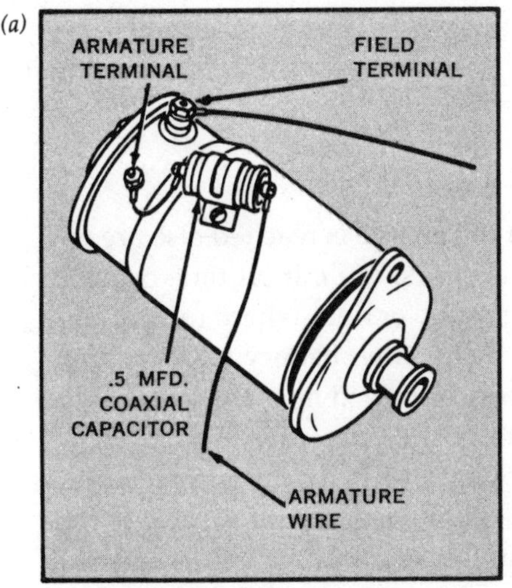

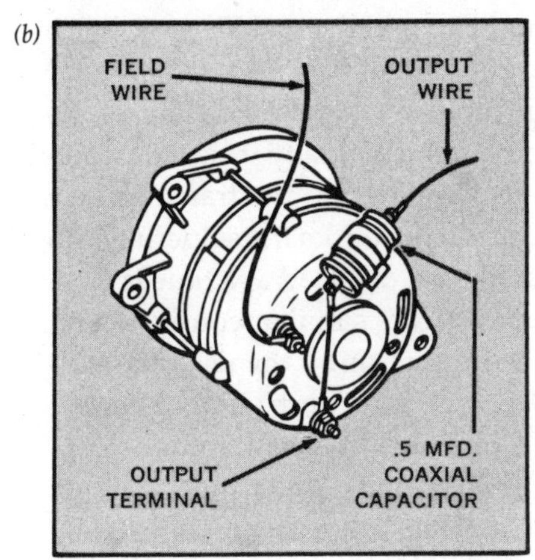

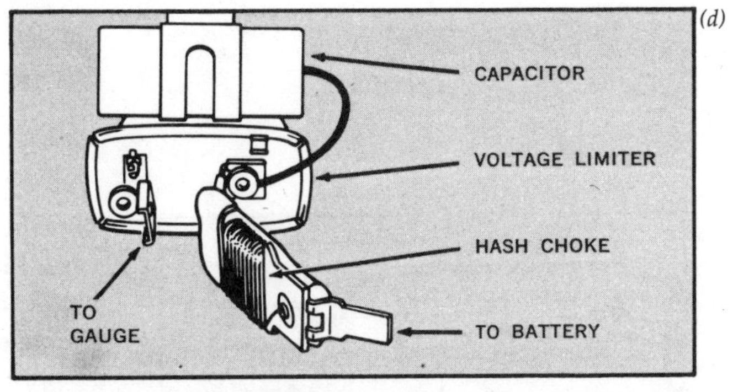

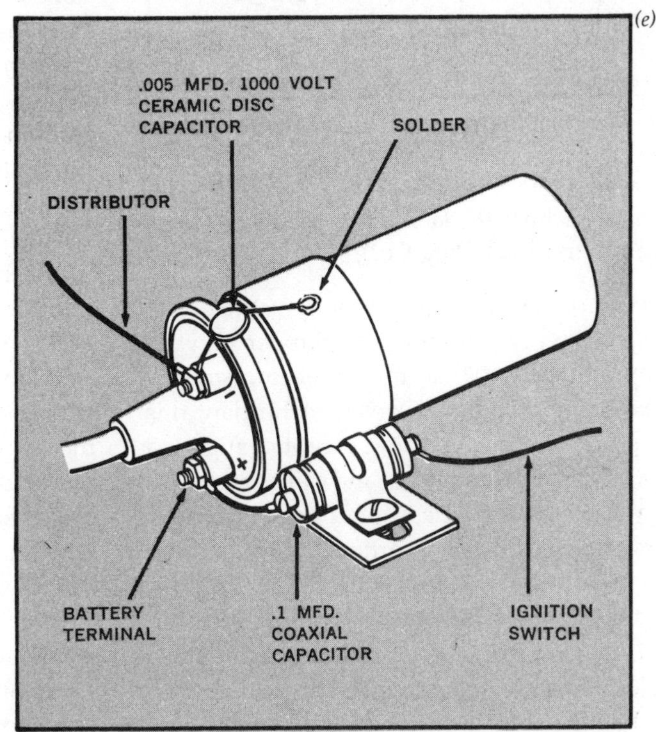

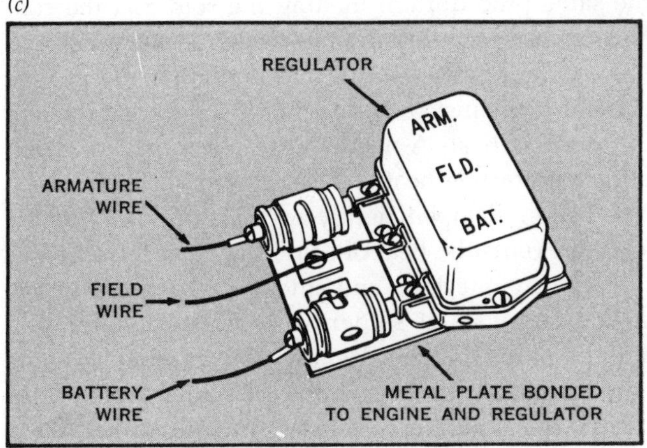

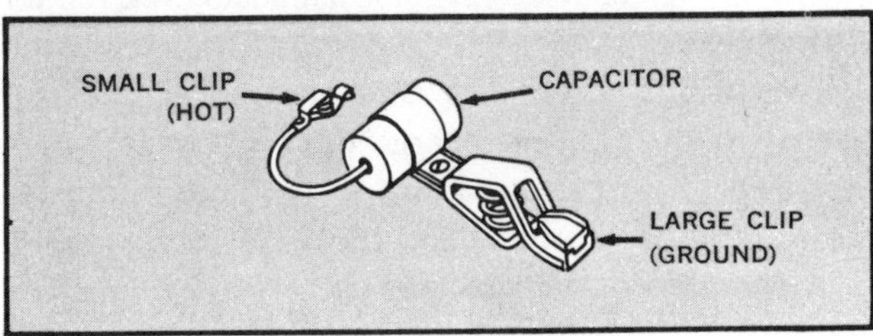

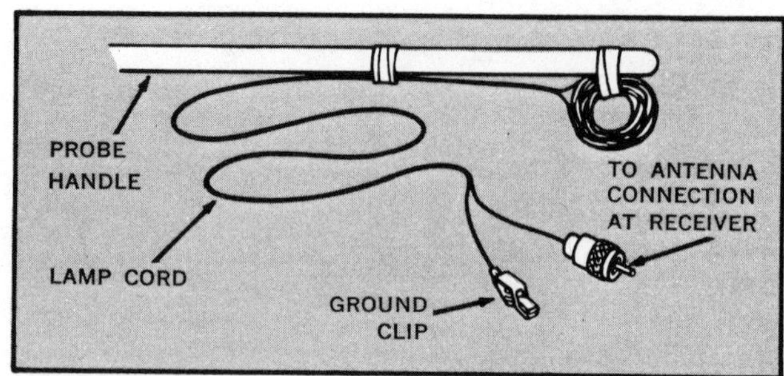

Figure 28-5. Often the source of radio interference is elusive and puzzling. A search coil like the one shown above, connected to the radio receiver, is held near all possible offenders. The increased noise from the loudspeaker identifies the offender. (Champion Spark Plug Co.)

Maintenance on radio transmitters is restricted by law, thus leaving very little that the skipper may do himself. Any adjustment that affects the transmitted wave may be made only by a licensed technician; this encompasses just about everything except changing a fuse, seeing to it that all feed wires and cables are tight, and keeping the instrument clean and free of moisture.

Insulators exposed to the weather quickly become coated with a salt film and forget that they are supposed to insulate. The film, especially when wet, becomes a fairly good conductor and can simulate a difficult-to-locate partial short circuit. This applies particularly to the insulators at the ends of bare long-wire antennas such as used on some sailboats. The remedy is obvious.

The output power of a radio transmitter is closely related to the state of charge of the storage battery feeding it. Low charge, low voltage, and diminished radio output follow in that order. Electronic maintenance therefore must include battery maintenance—and luckily the battery makes very simple demands.

A storage battery that is being used properly really needs nothing more than occasional additions of water—and some of the latest battery models claim not to need even this. Engine batteries get automatic good attention if the voltage regulator is in good shape. Separate batteries under the care of a charger that shuts itself off when full charge is reached also are well cared for. In both instances, the critical time comes at the end of the warranty period when replacement must be considered. Somehow, battery manufacturers hypnotize their products into dying almost simultaneously with the end of the guarantee.

Many of the stoppages of earlier electronic equipment powered by vacuum tubes were traceable to tube failure. The quick and easy cure was replacement of the tube. Pulling a vacuum tube out of its socket and pushing another one in required no skill, took but a minute, and put the equipment back in service.

A similar failure of a transistor in a solid state device is not that readily corrected, especially if the construction be printed circuit. Unsoldering the transistor is a trick in itself, and too much heat from the soldering iron ruins the board. Unsoldering one lead at a time is frustrating unless the solder can simultaneously be removed with a "solder sucker" or by the use of a "solder wick." The solder wick is a short piece of copper braid that sucks the molten solder away from the connection by capillary attraction; its use is shown in Figure 28–6.

This difficulty in removing a component from a printed circuit board is not unique to transistors, but is common to everything on the board. Taking off a capacitor, a resistor, an inductor, or whatever entails the same procedure of melting the solder at the connection and then sucking or wicking it away.

Resoldering a substitute component to the printed circuit board must be done without subjecting the part to excessive heat. A needle-nose pliers acts as a temporary protective heat sink, as shown in Figure 28–7. The pliers, clamped tightly to the upper end of the lead, prevents the heat of soldering from travelling all the way up the lead and into the sensitive internal structure. A small soldering iron and a minimum amount of rosin-core solder are the rule when working with printed circuit boards; excess solder could bridge over to the adjoining conductor strip and cause a short circuit.

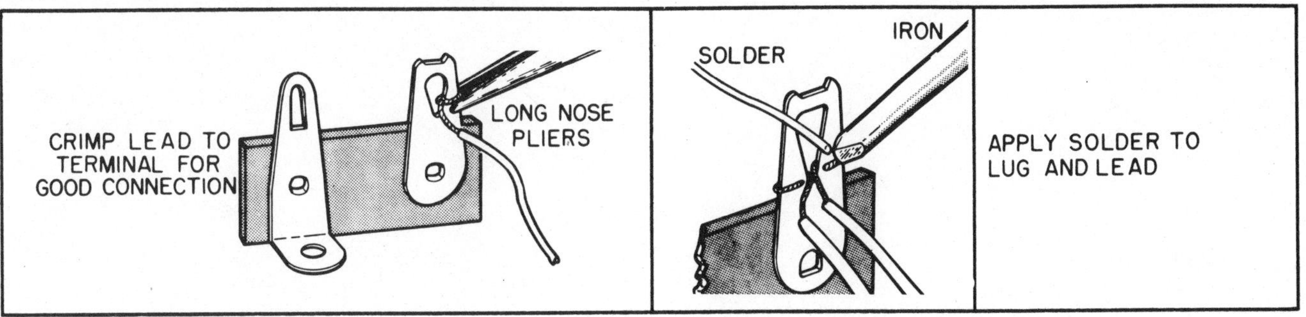

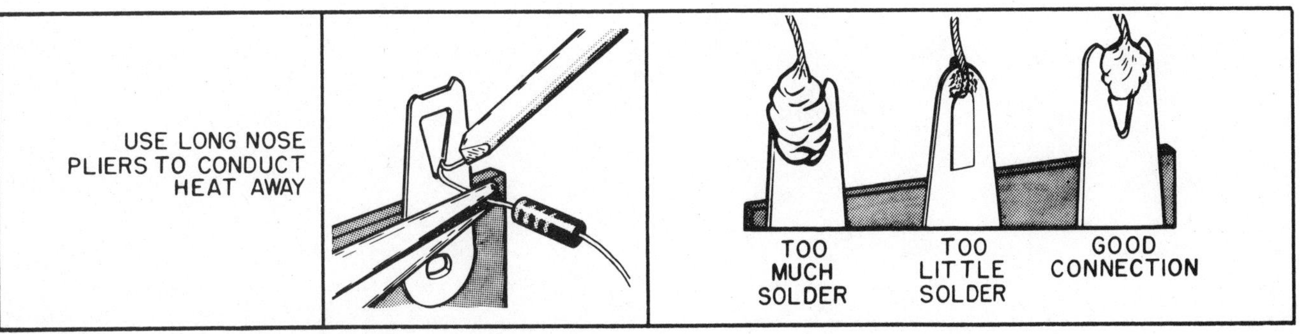

Figure 28-6. Maintenance and repair of electronic instruments often entails soldering wire connections to terminals. The illustrations above show how to do this correctly and also some mistakes to avoid.

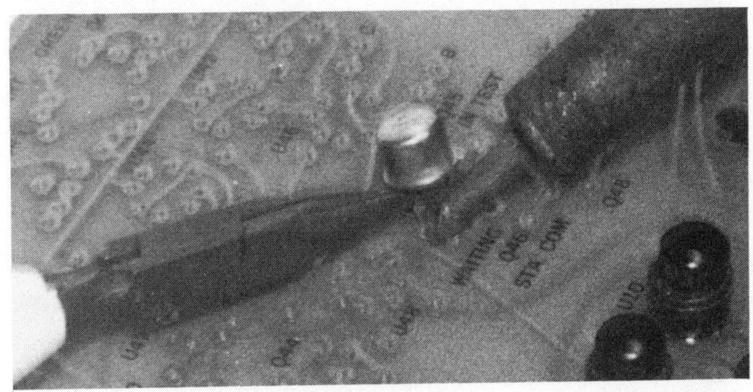

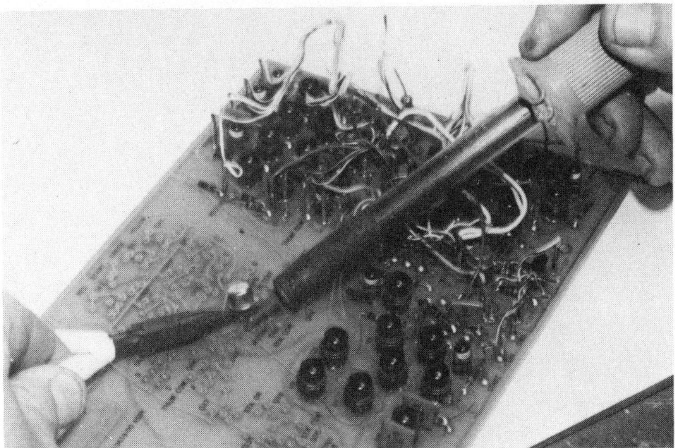

Figure 28-7. The pliers, held close to the component, absorbs the heat of soldering and prevents damage to delicate units when soldering them into a circuit.

Manufacturers pack printed instructions with all electronic equipment. This printed matter could be merely a single page for a simple device like a hailer, or it could take the form of an entire small book to accompany a radar set. By saving all these instructions and keeping them handy on board, the skipper has a ready-made mini-library that provides specific information about his electronics and guides him in its use.

Repairs or substitutions on a modern electronic device usually cannot be completed without the use of a soldering iron. Proper use of this tool therefore becomes an important skill.

Good soldering becomes easy if two admonitions are borne in mind: sufficient heat and absolute cleanliness. The working temperature and the thermal capacity of the soldering iron must be great enough quickly to impart a solder-melting temperature to the parts to be soldered. The solder should melt by being touched to the heated part, not to the iron.

Cleanliness means being clean in the chemical sense, devoid of all surface oxides. This condition is arrived at by first abrading the surface with abrasive paper or by filing. The clean condition is preserved during soldering by use of a proper flux and its chemical action. For electronic work only one flux is acceptable and that one is rosin; it forms the core in the universally employed rosin-core solder.

PART SIX

TESTING AND REPAIRING

29. *Trouble-Shooting*

Electronic trouble-shooting requires a clear understanding of the function of the particular portion of equipment under inspection and a knowledge of what normally should take place at each progressive point in the circuit. Thus armed, proceeding logically from the input to the output discloses the point at which the ensuing signal differs from the expected norm and thereby pinpoints the location of trouble. Corrective action then can be concentrated where it will be immediately effective.

The instruments for trouble-shooting may vary from the single volt-ohm meter in the hands of a knowledgeable skipper to the complete testing laboratory in the shoreside service shop. The volt-ohm meter (VOM) is perhaps the handiest single tool of them all. It can measure voltage and current, check the continuity of a circuit, and compare the actual resistance of a path with what the designer intended it to be in order to reveal any telltale discrepancy. (See Figure 29–1.)

Test lamps are worth their weight in gold and are easily made up at minor cost. One should be a plain 12-volt light bulb with long leads either coming from its socket or else soldered directly to the bulb base. This will be for all battery circuits. The other consists of two 120 volt light bulbs of any size connected in series and with two leads as shown in Figure 29–2. This tester works equally well at 120 volts (dim light) and 240 volts (bright light).

In some situations the test lamp will give a positive answer, whereas the VOM will be misleading. Consider two terminals, intended to provide full voltage, that somehow are not functioning. A VOM reading shows full voltage but the test lamp does not light, thus negating what the VOM is indicating. The explanation? The tiny and useless current available at the terminals because of some circuit fault is sufficient to operate the extremely sensitive VOM but far from enough to do useful work such as lighting the lamp.

A basic function of the VOM in electronic trouble-shooting is checking continuity—another way of say-

Figure 29-1. The VOM (volt-ohm meter) is one of the handiest diagnostic tools in the hands of the technician—and even when used by the skipper. It measures current in amperes, potential in volts and resistance in ohms. A special use is checking continuity. (Eico Electronics)

Figure 29-2. Test lamps should be in the trouble-shooting kit of every skipper. They are made easily as shown. One is for 12 volt battery circuits. The other checks both 120 volt and 240 volt circuits. The bulbs are dim for 120 volts and bright for 240 volts.

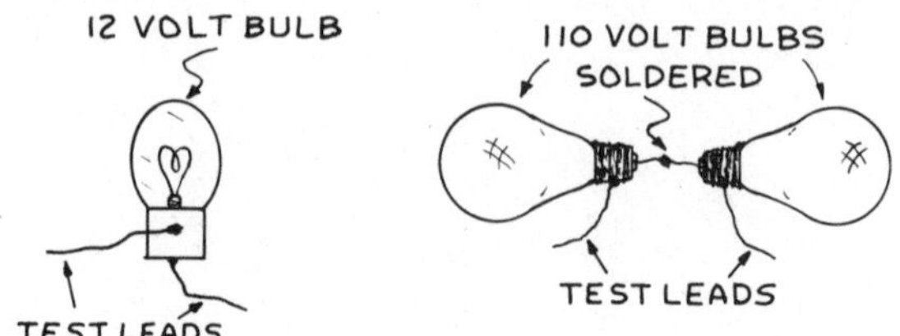

ing that resistances will be measured. The wire itself that forms a connection between two points should have practically zero resistance and should show on the VOM as a short circuit if everything be in order. A break in the wire path will be indicated as infinite resistance. These are the two opposite conditions in which the wire may be found when tested. Conditions in between, such as a corroded connection, will cause the VOM to read values much above zero. A resistor in series with the wire circuit will cause the meter to give the resistor's value if nothing be amiss.

A good capacitor will indicate infinite resistance, since it is a barrier to the direct current supplied by the meter's battery. A bit of ambiguity enters here because a poor capacitor often may not reveal itself on the meter unless it is an absolute short circuit. The reason is that it may be "good" at the very low voltage of the VOM and bad at the higher voltages it encounters in service. Note that all continuity tests of this nature are made with power off.

One little trick comes in handy when testing capacitors. The unit first is checked for continuity; this charges the capacitor up to the voltage of the meter's battery. The VOM then is switched to a low voltage scale and connected to the capacitor. If the capacitor be good enough to hold a charge, the meter needle will flick.

Novice users of the VOM often are puzzled when checking the coil of a dynamic loudspeaker. The unit is marked eight ohms, yet the meter shows it to be almost zero ohms, a short circuit. The discrepancy arises because the speaker is rated eight ohms only when fed with a high frequency alternating current; actually, this is its "reactance." Its resistance to direct current is truly close to zero.

Trouble-shooting a complicated electronic circuit without a wiring diagram is difficult and time-consuming and most technicians will avoid doing it. The wiring diagram is a road map that gives clear point-to-point instructions and shows what components lie in between. The nature of these components and an understanding of what they do predict the change the signal should undergo in a normally functioning circuit. Most owner's manuals include a wiring diagram, and it should always be retained for future reference.

The easiest trouble-shooting may be accomplished when the manual contains a complete table of resistances and voltages. Now the VOM needs only to be connected to each listed point for a reading of resistance, with power off, and voltage with power on. Comparisons with the design parameters in the table quickly lead to the location of trouble. In most cases, readings within ten percent are considered to comply with the table.

In another helpful system various "test points" along the circuit are indicated on the wiring diagram. The instructions describe what is to be expected, in the way of voltage or signal, at each test point when everything is operating normally. The ensuing voltages are measured with a VOM; the signals are sampled either with a signal tracer or with an oscilloscope if the shape of the waveform be important. (See Chapter 30.)

Many apparently abrupt failures of electronic instruments aboard may be traced to basic causes that are external to the unit that has conked out. The first thing to suspect is power. A discharged battery, a broken connection, a blown fuse, an open circuit breaker—any of these will stop operations. Trouble-shooting here is best done with the test lamp rather than with the VOM for reasons already stated. Check the instrument terminals first to verify the lack of power. Then start at the battery and check progressively forward from there.

Blown fuses, inoperative circuit breakers, and faulty switches are quickly found with the aid of the VOM set to its voltage measuring function. The instrument is turned on. The two probes from the VOM then are placed on the two ends of the suspected fuse; any indication on the meter condemns the fuse as blown. The same method, with the probes astride the circuit breaker or the switch that supposedly is "on," reveals a defective unit if the meter reads any voltage.

One caution in the general use of the VOM around the boat: Make certain that the dial is set for the intended test. The correct range should be set when measuring voltage. If the tested voltage is unknown, start with the highest range of the meter and then move down to the most appropriate scale. Never connect the VOM to any source of voltage when its dial is set for resistance readings. Carelessness in range setting could result in a burned out meter, although some of the more expensive instruments are protected against such misuse by internal diodes.

Electronic instruments that use vacuum tubes maintain ground negative. The voltage from the power supply that feeds plates and screens of the tubes is highly positive. Things are opposite in most transistor circuits. Ground is positive, and the power supply (or battery) delivers low negative voltage. This means that with vacuum tube devices the negative probe of the VOM is touched to ground or chassis, while with transistor units it is the positive probe.

Since a transistor consists essentially of two diodes back to back, a VOM set to measure resistance can give a general idea of whether a transistor is still among the living. It does this by taking advantage of the fact that a diode allows current to pass in only one direction. (See Chapter 3.) With one probe on the base of the transistor, touching the other probe to emitter and to collector should show either very high or very low resistance—and the exact opposite when the probes are reversed. A transistor that passes this test may be expected to work in a circuit, although how well it will do so cannot be judged by this simple means. A true calibration of the transistor's functional ability requires a transistor tester or a transistor curve tracer, both of which are described in Chapter 30.

A novice trouble-shooter may find difficulty in correlating a wiring diagram with the apparently random tracks on the printed circuit board. Such correlation was easy with the old-time discretely wired chassis because each wire could be traced from origin to destination in accordance with the drawing. Some circuit boards help a little by identifying each component, providing various points of reference that may be referred to the diagram. While there is no substitute for experience, a few hints may help.

Begin by finding the input to the circuit board under observation, whether a voltage from the antenna, a signal from a previous amplifier, or whatever. Check the wiring diagram to learn what component is next in the path. If it be an inductor, for instance, locate the nearest inductor and verify the connection by tracing the copper track. Locate the next component, verify the connection, and continue until the output terminals of the circuit board are reached. Now that the eitire path is clearly defined, check for the presence of the correct signal at each point.

Keeping in mind some generalizations of what each sub-circuit of an instrument does to the signal also aids in trouble-shooting. For example, the "front end" of a communications receiver tunes the circuit to accept the desired frequency and usually also amplifies the incoming signal before passing it on to the mixer. The amplifying is done at the original received frequency by so-called radio frequency (RF) amplifiers. Thus the front end comprises tuning means, either a variable inductor or a variable capacitor, plus an active amplifier, and may be recognized as a distinctive unit. Any testing in this area reveals the original signal frequency, and it should be stronger at the output than at the input if things are working correctly.

The mixer is next in line to process the received signal. This mixer receives not only the output of the front end, but also an entirely separate frequency generated by an internal tuned oscillator. (See Chapter 16.) The mixer has the ability to combine (heterodyne) these two inputs into a fixed lower intermediate frequency (IF); therefore trouble-shooting at this point

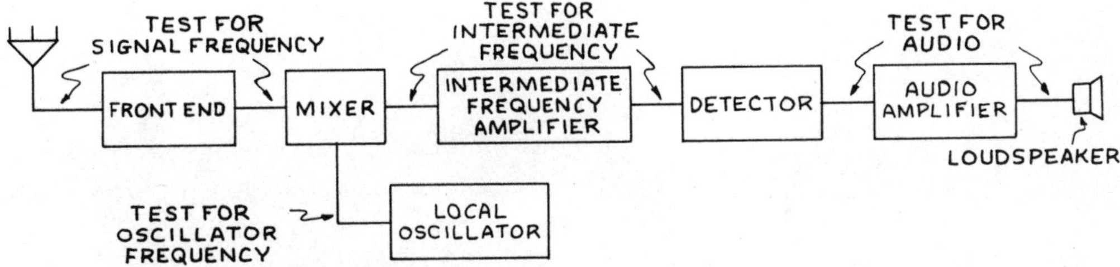

Figure 29-3. An inoperative radio receiver is tested progressively for the presence or absence of a signal of the frequency noted in the above diagram. The output at any point always should be of greater amplitude than the input at the preceding point.

concerns itself with verifying that the IF exists in suitable amount.

The intermediate frequency amplifier, as its name testifies, has the job of strengthening the output from the mixer. Trouble-shooting this unit must verify that a signal of the new lower frequency is there and that it is stronger at the output than at the input. All testing so far has of necessity been at radio frequencies because the audio intelligence has not yet been developed.

The function of the detector is to develop this audio intelligence. Trouble-shooting now brings out actual sound, and the testing concerns itself with its clarity and freedom from unwanted noises. Checking the following audio-frequency amplifier (AF) again verifies that the output is a faithful, stronger version of the input.

The foregoing should illustrate the statement that trouble-shooting is nothing more than a concise, logical, sequential application of basic circuit knowledge. It boils down to expecting something at a certain point in a circuit and deciding whether or not it is there in adequate quality and amount. (See Figure 29–3.)

The most modern electronic instrument designs separate the various subcircuits into separate modules, each on its own circuit board. This brings a totally new facet to trouble shooting. Faults may now be found by simple substitution. When trouble develops, the modules are slipped out, one at a time, and a known good module is inserted instead. If one of these substitutions restores the instrument to operating condition, the fault obviously has been located. This substitution technique may be combined with the point-to-point system already described, and thereby the trouble may be localized still more quickly. (See Figure 29–4.)

Trouble-shooting often singles out a single component as being defective and then this component must be repaired or replaced. Very few electronic parts can be repaired, except for superficial damage, and replacement is the general procedure. Must the new part be procured from the original manufacturer of the instrument or will something from an electronics store be satisfactory? It depends. An inductor or a transformer may be specially wound and that restricts the choice to the manufacturer. But a capacitor with an exactly equal specification is the same wherever bought. Transistors and vacuum tubes of reliable make and bearing identical numbers are identical in operation regardless of trademarks. A resistor is a resistor. A meter is a meter, although quality and sensitivity levels may vary greatly. As in so many fields, common sense eventually must govern the decision.

The instruments employed in trouble-shooting must be capable of greater accuracy than the limits prescribed for the device under test. When the specifications call for a reading within plus or minus one percent, for instance, a meter that can only come to

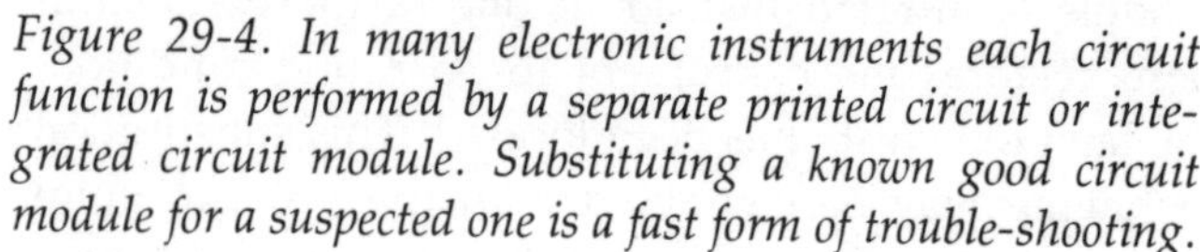

Figure 29-4. In many electronic instruments each circuit function is performed by a separate printed circuit or integrated circuit module. Substituting a known good circuit module for a suspected one is a fast form of trouble-shooting.

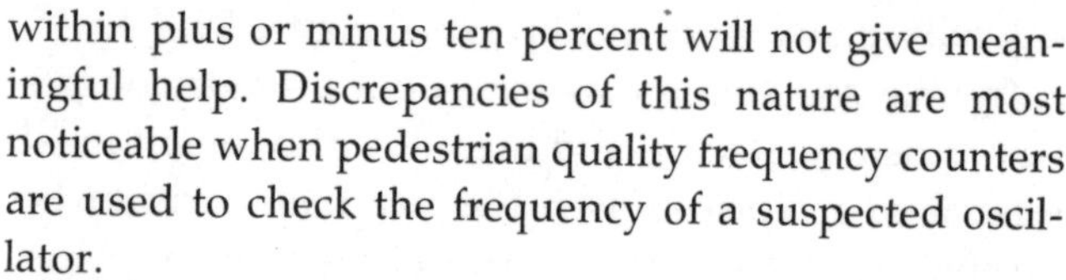

within plus or minus ten percent will not give meaningful help. Discrepancies of this nature are most noticeable when pedestrian quality frequency counters are used to check the frequency of a suspected oscillator.

Trouble-shooting a radio receiver may be done by tuning in signals off the air and checking their progress through the circuits. A more reliable method is to feed the output of a signal generator into the antenna terminals. In this manner, the input is constant, does not vary as a signal would, and consequently allows more accurate comparison of the various circuit outputs with the norm. The idea is to inject from the signal generator the lowest amplitude that will produce results at the other end.

As electronic navigational instruments become more and more compact, trouble-shooting, in turn, requires an ever more delicate approach. The conductive lines on a printed circuit board are so closely spaced that a carelessly placed probe could cause a disastrous short circuit. Any but the finest point soldering iron could cause bridging from one track to an adjacent one. Sensitive transistors could be damaged by the voltage and current from the battery in a VOM applied through an incorrectly placed test probe.

The "trouble-shooting road maps" that follow place the gist of this chapter into diagrammatic form so that the eye may verify and help apply the information gained. (See Figure 29–5.)

Figure 29-5. Trouble shooting an ailing electronic instrument should proceed logically from point to point as shown for a radio receiver.

TROUBLE CHART

ACTION	QUESTION	ANSWER	SEE
TURN INSTRU-MENT ON	DOES PILOT LIGHT?	YES NO	1 2
CONNECT TROUBLE LAMP TO INSTRU-MENT TERMINALS	DOES TROUBLE LAMP LIGHT?	YES NO	3 4
CHECK FUSE	CONTINUITY?	YES NO	5 6
CHECK BATTERY	SPECIFIC GRAVITY?	AROUND 1200 MUCH LOWER	7 8
CHECK SWITCH	CONTACT?	GOOD NO	9 10

1. Power is reaching the input of the instrument and the trouble lies beyond in the circuit. Considering the complexity of modern electronic circuits, trouble in this area is best left to trained technicians and well-equipped electronic laboratories. Working on the circuits of transmitters and radars is illegal except for properly licensed personnel.

2. The pilot light may be defective, the switch may be defective and power may not even be reaching the instrument terminals. See next item.

3. Connect a trouble lamp such as shown in Figure 29-5 to the input terminals of the instrument while the instrument is connected to the source. If the trouble lamp lights, power is being received and the trouble lies in the switch, in the pilot light or in the circuit. This test should work with instrument switch on or off.

4. The trouble lamp not lighting indicates that no power is reaching the instrument. Check the feed line progressively from battery through main fuse or breaker to instrument. (Use trouble lamp instead of multimeter. See Text.)

5. If trouble lamp lights when touched to instrument terminals, fuse is all right; see #1.

6. If trouble lamp touched to instrument terminals does not light, momentarily short the instrument fuse. If trouble lamp lights, replace fuse.

7. Check each cell of the battery with a hydrometer. If the readings be fairly uniform and near 1200, battery is all right.

8. Hydrometer readings close to the 1100 mark indicate a battery unfit for normal service. Be alert for one very low-reading dead cell. If only trouble is a discharged battery, place the battery on charge until the cells gas uniformly.

9. With instrument connected but switch off, touch trouble lamp terminals to the two connections of the switch. The trouble lamp should glow (provided it has been made with a very low candlepower bulb. Turn switch on; trouble lamp should go out with a normally functioning switch.

10. If trouble lamp does not go out when switch is closed in #9, switch is defective. Sometimes, with marine equipment, the trouble is only corrosion and may be corrected by cleaning the contacts with very fine abrasive paper.

30. *Test Equipment and Its Use*

Reliable, accurate test equipment is a necessity if meaningful trouble-shooting and testing are to be performed on marine electronic devices. Navigational electronic instruments have become so highly sophisticated that random hunt and peck methods are inadequate except for the simplest malfunctions. All electronics aboard have manufacturer-dictated specifications to which they conform when new, and it is the duty of a competent trouble-shooter to return them to that level when they fail.

The volt-ohm meter and the test lamp, already described, have a rightful place in testing, but they are at the elementary level of test equipment. From that point up, electronic testing instruments become increasingly complicated, perhaps reaching a zenith in an all-purpose oscilloscope. Like the navigational devices they are employed to check, today's test instruments are also solid state—more reliable, more compact, and more miserly with current than their predecessors. The instruments seek information along several channels: the intensity and polarity of a voltage, the shape of a wave form, the clarity and strength of an output, and the frequency of a locally generated signal.

The "signal tracer" is one of the simpler test instruments, yet it has wide usefulness. Essentially, this is an amplifier that is able to process both radio frequency and audio-frequency signals applied to its input. The instrument includes a meter that indicates relative strength and also a loudspeaker to produce an audio output. Two probes are supplied, one for radio-frequency input and the other for audio frequency checking. (See Figure 30–1.)

The radio-frequency probe contains a diode and is connected to the instrument via a shielded cable that excludes extraneous pickup of electrical noise. The diode rectifies the radio frequency alternating current into a pulsating direct current suitable for the amplifier; what actually takes place is a demodulation that recovers the intelligence in the radio wave. The audio probe is a simple direct connection to the amplifier.

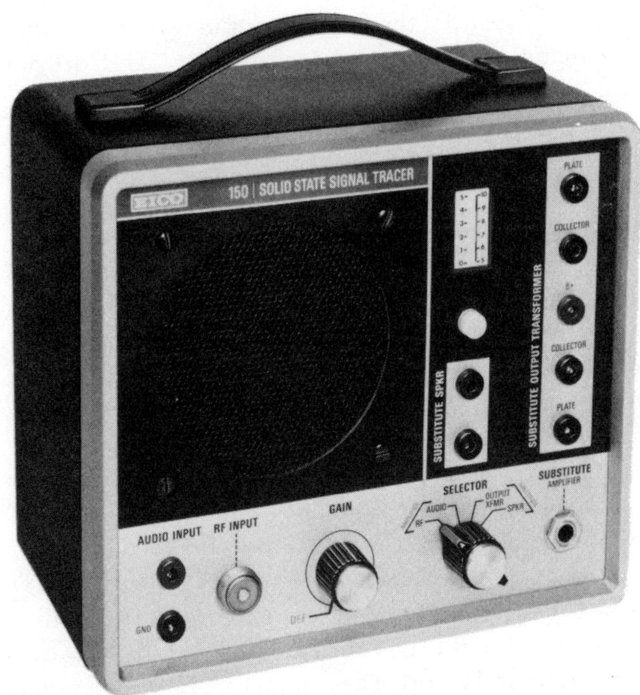

Figure 30-1. The progress of a signal through an electronic circuit may be checked with a Signal Tracer. The metal tells whether there has been gain or loss. (Eico Electronic Instrument Co.)

The rationale of the signal tracer is to sample the circuit under test at a number of progressive points and to note whether the sampled signal becomes stronger as one goes from input to output. Indicating these relative strengths is the duty of the meter on the panel. The radio probe is used wherever radio frequency current is presumed to be present, and the audio probe comes into play at all other points. On a communication receiver, for instance, the radio probe would start at the antenna terminals, and would continue on to the radio frequency amplifier, the mixer, and the output of the intermediate frequency amplifier. The audio probe would take over at the detector and go from there to the output. Remember that the meter readings are purely relative and do not necessarily denote any actual power level.

In the receiver test just described, it was presumed that a signal was being received from some distant transmitter. This may become a chancy method if the transmission be intermittent or if fading takes place.

A preferable procedure is to feed a constant signal into the receiver from a signal generator. Subsequent measurements then become truly relative and more meaningful. The signal generator is modulated with a fixed audio tone. The only drawback with this way is that distortion may be difficult to detect in a single tone; distortion in music or voice is more apparent.

An additional hint for point-to-point signal tracing: Active elements such as transistors and vacuum tubes are the sole sources of all gains in signal strength, so it is wise to check each of them separately to learn whether they are doing the job. With the probe at the base of a transistor (or the grid of a vacuum tube) a certain signal level is registered. Moving this probe to the collector of the transistor (or the plate of the vacuum tube) will raise the level on a functioning unit. A critically tuned circuit may be detuned by the capacitance of the probe touched to it, but this should not interfere too greatly with tracing.

The signal tracer may be metamorphosed into a field strength meter by attaching a piece of wire to the radio frequency probe to act as an antenna. The overall operation of a transmitter may thus be checked. The probe antenna is placed a short distance from the transmitting antenna and the reception is monitored on the meter and on the loudspeaker of the signal tracer. Full, clear reception will be had only from amplitude-modulated transmitters, but frequency-modulated transmitters still may be checked relatively.

The signal tracer may occasionally indicate the presence of a signal where no signal should be, and this obviously is a sign of trouble. Such a case would be the fully bypassed emitter of a transistor or cathode of a vacuum tube. A signal here indicates a defective bypass capacitor.

The signal generator is an all-around workhorse and, as described above, enables the testing of radio receivers and other circuits without the need for a distant transmitter. The signal generator is a high quality variable oscillator (see Chapter 7) that is switchable to a number of frequency bands and has provision for being modulated by audio tones. The output power is minimal, but the quality and frequency stability of the output meet laboratory standards. (A typical signal generator is shown in Figure 30–2.)

The broad usage of a signal generator in circuit test procedures may be better understood by remembering that this instrument is essentially a means for establishing a numerically exact frequency. This frequency then may become a prime source (as when it is fed to a

Figure 30-2. The RF Signal Generator produces a desired radio frequency for injecting into a circuit under test. (Eico Electronic Instrument Co.)

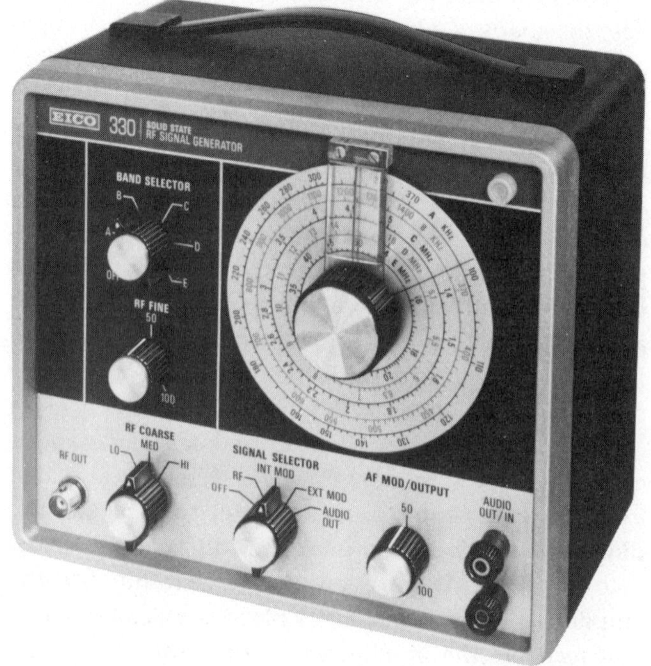

receiver under test) or it may become a standard against which others are compared for purposes of calibration.

A major use of the signal generator is in the "alignment" of the various tuned circuits of a radio receiver. (Alignment, in this sense, means setting each tuned circuit to the frequency specified for it.) As an illustration: The intermediate-frequency amplifier of most broadcast radio receivers is designed to be tuned to 455 kilohertz. Therefore a 455 kilohertz signal is fed into the receiver by the signal generator, and the intermediate frequency transformers are adjusted until the maximum signal is fed through. (The alignment procedure is slightly more complex with some high fidelity receivers that require tuning to two closely related frequencies in the interest of more true-to-life sound.) A VHF marine receiver requires substantially the same treatment except that the injected frequency is usually 10.7 megahertz.

Calibration by comparison works two ways with the signal generator. The signal generator itself may be calibrated (or verified) by comparison with a known-frequency broadcaster; or the signal generator may be employed to calibrate the tuning dial of a radio receiver. In these operations heterodyning methods, in which two signals "beat" against each other to produce silence when they are exactly alike, are commonly resorted to.

The radio frequency signal generator has a counterpart at audio frequencies in the sine-wave and square-wave generator. This instrument is used in the testing of audio amplifiers, such as hailers, and the audio amplifier portions of radio transmitters and receivers. This, again, is basically a variable oscillator whose output frequencies may be varied from the sub-sonic to the ultra-sonic. Its output may be chosen either as the smooth sine wave form or as the sharp square wave form. The square wave accentuates minor faults in an audio amplifier that the sine wave may gloss over. (See Figure 30–3.)

Figure 30-3. The output of the Sine and Square Wave Generator makes it possible to test an audio amplifier for amplifying ability and for its fidelity. (Eico Electronic Instrument Co.)

When the audio frequency generator is employed in connection with an oscilloscope, the entire overall operation of an amplifier may be gauged visually within a few minutes. The wave form displayed on the scope reveals an internal diagnosis of the amplifier under test. The perfect wave form injected into the amplifier by the generator emerges at the output either in its original form or in a distorted version. The amount of this distortion indicates how far the amplifier deviates from true high fidelity in its operation. Seldom is an amplifier perfect, but a good one does very little mayhem on the square wave going through it.

Reproductions of the wave shapes produced on the oscilloscope by various amplifiers are shown in Figure 30–4. At "a" is the shape of the square wave fed into the amplifier; if anything very close to this appears at

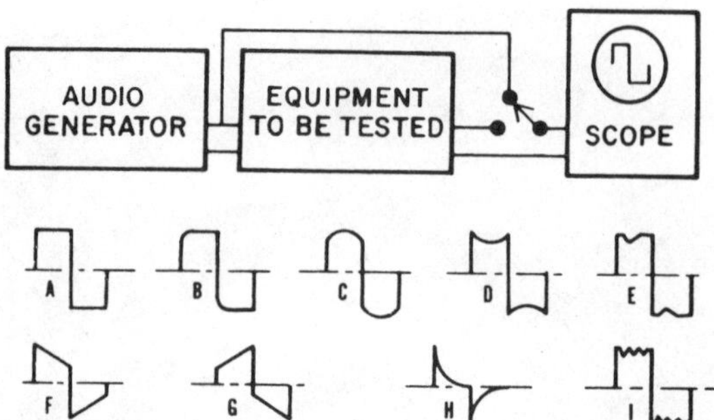

Figure 30-4. The drawing shows how the square wave generator and the oscilloscope are connected to the audio amplifier under test. The square wave fed into the amplifier is depicted at (a). The meaning of the resulting outputs is as follows: (b) drop in gain at high frequencies, (c) increased low frequency gain, (d) drop in low frequency gain, (e) less gain over narrow band of frequencies, (f) and (g) phase shift at low and high frequencies, (h) differentiation is taking place, (i) amplifier is oscillating.

the output, the amplifier is beyond reproach. The distorted diagrams that follow are the result of both phase and frequency discrimination within the tested amplifier, as noted in the caption. An experienced technician is able to read these oscilloscope patterns the way a physician deciphers an electrocardiogram.

The oscilloscope, as noted above, visually reproduces the shape and amplitude of any electrical signal presented to its input terminals. It does this on the face of a tube that is a smaller version of the familiar TV picture tube. (An oscilloscope is shown in Figure 30–5.) This instrument, perhaps more than any other, provides a detailed look into the internal workings of electronic devices.

In the days before transistors, when vacuum tubes were the only active units on the electronic chassis, a tube tester was a basic tool in all electronic troubleshooting. The tubes were always the prime suspect as the cause of trouble, and the first move always was to take each one out and test it. For old times' sake only, a typical tube tester is shown in Figure 30–6. This one is the more advanced type that measures the dynamic mutual conductance of the tube under test; simpler testers measure only the emission.

The transistor analyzer, shown in Figure 30–7, is the modern counterpart of the tube tester. Transistors may be N type or P type (see Chapter 3) and the analyzer is switchable to take care of either. Transistors may also be "bipolar" or "field effect," and terminals are placed on the front panel to accommodate both.

An important parameter of the junction transistor is its "beta," and this is measured and rendered numerically by the analyzer. The higher the beta, the better the transistor, because beta is an indicator of the transistor's amplifying ability. The range of this beta is unique to the type of transistor, whether it be designed for small signal, large signal, or power amplification use, and the analyzer allows these separate classifica-

Figure 30-5. The actual shape of an audio or radio wave may be shown on the screen of an Oscilloscope. (Eico Electronic Instrument Co.)

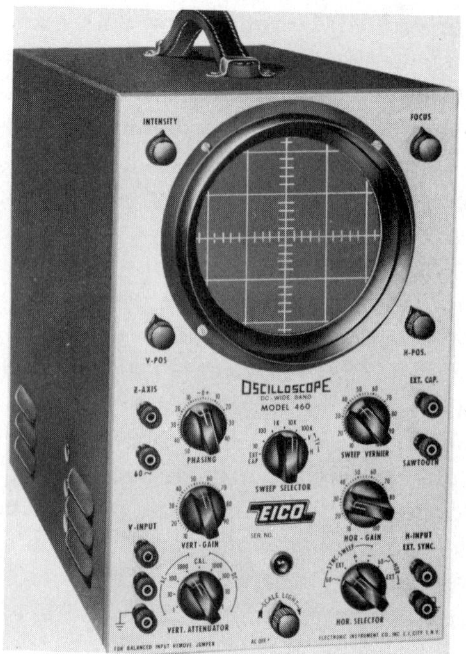

tions of measurement. Similar analysis is possible for the field-effect transistors.

The analyzer gives relative numerical readings, but more may be learned by observing the actual characteristics of the transistor in operation. This feat is accomplished by the semiconductor curve tracer (shown in Figure 30–8) when used in conjunction with an oscilloscope. Some of the curves that are seen on the face of the scope tube during testing are shown in Figure 30–9. Here, too, a technician can evaluate a transistor by looking at its curve. What is equally valuable, a defective transistor is shown up instantly without further ado or testing.

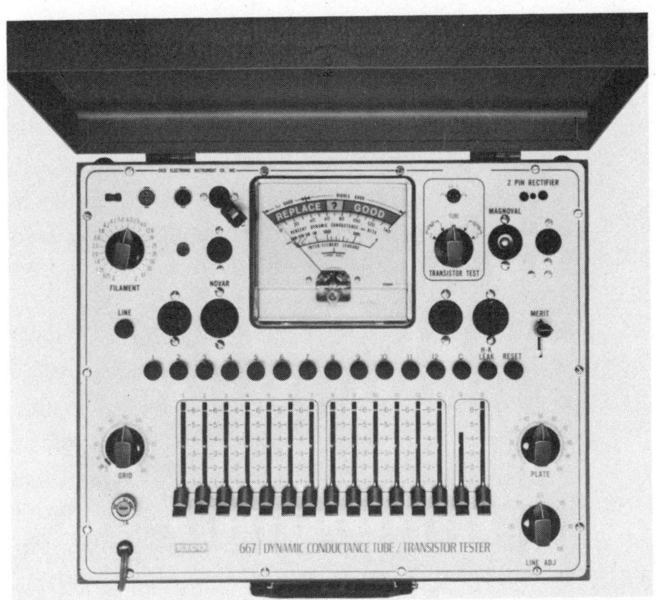

Figure 30-6. Although vacuum tubes are disappearing from electronic devices, there are still enough of them around to require the presence in a shop of a Tube Tester. (Eico Electronic Instrument Co.)

Figure 30-7. The Transistor Analyzer quickly determines the condition of transistors and diodes. (Eico Electronic Instrument Co.)

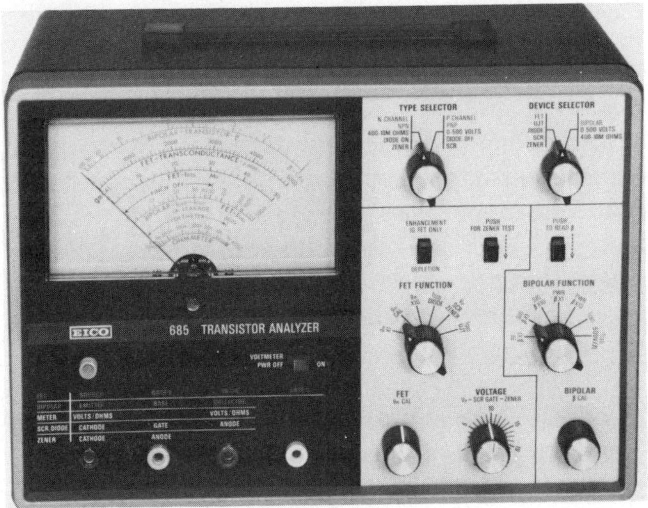

Figure 30-8. A Transistor Curve Tracer (shown here with an oscilloscope) develops a family of curves for the transistor under test that give a technician the inside story. (Eico Electronic Instrument Co.)

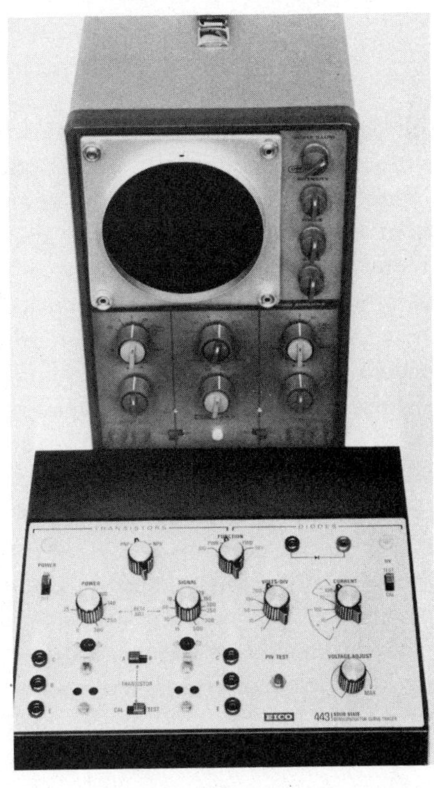

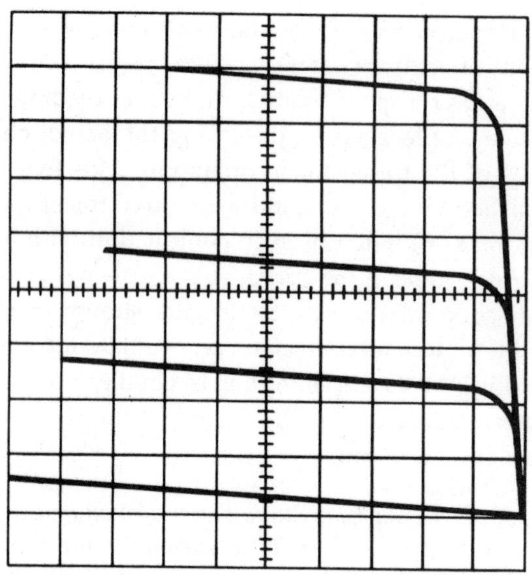

Figure 30-9. This is the response curve drawn by a transistor curve tracer for a functioning transistor.

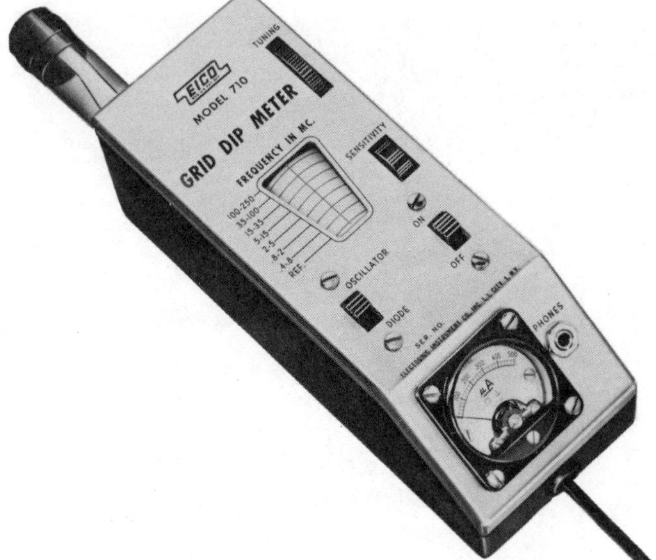

Figure 30-10. The Grid Dip Meter, held near an oscillating circuit, reads frequency and resonance. (Eico Electronic Instrument Co.)

Often during trouble shooting it becomes desirable to find the exact radio frequency to which a tuned circuit is resonant. This may be done without any actual connection to the questioned circuit by means of a "grid dip meter," such as shown in Figure 30–10. This instrument is held in close proximity and its knob is turned slowly to the maximum indication on the meter; the frequency then is read from the calibrated dial. Several coils are supplied and are plugged in for the various ranges of frequency. (The meter reading is the result of a transfer of radio energy from the grid dip meter to the circuit under test when the two are resonant.)

The increased public awareness of high fidelity sound reproduction has resulted in more attention being paid to the quality of the voice that emanates from marine radio transceivers. Quite naturally, as a result, instruments for checking audio quality have been added to the technical shelves of technicians. The measurements of the audio condition are made by noting the amount of distortion that is added to the original input rather than by trying to decide how pure a certain tone may be.

The distortion encountered may be either harmonic or intermodulation in character, and an instrument such as the one shown in Figure 30–11 measures both. Harmonic distortion occurs when an amplifier adds undesired harmonics to a fundamental frequency that is being passed through it. In intermodulation distortion, two frequencies passing through an amplifier react with each other to produce unwanted additional frequencies because of deficiencies in the circuit. Obviously, neither form of distortion is desired in high fidelity equipment.

The distortion meter measures harmonic distortion in a straightforward manner. For example, an amplifier is to be tested to learn at what level of fidelity it can handle a 1,000 hertz tone. Such a tone is fed to it from

Figure 30-11. The Harmonic Distortion Meter measures the percentage of undesired harmonics occurring in an audio amplifier under various conditions. (Eico Electronic Instrument Co.)

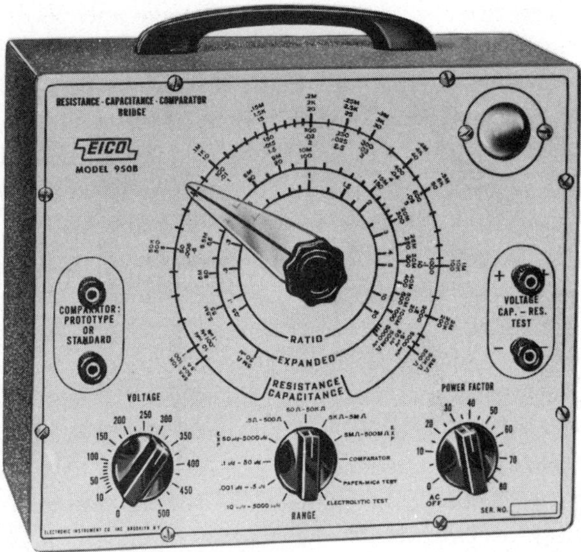

Figure 30-12. The RC Bridge accurately measures the values of resistance and capacitance. (Eico Electronic Instrument Co.)

an audio generator, and the output of the amplifier is delivered to the distortion meter. Now the dial on the distortion meter is turned slowly until the basic 1,000 hertz note is completely absorbed internally. What remains are the various undesired harmonics, and the amount of these is read on the meter as distortion.

Two distinct and widely separated frequencies going through an amplifier simultaneously may react with each other under certain circuit deficiencies. This will produce frequencies that are sums and differences of the original two and the result will be intermodulation distortion. The distortion meter generates its own 60 hertz and 7,000 hertz frequencies for simultaneous injection into the amplifier under test, and their relative strengths may be varied to suit. The output of the amplifier then is returned to the instrument for a measurement of the intermodulation distortion. (The general classifications for loss of fidelity are amplitude distortion, frequency distortion, and phase distortion.)

Inductance, capacitance, and resistance are the parameters of passive elements such as coils, capacitors, and resistors in an electronic circuit, and it is often necessary to measure their exact values. An instrument such as shown in Figure 30–12, called a bridge analyzer, is employed for this purpose. The term "bridge" applies to an internal circuit, shaped like a baseball diamond, that balances each parameter against a reference in order to arrive at a reading. Measuring resistance in this manner is more accurate than the results obtained by using the volt-ohm meter mentioned earlier. The instrument supplies high voltages for checking the integrity of capacitors at their commercial ratings.

Theoretically, a capacitor should be a total barrier to direct current and therefore its resistance should be infinite. High grade mica dielectric capacitors are true to this specification, but electrolytic capacitors are not; they allow considerable leakage. The analyzer measures this leakage at the working voltage and permits the technician to decide whether or not it is within tolerance. Measuring the direct current resistance of an inductor is merely a ballpark way to classify it; its reactance at any given frequency will be totally different.

The wide-spread use of frequency modulated (FM) transceivers on the marine VHF band has brought the need for more sophisticated methods of receiver alignment. Amplifiers and detectors in these transceivers must be able to handle without distortion the range of frequencies that are generated by talking into the microphone. A test instrument that reproduces these frequencies for the purpose of testing and alignment is called a sweep oscillator or, in the technical lingo, a "wobbulated oscillator." Used in conjunction with an oscilloscope, this instrument "draws" a picture of the response curve and identifies the frequencies of concern to the technician. It also enables setting the detector to be responsive only to frequency modulation and not to amplitude modulation. (See Figure 30–13.)

The original method of measuring the frequency of a radio transmitter was with either a wavemeter or a signal generator. The wavemeter is a tunable tank circuit consisting of a capacitor, an inductance, and an indicator light. The dial is turned slowly until the light becomes bright to show resonance between the meter and the nearby transmitter; the transmitter frequency then is read from the dial.

When the signal generator is employed, its output is combined in a receiver with that of the transmitter. This heterodyning action results in "beats" in the receiver. When zero beat is reached, the transmitter frequency is read from the signal generator dial.

The modern method, considerably simpler and faster, makes use of a frequency counter similar to the one shown in Figure 30–14. The internal circuits of the counter change the transmitter frequency into pulses. These pulses go through gates that are open for extremely accurate short periods of time controlled by quartz crystals. The instrument combines the number of pulses in the given length of time into a digital reading of the transmitter frequency.

Figure 30-13. The FM Sweep Marker Generator finds use in aligning wide-band radio frequency amplifiers such as those in television receivers. (Eico Electronic Instrument Co.)

Figure 30-14. The Frequency Counter has replaced older methods of measuring the frequency of oscillating circuits. (Eico Electronic Instrument Co.)

31. *Operating Procedures*

Radio conversations take place on the equivalent of a huge party line—everybody listening, everybody ready at any instant to jump in and talk. Courtesy and regulations are necessary to prevent utter chaos; yet any knowledgeable listener can verify that both are often most noticeable by their absence. Courtesy is beyond the province of enforcement, but the regulations are federal law, and fines are specified for lack of compliance.

Ignorance is probably the greatest factor in failure to comply, but it is a well-known principle of law that ignorance is no excuse. Another factor is the vulgarity brought to marine radio from the channels of citizen's band. The citizen's band greeting, "Are your ears up?" or the peremptory "breaker, breaker," so dear to CB users, has made many a skipper wince.

The procedure for instituting and answering marine radio contacts is simple enough. The Coast Guard Auxiliary, the Power Squadron, and numerous publications have tried to pound the procedure into the heads of neophytes, but in too many instances the soil has not been receptive.

The ground rules that apply to all shipborne radio equipment specify that the transmitter must be covered by a valid license from the Federal Communications Commission (FCC). In addition, Part 83 of Volume V of the FCC Rules and Regulations must be on board, as well as a suitable radio logbook. The skipper or other authorized operator of the transmitter must have a personal license that may be as simple as the Restricted Radiotelephone Operator Permit that requires no examination. (Note that this license does not permit the holder to make any technical adjustments.)

The possession of a radio logbook and the entry into it of certain radio transactions are mandatory—but there is probably no other rule in the regulations ignored more universally by pleasure boat skippers. The maintenance of a listening watch is not legally required on a pleasure boat underway, but if such a watch be maintained, its exact duration must be entered in the logbook. Incidentally, each page of the logbook shall show the vessel name and call sign and it shall not be of the loose-leaf type. Each entry shall be signed by whoever made it.

The importance of keeping the radio log stems mostly from the fact that it will form a record of all distress and safety calls heard or made. Also, any repairs or adjustments made to the transceiver or the radar by a licensed technician are recorded, detailed, and signed. The rules are sensible in that they do not require the listing of ordinary radio communications that do not refer to distress or safety.

The first step in initiating a radio call to a vessel or shore station is to listen on the channel to be used. Obviously, if the channel be in use, the call is not made until it is clear. Talking is done in the strict sequence outlined below.

With the microphone button held pressed in: "Calling the yacht (name and radio call), calling the yacht (name and radio call), calling the yacht (name and radio call)—this is the yacht (your yacht name and radio call), Over!" This call is sent out on the distress and calling frequency that is Channel 16 for VHF and 2,182 kHz for the marine band. The regulations permit 30 seconds for this preamble, although it should not take that long. If no reply be heard, two minutes must elapse before the call is repeated. The called station answers by saying "This is (yacht name and number)."

Once contact is established with the called vessel or shore station, further use of the calling frequency is restricted to agreeing on the channel to which the conversation is to be switched (in other words, a channel common to both stations). All of this is quick and easy once the correct habit has been formed. Skippers are also requested to switch to low power for nearby contacts in order to reduce interference, but the efficacy of this is questionable.

It has always been a saga of the sea that distress takes precedence over everything, and this has rightly carried over into radio. There is a standard procedure for sending out a distress call, but the regulations realize that conditions could be hectic when life is at stake at sea and wisely give the caller wide latitude. Anything goes on any channel that will attract attention and bring help quickly. All stations must remain silent except those directly involved.

The distress call is preceded by the word "mayday" repeated several times. Then all pertinent information is given, such as position, nature of emergency, number of persons aboard, identification, and appearance of the distressed vessel. Chances are that radio direction finders on shore will home in on it to triangulate and verify its position. (Mayday is a phonetic version of the French "help me!")

Mayday is the extreme request for help and is to be used only when the distressed condition warrants. The lesser classifications of priority messages are those with "urgent" and "safety" information for all listening vessels; the preambles to these (phonetically) are pawn and saycuritay.

Because of the difficulty in distinguishing certain letters over the phone, it is preferable to use the phonetic equivalents. These are listed in Figure 31–1. Thus a station whose call letters begin with W and B would be called as "whiskey bravo." To eliminate the confusion often encountered between voice-transmitted "five" and "nine," the nine may be pronounced as "niner." A general rule of secrecy applies to all conversations overheard on radio.

The depth sounder supplies information that may be used in several ways by the skipper. First, of course, is the simple fact of how much water there is under the boat. Also, a knowledge of depth may assist in position finding, although it may prove only a rather approximate tool.

It must be remembered that the depth sounder reading is not expressible as absolute until the location of the transducer on the hull is known. If the transducer be at the garboard strake, the depth of the keel must be subtracted from the reading in order to ascertain the depth of water left for navigation. Nor does the depth sounder look ahead to warn of immediate peril.

Standard Phonetic Spelling Alphabet

A	ALFA	**N**	NOVEMBER
B	BRAVO	**O**	OSCAR
C	CHARLIE	**P**	PAPA
D	DELTA	**Q**	QUEBEC
E	ECHO	**R**	ROMEO
F	FOXTROT	**S**	SIERRA
G	GOLF	**T**	TANGO
H	HOTEL	**U**	UNIFORM
I	INDIA	**V**	VICTOR
J	JULIETT	**W**	WHISKEY
K	KILO	**X**	X-RAY
L	LIMA	**Y**	YANKEE
M	MIKE	**Z**	ZULU

Figure 31-1. The phonetic alphabet listed in the table above are used to remove the ambiguities of speech. Some of the phonetics sound peculiarly American but, presumably, they are recognized around the world.

To employ the depth sounder in position finding, its reading usually is combined with some other line of position. For instance, an azimuth line that has been established may be inspected along its length to find the charted depth that coincides with the depth sounder reading. That point would be the fix. (See Figure 31–2.)

Another scheme, wherein the depth sounder readings are used alone to establish position, is admittedly more viable in theory than in practice. It may prove

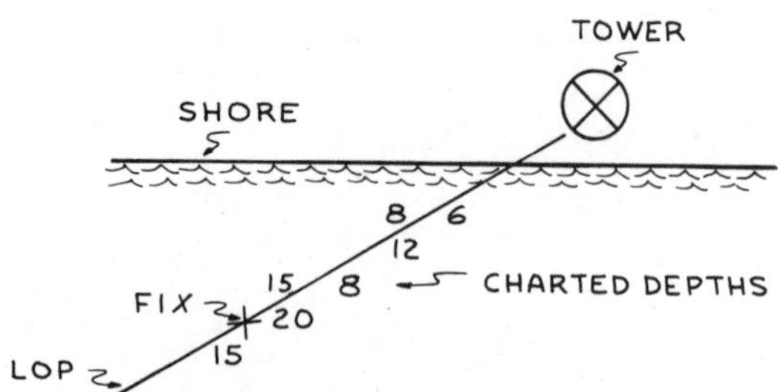

Figure 31-2. This skipper has drawn the above LOP on his chart. His depth sounder reads 20 feet. The coincidence of these two facts marks his fix. It must be remembered that this procedure is approximate at best.

helpful when the skipper knows his approximate location. A series of depth sounder readings is taken at constant speed and at exact intervals. These are laid out on a piece of transparent plastic or tissue paper to the scale of the chart. Moving the layout about on the chart until it coincides with printed depths should (theoretically) show the skipper where he is.

The best secondary use of the depth sounder is to verify a fix obtained by established procedures. The charted depth and the sounder depth at the fix should be close to agreement; a discrepancy indicates that someone has goofed. (This time the distance from the transducer to the waterline must be added to the reading and allowance must be made for tide.)

The radio direction finder may be visualized as an active pelorus. Whereas the pelorus is used as a passive instrument for sighting, the RDF does its own "sighting" by receiving signals from a known point. Both instruments arrive at the same bearing from the same source when properly used and in good condition. Any difference in bearing readings (deviation) is a sign that the radio waves have been reflected or refracted by something in their path or aboard.

If the radio direction finder is to be used at only one location aboard, a deviation table for it is not difficult to make. The procedure is somewhat like "swinging ship" for compass deviation. The RDF and the pelorus are employed simultaneously and the deviation on each heading is tabled. The target should be a prominent object that is charted so that the chart may become an added reference.

Each corrected bearing from the RDF is a line of position and, as is standard practice, where the lines cross is the supposed fix. One caution when using broadcast radio stations as targets is to consider only those whose transmitter is charted. A station listed only by city may have its transmitter some distance away.

Position finding by radar may also be likened to the use of an "active" pelorus, but now there is one great advantage: The distance to the target is measured as well. Now one bearing is able to determine a fix. However, the main job of radar aboard is collision avoidance.

All pips on the radar screen look alike to the novice, but the trained eye can distinguish between a stationary navigational aid and a moving boat. This training is acquired by repetitive practice during bright weather when what the eye sees and what the radar "sees" may be correlated.

Although the bearings derived from radar observation are plotted as lines, the actual radar beams are slices of pie, increasingly divergent as the distance to the target increases. This denigrates the accuracy of the bearings taken and must be considered when laying down lines of position. The radar's automatic measuring of distance generally is more accurate and making a radar fix by crossing two circles of position based on distance from two targets is a preferable procedure. (See Figure 31–3.)

The vessel without radar may yet take some benefit from the fact that so many ships are radar-equipped. By hoisting a radar reflector to its masthead or yardarm, it increases the chance of its being "seen" by an approaching boat whose radar is in operation. Hopefully, the oncoming skipper will pay heed.

Radar reflectors for boats take many shapes, some large, some small, but all claiming efficacy. The metal faces of the reflector are far better at returning imping-

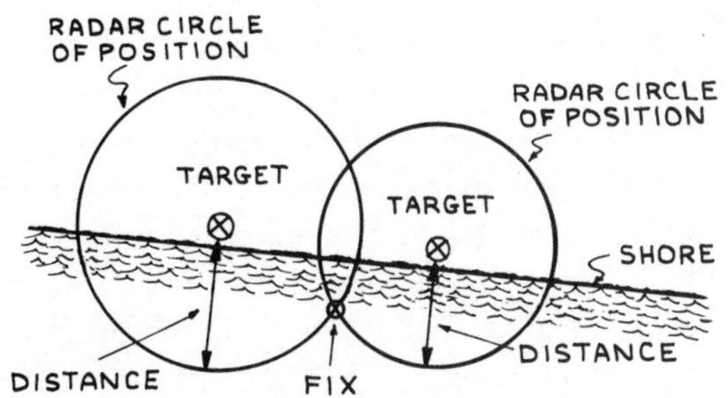

Figure 31-3. Two distance measurements by radar on two convenient targets provide circles of position on the chart that can result in a fix. One of the two intersections of the circle becomes obviously incorrect.

ing radar waves, mirror fashion, than the wood or fiberglass of pleasure boats. Radar waves are so small that the faces of the reflector need not be large to be effective. The echo displayed as a pip on the oncoming radar screen is much stronger from the reflector than it would be from the boat alone. (One form of radar reflector is shown in Figure 31–4.)

In the earlier discussion of transmission lines in Chapter 24 it is revealed that a theoretically perfect line, with no resistive losses and perfectly matched at both ends, transmits to the load all the power fed into it. But, since no actual transmission line is perfect and therefore power losses occur in real life, it is helpful to know how much loss is taking place every time the push-to-talk button is pressed.

The key to this situation is the standing wave ratio, the SWR, and this may be measured with an SWR meter, such as shown in Figure 31–5. The meter is inserted in series between the transmitter and the transmission line that, in most marine installations, is a coaxial cable. Since all the power in a perfect line is delivered to the load, there is no rejected or reflected power; this would be expressed by a SWR of 1:1, hardly attainable. The less than perfect line reflects back some of the power it should have delivered and its SWR would be greater than unity, perhaps 2:1 or similar. This ratio is read directly on the meter. Technically, the SWR is the ratio of the characteristic impedance of the line to the impedance of the load. Practically, the idea is to get the reading of the SWR meter as low as possible; then the maximum transmitter power goes out where it can do its rightful job.

The Omega navigation system divides the surface of the earth into invisible radio lanes, and under certain conditions there is a chance for ambiguity in determining the particular lane in which the vessel finds itself. A further error may result from failure to use the propagation tables or from incorrectly reading them.

Since the eight Omega transmitters that power the system are distributed world-wide and each may be heard up to 10,000 miles, an error in lane recognition is not that hard to come by. At the very low frequency used (10.2 kilohertz) each lane is approximately eight

Figure 31-4. A radar reflector is good insurance for vessels traveling on or near commercial lanes. The reflector provides a pip on the radar scope of the oncoming vessel. One form of radar reflector is depicted. (Jay Stuart Haft Co.)

Figure 31-5. The standing wave ratio meter (SWR) reads the relative efficiency of the antenna installation.

miles wide on the baseline and that distance would constitute the positioning error that accrues per mistaken lane. Thus a navigator setting his Omega into operation should know his dead reckoning position within four miles—the instrument will do the rest.

The propagation tables are necessary because the rate of propagation of Omega's very low frequency waves varies with the time of day and season. The waves travel between the earth and the ionosphere. The variations in height of the ionosphere during the 24-hour day are causes of the discrepancy and the tables supply correction factors. The drawing in Figure 31–6 shows how the surface of the earth and the bottom of the ionosphere in effect form a duct or psuedo wave guide through which the Omega waves travel.

Earlier Omega receivers depended on the navigator to make the propagation corrections. The most modern instruments insert the corrections automatically; they keep the propagation factors in memory to be called forth when needed. Failure to employ propagation corrections could result in a position error of more than 30 miles.

The early Conelrad system, ineffective and now long since abandoned, had the same problem of lane ambiguity mentioned earlier as a caution in the use of Omega. The problem is ameliorated in some of the most advanced Omega instrument designs; an automatic lane check using three frequencies is provided. The technical status of the entire Omega system is announced sixteen minutes after each hour on six frequencies by WWV, the Bureau of Standards station that broadcasts exact time continuously.

The very high radio frequencies that make up the VHF-FM marine band bring line-of-sight problems. Theoretically, these transmissions act like rays of light and are projected in straight lines, although occasionally local refractions may bend them slightly around the curvature of the earth. The distance that may be covered effectively with VHF-FM communications thus resolves itself to the heights of the antennas.

Antenna height on a pleasure boat obviously is restricted, so the height must be obtained at the other

Figure 31-6. The space between the surface of the earth and the inside of the ionosphere functions like a waveguide for radio waves.

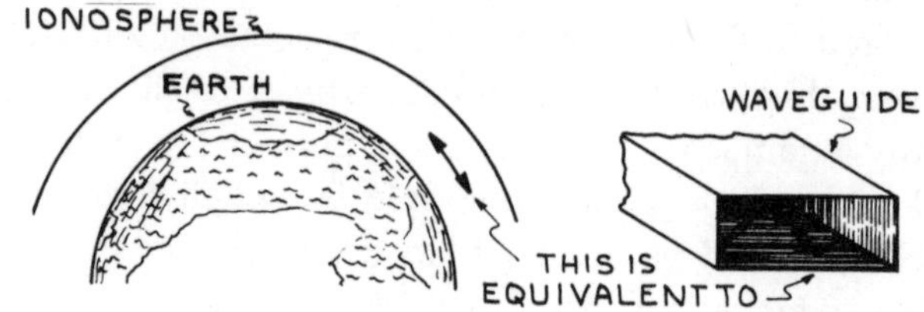

Figure 31–7. These are the locations of Coast Guard antennas and their height and range.

	VHF-FM		
	Location	Antenna Height	Minimum Range
	Steinhatchee, FL	150	17
	Crystal River, FL	450	30
	Tarpon Springs, FL	450	30
	Seminole, FL	300	31
	Venice, FL	450	33
R	Pine Island, FL	200	25
	Naples, FL	500	34
	Sugar Loaf Key, FL	330	30
R	Key West, FL	200	25
	Marathon, FL	200	32
	Islamorada, FL	300	34
	Princeton, FL	927	39
	Miami Beach, FL	225	25
	Delray Beach, FL	285	30
	Jupiter, FL	300	40
	Ft. Pierce, FL	400	39
	Cape Kennedy, FL	400	28
	Flagler Beach, FL	237	22
	Jacksonville Beach, FL	257	22
	Jekyll Island, GA	300	24
R	Tybee Island, GA	145	20
	Parris Island, SC	300	23
	Mt. Pleasant, SC	450	29
R	South Island, SC	145	20
	Nyrtle Beach, SC	230	22

NOTE: Minimum coverage was determined by a survey utilizing a 1-watt handheld transceiver and is based on the MAXIMUM distance the site was able to work/receive a low power transmission satisfactorily.

R—Channel 16 Guard Receiver Only.

end, in this case the Coast Guard. The sites and antenna heights chosen by the Coast Guard are listed in Figure 31–7 together with the minimum distances over which reliable communication could be maintained with a very low power transmitter on the boat. A valid assumption is that more transmit power on the boat could increase the distances.

The various electronic navigation systems have as their goal more accurate position fixes and less work on the part of the navigator to achieve them. It is interesting to learn how well this goal has been attained and some comparisons are in order. Comparison with celestial navigation, as the oldest and most universal method, puts the matter into perspective.

The average accuracy of a celestially obtained fix is approximately two miles under good conditions and when made by an experienced navigator. Omega, with its waves travelling in daylight, would lower this to one mile, but at night the error would rise to at least two miles. Loran-C could provide an accuracy of about one-eighth of a mile with its ground waves, but its sky waves could prove very much worse. The Transit satellite would beat them all with a standing accuracy of fix of at least one-twentieth of a mile.

Comparing the various systems weatherwise immediately relegates the celestial observations to clear skies. The electronic systems are not affected by weather, although severe ionosphere conditions and heavy nearby electrical storms could interfere temporarily. All of the foregoing is based on the presumption that everything is functioning perfectly, a condition that is not normal despite navigators' prayers.

Racons provide a further navigational assistance to the skipper whose vessel is equipped with radar. A "racon" is an aid to navigation that has a radar transponder tuned to the X-band of frequencies from 9,300 megahertz to 9,500 megahertz. Racons are being installed in increasing numbers.

The transponder functions as a receiver and automatic answering transmitter. The radar beams from an approaching vessel activate it. Each racon is identified by an assigned letter in Morse code. No special transmission is demanded of the vessel. The racon response appears on the radar scope of the activating vessel as a short thin line extending radially from the associated blip.

Under many operating conditions, the identifying line on the blip will be received only intermittently, but with enough repetition to be helpful. When the racon is close aboard, the racon identification will be in evidence almost continuously.

Whether or not the racon identification is received depends to some extent on the adjustment of the radar scope controls. When the adjustments are set to eliminate interference from sea returns, the chances are that the racon will be eliminated too. It is advisable to keep the radar scope at maximum sensitivity when racon reception is desired.

32. *Television Afloat*

Television is not an instrument for electronic navigation (at least not yet) but its prevalence on pleasure boats entitles it to a place in a discussion of marine electronics. Television occupies its own niche aboard as the electronic opiate that gently lulls the crew to sleep, drink in hand, when the craft is tied up safely after a hard day's run. It seems that every boat, even one with only a modicum of cabin, has a hallowed space reserved for television. The standard home scene of an enchanted audience hypnotized by the luminous screen has become a common scene afloat.

The television receivers that go to sea are the standard home styles, even though these are ill-equipped to withstand the marine environment. They are designed for operation on 120 volt alternating current, but this is no obstacle because they will be used dockside on pier power. Some receivers are available that will accept 12 volt direct current in addition to the AC. These are a boon to vessels, generally sailboats, that often anchor out but have no onboard source of power other than a storage battery. Screen sizes vary from the miniature to the ultra-grand and, as at home, screen sizes are a function of the pocketbook.

A fortunate peculiarity of the human eye makes television and motion pictures possible. If a series of pictures be flashed before the eye at a rate of approximately eighteen per second or faster, the interval between the presentations will not be noticed and the appearance of continuous motion will result. This characteristic is called "persistence of vision." The movies project 24 still pictures (or frames) per second, in each of which a character is slightly more advanced, and the eye sees a man walking. Television accomplishes its purpose in much the same way, except that 60 "fields" are shown per second in order to smooth the apparent motion still further.

A basic building block of television is that any illustration may be constructed of varying sizes and intensities of dots. A close examination of a newspaper picture under a magnifying glass will verify this. If the dots be kept numerous enough and small enough, the unaided eye will not be able to see them individually but will advise the brain only of the total effect. The television picture is such a combination of luminous dots and successive lines.

How a scene from life becomes a miniature replica on the TV screen is explained by the block diagram of Figure 32–1. A lens projects the scene onto a camera tube that electronically scans it and reduces it to a series of voltage variations that are then greatly amplified before being passed to an amplitude modulator. Meanwhile a synchronizing generator is delivering its pulses to both the camera and the modulator.

A master oscillator and its multipliers set the frequency at which the TV station broadcasts, and this is amplified to a suitable level. The modulator controls this level in accordance with the variations from the camera and the result goes on the air to the viewers via the antenna. The synchronizing pulses contained in the TV broadcast keep the viewer's screen and the

Figure 32-1. The scene is focused by a camera and the resulting voltage variations are amplified and applied to the modulator that controls the station output. A separate FM transmitter adds its audio output to the antenna.

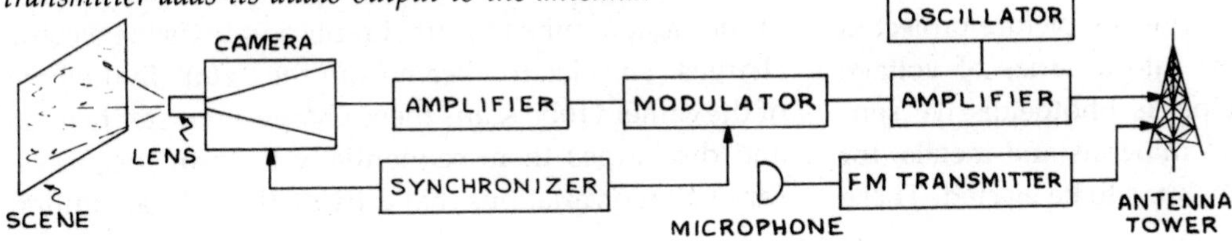

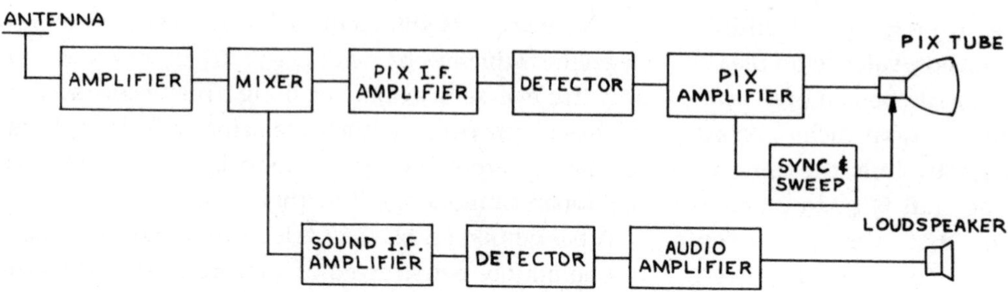

Figure 32-2. Television reception is accomplished in the stages shown in the above block diagram. Note that there are actually two receivers, a picture receiver and an FM sound receiver.

camera in exact relationship; without these, the picture would be a hodge-podge. The studio sound is picked up and transmitted in standard manner as part of the picture broadcast and reappears at the loudspeaker in the receiver.

The block diagram in Figure 32–2 summarizes what happens at the TV receiver when its antenna picks up the picture broadcast. The signal received usually is exceedingly weak, perhaps only a few one-thousandths of one volt. This is first amplified by the so-called "front end," the radio frequency amplifier. Heterodyning action at the mixer reduces the TV frequency to a much lower one better suited for intermediate frequency amplification. Now the detector recovers the picture intelligence which is amplified and then controls the brightness of the picture tube's spot. Coincidentally, oscillators move the spot in a raster pattern that is a duplicate of the camera's scanning to reproduce the televised picture. The sound has automatically traversed the audio circuits to the loudspeaker.

The heart of the camera is a tube, either a "vidicon" or an "orthicon," that transforms the picture projected upon it by the camera lens into a series of voltage variations. The basic principle is photoemissive sensitivity; light striking certain minerals and metals, for example caesium, causes electrons to be ejected. These electrons constitute a flow of electric current (see Chapter 2). Photos of these camera tubes are shown in Figure 32–3.

The cross-sectional drawing in Figure 32–4 helps explain how the orthicon functions. The forward, larger portion of the tube is the optical section. Here the televised scene is focused on the photosensitive cathode and causes the emission of electrons in accordance with the lights and shades of the picture. These electrons are directed and attracted to a target on which they form an invisible electronic replica of the scene. A beam from an electron "gun" in the base of the orthicon sweeps over the target in raster fashion and discharges the image to form equivalent voltage variations. An electron multiplier section in the tube amplifies these variations perhaps a thousand times before they are fed onto the TV transmitter.

The vidicon, shown in cross-section in Figure 32–5, achieves the voltage variations into which a scene is rendered in slightly different fashion, although still dependent upon photoemissivity. As before, the scene to be televised is focused on the photosensitive surface of the camera tube to form a replica in invisible electric charges. The electron beam from the "gun" in the base of the camera tube scans these charges in raster fashion and discharges them sequentially to form the necessary voltage variations that activate the TV transmitter.

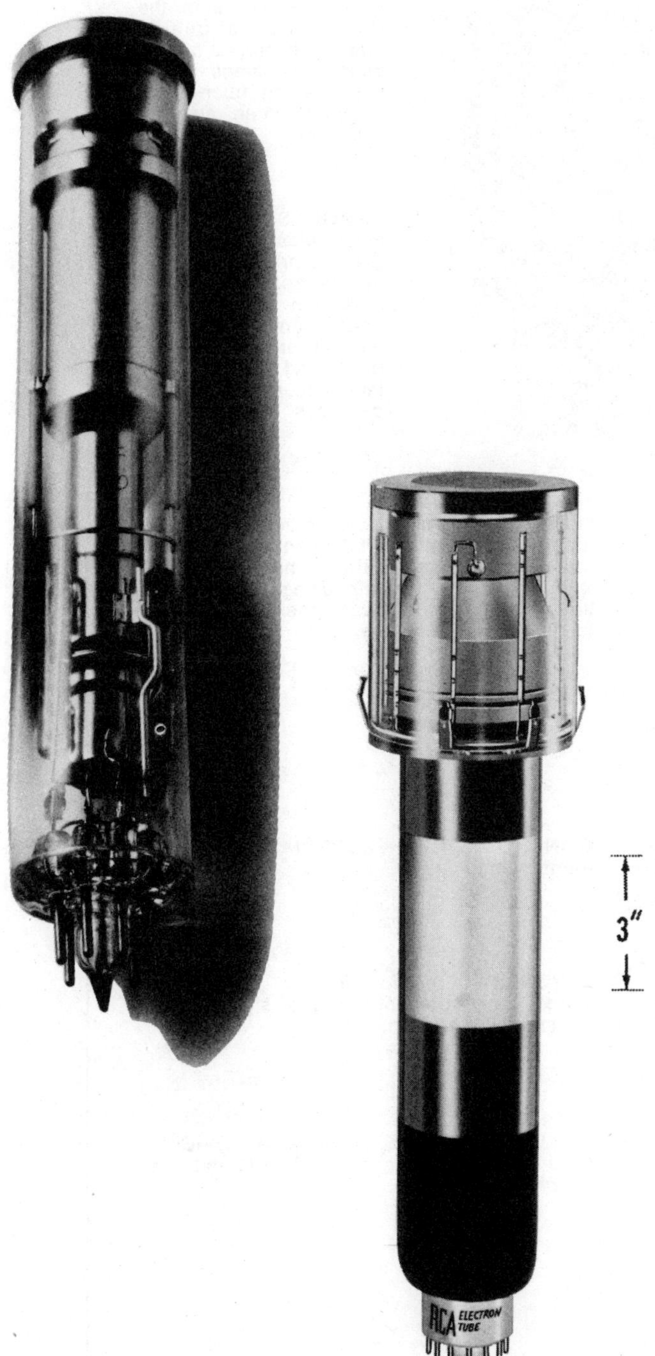

Figure 32-3. The Orthican and the vidicon camera tubes are the heart of television. (RCA)

The television picture tube (or CRT for cathode ray tube) in the home receiver changes the received voltage variations back into the original picture. (The cross-sectional drawing in Figure 32–6 shows how this occurs.)

The elements of the picture tube are an electron gun and a series of charged grids in the neck that direct and accelerate the electron beam. The final acceleration is provided by a conductive coating around the flared portion of the tube that is maintained at a high positive voltage. The magnetic field from a focusing coil around the neck of the tube brings the beam to a fine point when it strikes the inside of the picture face. The magnetic fields from horizontal and vertical deflection coils, also around the neck, cause the beam to scan the face in the raster pattern.

The inside of the picture tube face is coated with a phosphor compound that becomes luminous wherever the electron beam strikes it. The intensity of the luminous light depends on how hard the beam strikes, and this in turn is determined by the voltage variations sent on from the studio camera because these voltages control the electron gun. The coating has persistence; it remains luminous for a finite time after the beam leaves. This persistence combined with the persistence of vision obscures the repetitive action and causes the eye to see it as continuous.

The formation of the raster in the studio camera tube and in the picture tube is identical. The horizontal deflection coils in both tubes cause the electron beams to move from side to side at a rate of 15,750 times per second. At the same time, the vertical deflection coils are moving the beam up and down at a rate of 60 times per second. The result is a series of slightly canted luminous lines (262.5) to cover the face of the tube in one-sixtieth of one second. (Note that one-half line because it is the secret of "interlaced scanning" used in all receivers.) The next one-sixtieth of a second brings another 262.5 lines that are placed (interlaced) between the previous lines to make a complete raster of 525 lines, the standard in the United States.

Cut-away of Image Orthicon

"ANTI-GHOST" IMAGE SECTION—A specially designed image section used in certain RCA Image Orthicons. Provides proper geometry and suppresses highlight flare or "ghost" that occurs when field-mesh image-orthicon types are operated above the knee.

PRECISION CONSTRUCTION—Employed in the manufacture of certain RCA Image Orthicons. All tube parts including the envelope are precision made, precision spaced, and precision aligned. Provides tubes having excellent registration capability and uniformity of tube characteristics.

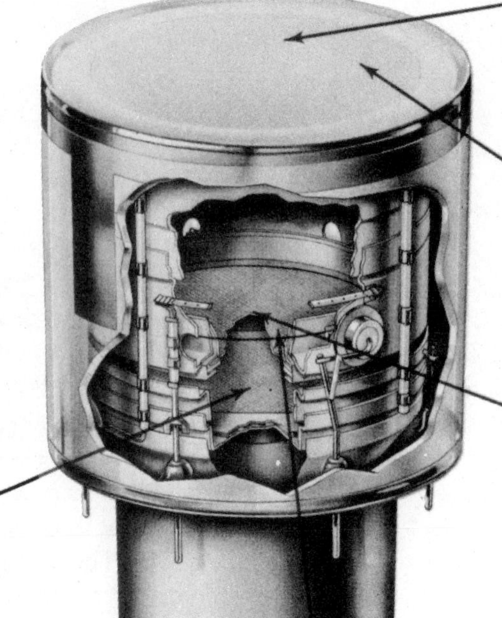

PHOTOCATHODES—Individually processed in each tube to provide maximum sensitivity and uniformity.

OPTICAL-GLASS FACEPLATES — Made of the finest optical-quality glass to eliminate optical distortion. Image Orthicons having fiber-optics faceplates can also be provided.

TARGETS — Made of electronically conducting glass to provide long life, stable sensitivity, and maximum resistance to raster burn in, semiconductive material for high gain, and specially selected blemish-free optical glass.

FIELD-MESH — A fine mesh screen employed in certain image orthicons which causes the scanning beam to approach the target perpendicularly at all points and prevents "beam-bending" due to charge pattern on the target. Provides a picture that is relatively free of unwanted bright edges or "overshoots" at the boundary of brightly illuminated portions of a scene. Improves dynamic registration in color pick-up applications and provides superior picture sharpness.

MICROMESH—A delicate, precision, electroformed mesh having 750 openings per linear inch. Prevents mesh-pattern and moiré effects without the need for defocusing. Improves picture detail when used with aperture-correction to provide full response.

SUPER DYNODE — The first dynode of the multiplier section in non-field mesh tubes designed to provide freedom from grainy background and to minimize dynode burn.

CONTROL GRIDS—Gold-plated to reduce thermionic emission.

DYNODES — Precision formed, spaced, and aligned to assure uniformity of signal gain in the multiplier section.

HIGH-GAIN DYNODE — The first dynode of the multiplier section in field-mesh tubes designed to increase the output signal-to-noise ratio and signal-output level.

X-RADIATION INSPECTED GUN ASSEMBLY—Assures accurate alignment of parts and spacing of electrodes.

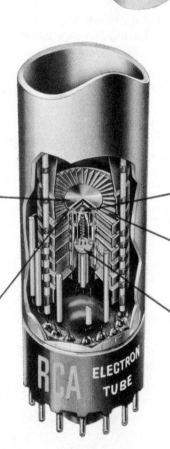

Figure 32-4. How the Orthican functions is shown above. (RCA)

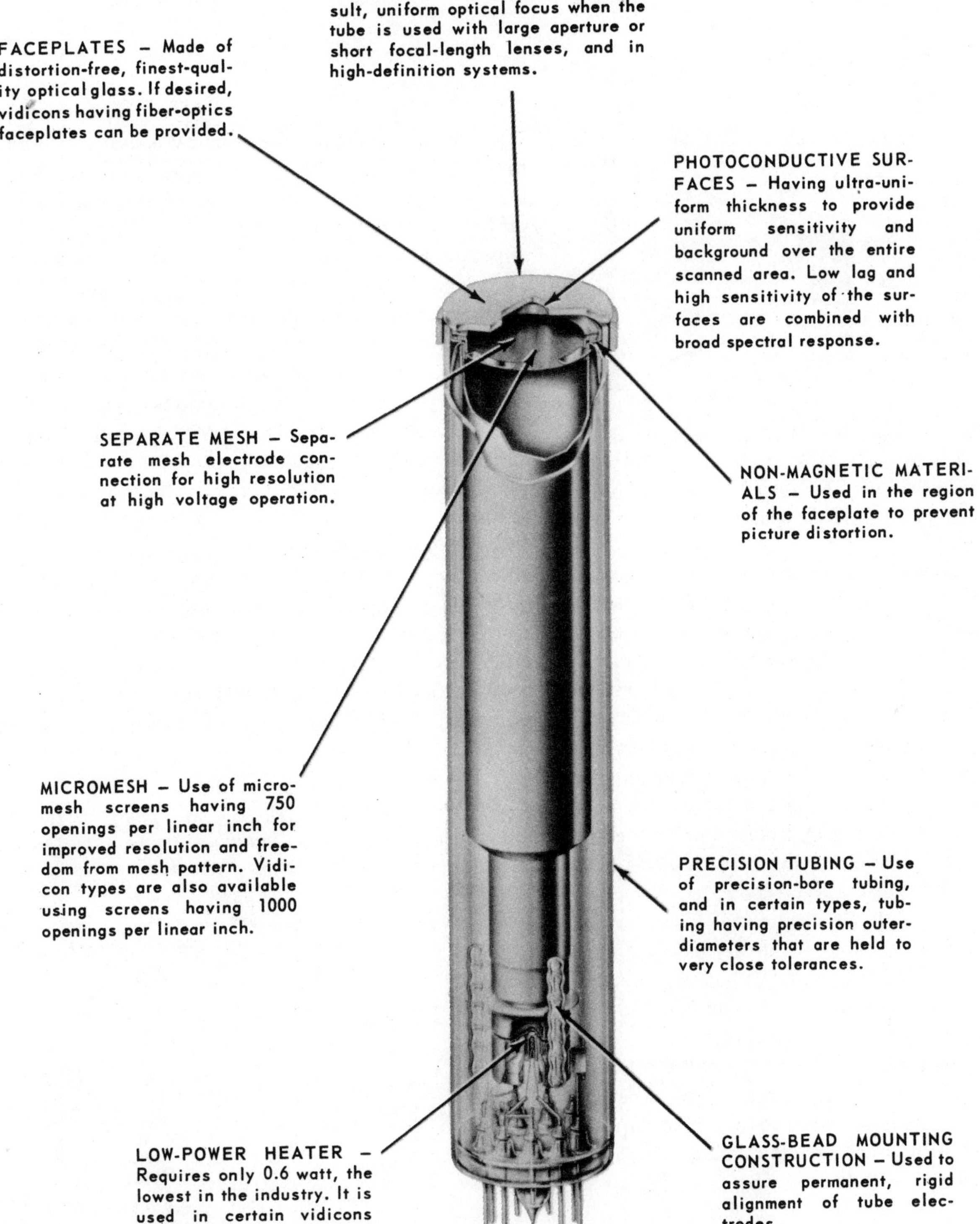
CUTAWAY OF A VIDICON
ALIGNMENT — Accurate alignment of tube faceplates such that the perpendicular through the tube axis and the faceplate is held within a tolerance of ± 1/4 degree. Net result, uniform optical focus when the tube is used with large aperture or short focal-length lenses, and in high-definition systems.
FACEPLATES — Made of distortion-free, finest-quality optical glass. If desired, vidicons having fiber-optics faceplates can be provided.
PHOTOCONDUCTIVE SURFACES — Having ultra-uniform thickness to provide uniform sensitivity and background over the entire scanned area. Low lag and high sensitivity of the surfaces are combined with broad spectral response.
SEPARATE MESH — Separate mesh electrode connection for high resolution at high voltage operation.
NON-MAGNETIC MATERIALS — Used in the region of the faceplate to prevent picture distortion.
MICROMESH — Use of micromesh screens having 750 openings per linear inch for improved resolution and freedom from mesh pattern. Vidicon types are also available using screens having 1000 openings per linear inch.
PRECISION TUBING — Use of precision-bore tubing, and in certain types, tubing having precision outer-diameters that are held to very close tolerances.
LOW-POWER HEATER — Requires only 0.6 watt, the lowest in the industry. It is used in certain vidicons designed for transistorized cameras.
GLASS-BEAD MOUNTING CONSTRUCTION — Used to assure permanent, rigid alignment of tube electrodes.

Figure 32-5. How the vidicon functions is shown above. (RCA)

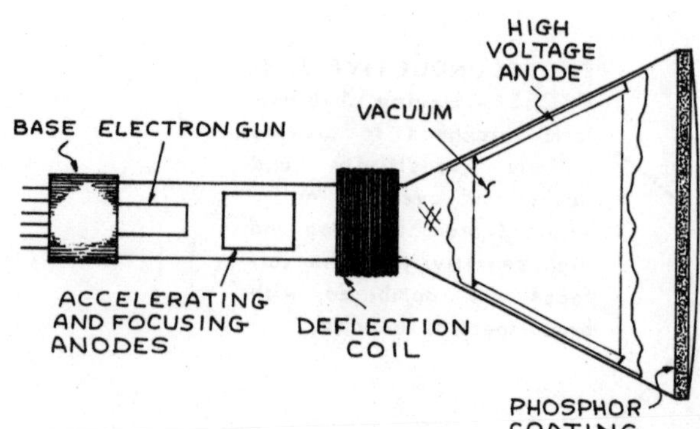

Figure 32-6. The elements of a TV picture tube are shown schematically. The various anodes focus and accelerate the beam from the electron gun. The coil about the neck of the tube deflects the beam to form the raster.

Each 262.5 presentation is called a "field" and the complete 525 line picture produced by two fields is a "frame." Thus 30 frames of motion are presented to the eye per second, more than enough to eliminate flicker, especially as there are actually 60 changes per second. The series of voltage pulses into which the studio scene was broken down has now been changed back into the original picture. The synchronization that made this possible was brought about by the synchronizing generator in the TV transmitter that fed its pulses to the camera and then sent them along with the picture via radio to the receiver.

The black and white (monochrome) television receiver that simulates colors with varying shades of gray has given way to the color TV that shows actual hues. Color TV has become a status symbol, and few skippers will permit monochrome aboard.

In essence, the internal functionings of the color TV and the black and white TV receiver are identical. Everything that has been described so far applies equally to both types. Additional circuits dealing exclusively with color have been added to the basic layout to achieve the rainbow display, and the picture tube is more complicated.

As shown in Figure 32–6, the black and white picture tube has one electron gun, and the phosphor coating on the inside face of the tube is all one kind. It gives off luminous light varying from white down to none, depending upon how hard the electron beam strikes. The color picture tube has three electron guns all focused onto one spot. The inside coating of the face of the color picture tube is composed of a myriad of globules, each consisting of three tiny dots of three different phosphors whose luminous colors are red, blue, and green, respectively. With the aid of masks and other means, one electron gun's beam will hit only the red dots, the other gun only the blue dots, and the third only the green dots. (The operating principle of a color picture tube is shown in Figure 32–7.)

Red, blue, and green are the primary additive projective colors and, either singly or in combination, they will reproduce all hues from white to black (although white is not a "hue," but the presence of all three, and black is the absence of all hues). For a color presentation each picture tube gun receives separate signals

Figure 32-7. The principle behind the color television tube is explained by this diagram. The red, blue, and green components of the scene are passed simultaneously to three equally colored projection electron guns whose beams strike fluorescent beads of like color on the picture tube screen. Mirrors and masks help in the orientation.

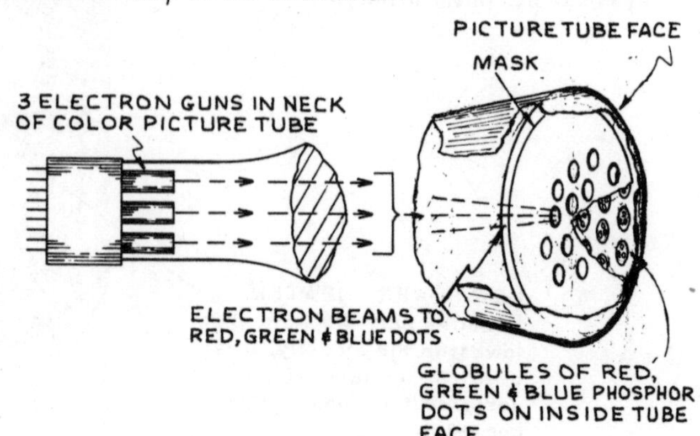

from the studio camera that now focuses the scene on its own red, blue, and green responsive orthicons or videcons.

Red, blue, and green projected in equal intensities on a screen will give the eye the sensation of white. That white light contains these primary colors may be proven easily enough by holding a glass prism in a beam of sunlight; all the hues from red to blue will be produced as a result of the refractive power of the glass. This feature of producing white by simultaneously projecting all three primary colors makes it possible for the color TV to receive and present a black and white program, as well as a color program.

The radio waves from the transmitter that are intercepted by the antenna of the color television receiver tell it whether the picture is to be in black and white or in color. A chroma signal in the complex video transmission permits the three color guns in the receiver to be controlled individually to produce the various hues on the tube screen. The absence of this chroma signal, as would be the case with a black and white transmission, locks the three guns together to produce white; a "color killer" takes over. Occasionally a color TV will show a color transmission in black and white—a sign that its color killer is out of whack.

The aim of television receiver designers is to make reception as automatic as possible, with a minimum of viewer-operated controls. Although the most modern deluxe receivers make color reception practically automatic, some control knobs still remain on the front of the receiver for the viewer and on the back for the service man. Often these controls can make the difference between an operative and an inoperative receiver and a familiarity with them may remove the need for a service call.

The received signal from a color transmitter contains four types of information that are sorted out by the detector in the receiver. They are the overall picture modulation, the synchronizing pulses, the chroma control, and the sound. The controls can affect any or all of these, and the following discussion concentrates on each. Figure 32–8 shows one method of beam orientation.

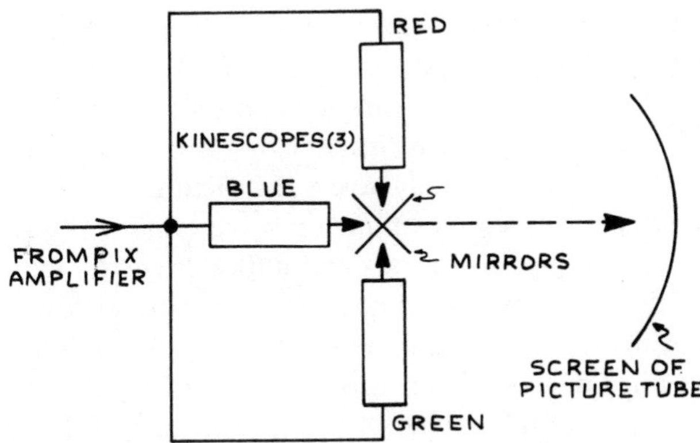

Figure 32–8. Special dichroic mirrors may be used for accomplishing beam orientation.

The sound in a television receiver may be considered a separate entity that differs little from what takes place in a standard frequency modulated (FM) radio. Two systems have been used. In one (now obsolete) the sound is taken off the incoming video signal directly after the tuner on front end and thence has its own amplifier-detector-amplifier path to the loudspeaker. In the other, known as the "intercarrier" system, the sound remains with the video until after the video detector and is then amplified, demodulated, and further amplified for the loudspeaker. Even though the picture is amplitude modulated, the sound in all television receivers is frequency modulated in the interest of higher fidelity.

The only adjustment of sound available to the viewer is the volume control that regulates loudness. Under certain conditions some of the picture controls will affect sound indirectly.

The "vertical hold control" keeps the picture on the tube in synchronization with the studio camera. Incorrect adjustment will allow the picture to roll either upward or downward across the screen. A simple manner of adjustment is to turn the control until rolling is induced and then to turn it slowly back until a black bar appears horizontally across the picture. This bar is made to move slowly to the top by careful knob turning until it just disappears and indicates correct sync.

The "horizontal hold control" keeps the picture in synchronization from left to right. Being turned too far in either direction from its correct position will cause the raster to tilt instead of being horizontal. Extreme maladjustment will crowd several partial pictures onto the screen from left to right. Horizontal synchronization depends upon the horizontal oscillator maintaining its correct frequency, and most TV receivers assure this with a circuit called "automatic frequency control" or AFC. When the horizontal frequency is brought close to normal by the manual hold control, the AFC takes over and keeps the horizontal oscillator on the hairline.

The "fine tuning adjustment" does what its name implies; it sharpens the image by bringing the front-end circuits into absolute resonance. On some receivers the fine-tuning adjustment is incorporated into the tuning mechanism and becomes a service check at the time of installation.

The "contrast control" determines how much the intermediate frequency amplifier will boost the received signal. Too little amplification (contrast control maximum counterclockwise) will result in a hardly visible picture or perhaps none at all. With the contrast control maximum clockwise the picture will be too dense because of over-amplification. The correct position is largely a matter of personal preference.

A standard, nonautomatic color television receiver will add two viewer-operated controls to those normal for a black and white receiver. These are the "color control" and the "hue control." Both affect the true-to-life appearance of the color picture. The color control is an overall adjustment for the total amount of color generated. The hue control might be considered to "fine tune" the naturalness of the scene, especially insofar as the appearance of the human complexion is concerned. (See Figure 32–9.)

Adequate signal strength is the main problem with onboard television. This is not a problem at home; a weak signal situation may always be countered ashore with more elaborate antennas or, with that final solu-

Figure 32-9. Color TV receivers often add additional viewer-operated controls, although the most elaborate modern receivers have eliminated most controls and have made operation almost automatic; in the better receivers it is completely automatic.

tion, the cable. Obviously, the multi-element Yagi antenna, complete with rotator, that comes to the rescue at home, is not for a boat. "Snow" (white specks covering the screen) in an otherwise correctly functioning television receiver is an indication of a very weak signal.

Skippers who dock or anchor within a few miles of a television transmitter usually find that satisfactory reception is attainable with the adjustable dipole antenna popularly called "rabbit ears." A simple alternative often is a folded dipole antenna made from twin-lead, a flat plastic ribbon that encases two wires and serves as a lead-in. (See Figure 32–10 for the details of construction.)

Several commercial antennas are offered on the market to alleviate the more difficult cases of weak signal strength. These are tailored to pleasure boat use and do not look too outlandish. Some have an electronic radio frequency amplifier built right in, and this is the preferable way to go. The amplifier derives its power from below via the down lead.

Whereas the ship's communications antenna should be sensitive in all directions (see Chapter 24), an antenna with such omni-directional characteristics is not the correct choice for television viewing. In most locations it would be productive of "ghosts" on the screen. The reason for the production of ghosts is explained in Figure 32–11.

Assume the boat with the receiver to be five miles distant from the transmitter in a straight unobstructed line. Assume further that a large, flat-surfaced building is also five miles from the transmitter at an angle to the right and that the distance from this building to the vessel is five miles. It can be seen that the direct wave travels five miles while the reflected wave must travel ten miles. If the two waves go through space at the same speed (and by natural law they must) then the reflected wave arrives later and its picture will be a duplicate slightly to the right of the original on the screen. A ghost has been born. A sharply directive antenna would discriminate against this reflected wave.

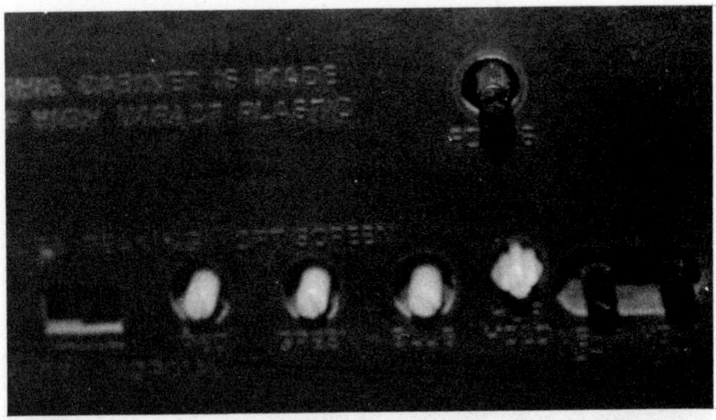

Figure 32-10. An antenna that may prove sufficient in strong signal areas may be made from twin lead, connected together as shown.

One form of television antenna seen quite often on pleasure boats is a folded dipole, constructed with wide metal bands instead of rods in order to increase the span of frequency coverage. A radio frequency amplifier is built into the support and a pilot light glows when power is received from below. Coaxial cable connects with the television receiver. (See Figure 32–12.) It seems that installers usually clamp this antenna to an available VHF or SSB antenna. This is not a good idea for two reasons: A dipole is sensitive only along a line at right angles to its width, and pressing

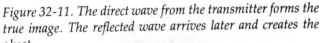

Figure 32-11. The direct wave from the transmitter forms the true image. The reflected wave arrives later and creates the ghost.

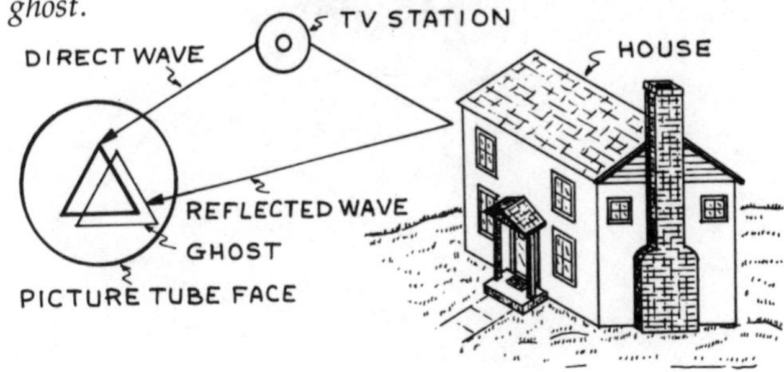

Figure 32–12. The unique dipole shown above is often seen on pleasure boats to capture television.

Figure 32–13. This large plastic "pie" contains a rotatable antenna controllable from below.

the microphone button can transfer sufficient energy to harm the TV set.

Another popular pleasure boat TV antenna is shown in Figure 32–13. It gives the appearance of a large plastic pie plate with a cover. Inside is a power-driven rotating dipole that is responsive to a remote control unit located at the receiver. Here also the connection from antenna to set is made with coaxial cable. (Note that coaxial cable is available with various impedances; the one to choose is the one that matches the input impedance of the receiver. In most instances the correct coaxial cable will be RG59U at 75 ohms.)

The down lead in most home television installations is twin lead wire. This could also be employed on a boat if both antenna and receiver will accept the much higher impedance. However, twin lead requires care in location; it should be away from sources of electrical noise and from large metal surfaces. By contrast, coaxial cable is self-shielding and may be run most anywhere except the bilge.

A shoddy antenna at bargain price will soon succumb to the tough marine environment. Ordinary steel invariably becomes a waste of money. The metal in the chosen antenna should be stainless steel or chrome-plated brass.

Tiny moulded transformers are available for matching the impedance of the down lead to that of the antenna or to that of the receiver when necessary.

33. *Hi-Fi on Board*

Music calms the savage breast, not only of landsmen but of skippers as well, and so the installation of entertainment radios, stereos, and tape decks has become almost as lavish on the boat as at home. Furthermore, the ears of the listeners are now more sophisticated and are no longer satisfied with the tinny tones from cheap reproducers. Life-like stereo sound is demanded and nothing less than what is popularly called "high fidelity" will do. Good entertainment sound on board consequently requires a substantial investment.

A definition of "high fidelity" runs into that old conundrum of "how high is high?" Reproduction that has absolute fidelity to the original is not yet technically attainable outside a laboratory. How close to perfection reproduction comes is relative, and so the term "high fidelity" itself is relative. Perhaps the least controversial definition is "sound that resembles the original closely and is pleasing to the listener."

Despite the exactitude possible with modern technical measurements, the concept of fidelity remains subjective and highly personal. The meters may agree on near perfection, and yet the listener is not pleased with what he hears; his ears do not consider the sound "natural." He will manipulate a tone control, introducing obvious technical distortion, and the resultant audio output will be "high fidelity" *for him*. Thus audio engineering must constantly be diluted with human response and compromise.

Human response is a vastly complicated thing because human ears are one of nature's marvels. The ear is not fooled so easily as is the eye. If you project pictures more rapidly than about twenty times per second, the eye loses count; persistence of vision blends everything into one. By contrast, the ear can recognize and distinguish up to twenty *thousand* vibrations per second. Such acuity of hearing means that even tiny imperfections in sound will be noticed. This newly critical listening attitude by the public is the reason for the almost universal switch to high fidelity equipment.

Sound is a wave motion that consists of alternate compressions and expansions in the medium through which it is travelling to the hearer. This medium may be almost any material from solids to liquids to gases to air with a different speed of propagation through each. Sound travels fastest in hardened steel and slowest in air, although it cannot traverse a vacuum. The velocity of sound in air, which is the parameter of interest in connection with high-fidelity audio, is approximately 1,100 feet per second regardless of pitch or frequency. Doubling the frequency of any sound raises the pitch by one octave in musical terms. There are interrelations within an octave that distinguish the various musical scales.

The human auditory system, in optimum condition, can hear the wide span of frequencies between 30 hertz and 20,000 hertz. (By comparison, the eye can respond only to the narrow band of frequencies lying between blue and red.) Sound waves enter the outer ear and impinge on the eardrum, vibrating it in unison with the arriving frequencies. The eardrum, in turn, actuates the bones of the middle ear, the hammer-anvil-stirrup, that convey the impulses to the inner ear and finally to the cochlea. The auditory nerve, attached to the cochlea, responds by sending a coded electrical message to the brain where identification of the sound is made. (See Figure 33–1.)

Figure 33-1. The functioning of the human ear is shown schematically. Sound is collected by the outer ear, led to the middle ear and actuates the ear drum. The auditory bones amplify the sound from the ear drum. The auditory nerves transform the vibrations into electric currents that go to the brain.

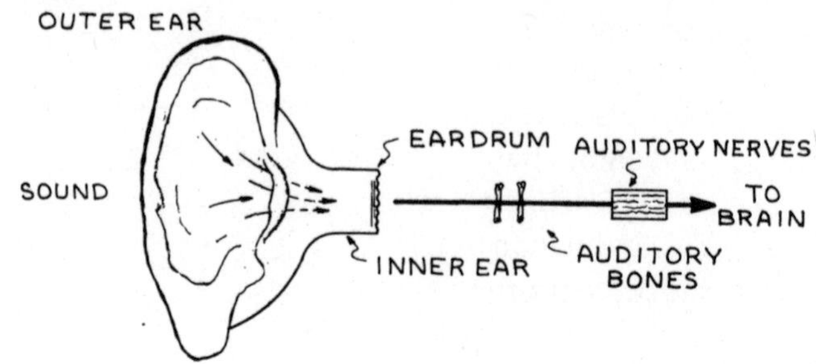

The ear is able to detect the tiny interval of time it takes a sound wave to pass from one ear to the other, in technical language, the difference in "phase" at the two ears. Although this time interval is only a fraction of one-thousandth of one second, it is part of the process that enables the ear to pinpoint the direction from which the sound is emanating. This ability of the ear to discern direction enables a listener in a concert audience to position each instrument on the stage even with his eyes closed; it provides spatial presence. The idea behind stereophonic reproduction of sound is to duplicate this feeling of presence.

We listen with two ears, and each ear hears the sound slightly differently because of the phase change just described. But when a listener sits in front of a single loudspeaker he hears only a point source of sound regardless of the size of the orchestra that is playing. Placing two loudspeakers before him, each fed by its own microphone, electrically transports both ears to the site of the performance and, in theory at least, restores his binaural hearing. Hence the two speakers of stereophonic sound reproduction and the four speakers of later improvements. (See Figure 33–2.)

Supplying the two channels of information required for stereo has evolved from very crude to highly technical. The first stereo broadcasts were accomplished by listening to two radios, one AM and one FM, simultaneously. The two microphones for the two broadcasting systems were side by side in the studio, one serving as the left ear and the other as the right. Judging from what is known about AM and FM transmissions, the result of this early striving could not have been a star performance. The breakthrough that has made modern stereo broadcasting possible was the elimination of two discrete transmissions by "multiplexing" the left and right channels into one broadcast.

The multiplexed signal is at present used only on frequency modulation stations. One channel is modulated directly onto the FM transmission. The second channel modulates a sub-carrier that in turn is modulated on the main carrier. Pilot signals also go along. The receiver decodes this into two discrete channels, each with separate amplifier and speaker. Phonograph records and tape decks also provide two channel stereophonic reproduction.

Figure 33-2. FM receiver, power amplifier and record player are stacked conveniently in a corner of the salon on this yacht. Stereo speakers are located in various locations on the boat.

The sources of music for high-fidelity reproduction aboard may be radio, records or tape. In this connection, the designation "radio" is restricted to frequency modulation (FM) exclusively. Amplitude modulation (AM) is inherently too noisy, and Federal Communication Commission rules permit AM broadcasters to use only a limited span of audio frequencies. By contrast, leading FM stations broadcast all the frequencies the human ear can hear.

Frequency modulation radio receivers are especially suitable for use on pleasure boats because their circuitry is reasonably deaf to ignition and other electrical noises, whether on board or atmospheric. Antenna requirements for FM receivers are easily met and FM broadcasters are now so numerous that a station is almost always near enough to produce a good signal. The larger FM stations all transmit stereophonically, but their emissions are "compatible" and monophonic receivers are able to use them fully without loss of the inherent high quality.

In areas of strong signal, a few feet of wire connected to one of the antenna posts of the FM receiver will capture enough voltage to produce full response. In lesser areas, a folded dipole constructed of twin-lead wire, as shown in Figure 32–10, will bring in sufficient signal. In the few remaining areas of poor signal more elaborate antenna provisions must be made—often in conjunction with the television antenna.

Phonograph records are an excellent source of musical program material for high fidelity systems. However, records are not at their best on boats and, in fact, may be used routinely only on the larger, more stable vessels. The reason is the feather weight the best phono cartridges impose on the record groove and the consequent ease with which even a slight roll of the boat will send the tone arm slithering across the platter. This is obviously harmful both to the record and to the delicate diamond point of the pickup. Incidentally, greater pressure by the tone arm on the groove is not the answer; it would ruin records quickly. Some deluxe record players have "anti-skating controls" that reduce the problem of a wildly sliding tone arm.

The hull, planking, and structural members of a boat are excellent transmitters of sound; engine and other vibrations are constantly traversing them. Unless the record player is mounted on sound-absorbing feet, these vibrations will find their way to the pickup cartridge and will emerge from the loudspeakers fully amplified and annoying. Simply mounting the player on pieces of rubber is not enough. It takes more sophisticated mountings to absorb the interfering noises and yet not so spineless that they wobble like a gelatin dessert.

It's a far cry from the wax cylinder of Edison to the longplaying record of today—and even from the brittle 78 rpm record that served during the interim. The improvement in the composition of the record material alone has reduced needle noise to a negligible level. The playing time of a record has been multiplied by ten. The range of frequencies that are reproduced has been extended to cover the span of hearing. Original course grooves have been replaced with microgrooves that are cut 250 to the inch. Turntable revolutions have been standardized at 33-1/3 per minute, greatly extending playing capacity. Most marvelous of all, each groove carries two distinct channels of information to make stereophonic reproduction possible.

Phonograph records and playback equipment are now produced to an internationally agreed upon standard and this makes any combination universally playable. A record pressed in Germany is as compatible with a player manufactured in Japan as if both came from the same factory. The standard specifies that the left and right channels shall be cut into the groove at right angles to each other and at 45° to the surface. Sounds to the left and to the right of the recording microphone are maintained on the record in their proper relationship so that the sense of correct directivity shall not be lost on reproduction.

The "needle" of a high quality pickup head is a tiny diamond whose angle and point are ground to exacting specifications. In a common type of pickup, the needle is an extension of a miniature iron armature that moves within two small coils in response to the sound track the needle is traversing. One coil develops the voltage that actuates the left channel amplifier; the other coil performs the same service for the right channel amplifier. (See Figure 33–3.)

Reproducing the desired high fidelity music from

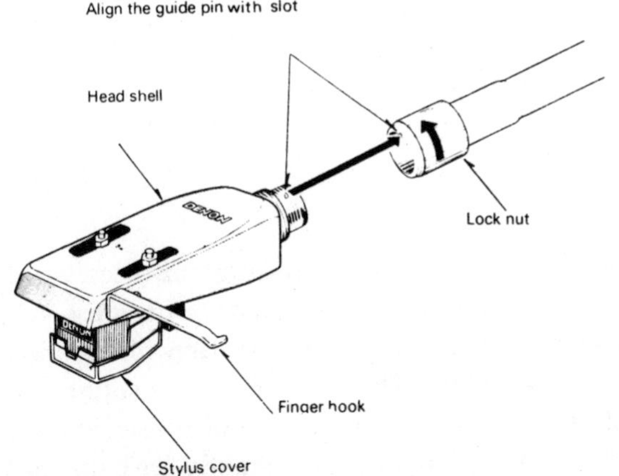

Figure 33-3. The phono pickup must be sensitive enough to respond to the slightest changes in the record groove without exerting undue pressure upon it. The "needle" or stylus is a carefully ground diamond. A stereo pickup simultaneously follows a right channel groove in one plane and a left channel groove in another.

magnetic tape is more suited to conditions on a boat because this method is immune to rolling and other normal marine motions. The original magnetic recordings were made on steel wire that was passed under recording and reproducing heads. The magnetic fields impressed on the wire during recording induced equivalent voltages in the reproducing head that activated amplifiers, headphones, or speakers. The system was crude and more a laboratory toy than a practical device. Magnetic tape took over easily and even in its initial stages was far superior.

Magnetic recording tape is a strip of plastic about one-quarter inch wide and usually less than two one-thousandths of one inch thick. It is coated with selected oxides of iron that form the magnetizable substance. The plastic base is chosen for its ability to resist loss of flatness, stretch, brittleness, and moisture absorption. Acetate was in use but present day choices are more sophisticated plastics. Unrecorded tape is commercially available in cassettes and in long lengths on large reels. Most recorded music tapes for home use are in cassette form.

The magnetic coating on the tape is, of course, the actual repository of the recorded sound. This coating must be extremely uniform to reduce background noise on replay. It must be homogeneous to avoid dead spots. It must adhere permanently to the tape and its surface must be polished to reduce friction when it slides by the recording and reproducing heads. One difficulty that may be encountered with tapes that are stored tightly reeled for long periods of time is the transfer of the recording to the adjacent layer.

Magnetic tape is "erased" in the same manner that steel objects are made to lose their magnetism—by being subjected to a strong alternating current magnetic field. An erasing head to do this is part of the tape equipment. By the same token, recorded tapes should not be placed in the vicinity of electrical appliances that may radiate stray fields.

The latest development in the ever-continuing search for more life-like sound reproduction and the feeling of being present at the performance is "four channel stereo." This adds two loudspeakers and two channels of information. Thus it has a speaker for the left front, one for the right front, one for the left back, and one for the right back. The listener is truly wrapped in sound.

True four channel stereo requires four distinct and separate channels fed by four microphones at the four opposite points of the stage. A matrixed form of four channel reception also has been advocated; it derives the two additional channels by electronically modulating the two normal stereo channels and thus is not a true four channel system.

Figure 33-4. The principle of the phono pickup is shown schematically. The needle following the sound pattern in the record groove causes the attached coil to move and "cut" the lines of force of the magnet. This generates the voltage that is sent on to the amplifier.

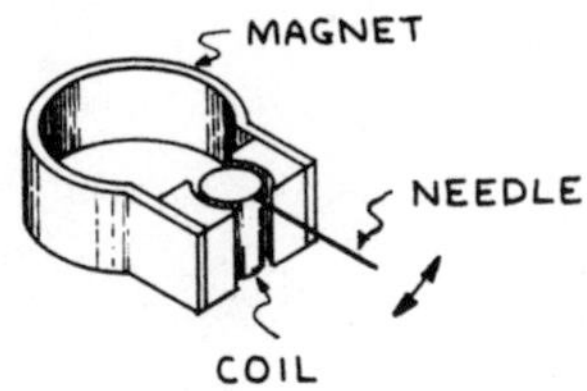

Magnetic tape is the recording medium most easily adapted to four channel sound. There is no problem in recording four channels on the tape, side by side; reproducing heads with four separate gaps and coils are on the market. Two amplifier/speaker systems are added to the two normally in use for standard stereo.

Putting four channels into each groove of the phonograph record is considerably more difficult, but it has been done via a high frequency carrier wave. The carrier is in the ultrasonic range, and it is modulated by the channel informations; in essence the two cuts now in the groove for stereo are each doubled up. The problem here is one of longevity: The ultra-fine undulations of the carrier wave are worn down quickly.

The four channel system does not fit into the present FM broadcasting scheme, even though the stations are equipped to transmit stereophonically. Additional equipment is required, and the question becomes one of business economics.

"How much amplifier power is needed for adequate sound reproduction?" That question often arises—and the answer is not as obvious as it may seem. The power required to fill spaces the size of the average pleasure boat cabins and salons is minimal. One watt would be more than sufficient. But that is not the whole story. The one watt of sound output would play the normal sections of a musical program adequately, but would fall down miserably when fortissimos and deep basses appear. These could take many watts. A safe choice for most boats would be amplifiers capable of twenty watts output played with the volume control at a comfortable level.

A listener sitting in an auditorium receives his sound from many directions, not only from the stage. Reflections of the original sound reach him from the walls and the ceiling. These reverberative sounds have a great deal to do with the feeling of presence the listener receives, and attempts constantly are being made by equipment manufacturers to introduce them into the home reproduction.

Each reflected sound arrives at the ears of the listener a tiny fraction of time later because it has further to travel. Thus it is characterized by delay and an equivalent delay may be introduced electronically in the reproduction to give the feeling of a reverberation. The reverberation time is such an important parameter that halls and auditoriums are rated by it.

The close confines of a boat are psychologically broadened when reverberation is added to the reproduced music. The electronical delay circuits for achieving reverberation are included in some deluxe equipment and may also be purchased separately as a later add-on. The amount of reverberation is controllable to suit the enclosure in which the music is being heard. Drapes, carpets, glass, cubic volume—all have an effect.

The placement of loudspeakers in the cabin or salon is an art and not a science. Trial and error is the word here. Outside of keeping the left channel on the left and the right channel on the right, not much more can be given in the way of directions. The speakers are moved around until the sound gives maximum pleasure.

GLOSSARY

A abbreviation for ampere, unit of current flow

AGC automatic gain control

ALTERNATION one-half of a cycle of alternating current

AMMETER instrument for measuring current flow in amperes

AMPLIFIER a circuit for increasing the strength of voltage or current

AM amplitude modulation

ANTENNA an exposed conductor for radiating or accepting radio energy

ANTENNA ARRAY a group of antennas

ANTENNUATE to decrease the strength

AUDIO the frequencies within human hearing ability

AUTO TRANSFORMER a transformer with only one winding

B abbreviation for Beta, designation for transistor current gain

BANDWIDTH the range of frequencies that a circuit will pass or reject

BEAT FREQUENCY the resultant of two dissimilar interacting frequencies

BIAS a voltage applied to a circuit to control its action

BRIDGE four elements arranged like the bases of a baseball diamond

BUFFER an amplifier placed between two circuits to prevent interaction

BYPASS a capacitor providing a "detour" for alternating currents

CAPACITOR two conductors separated by an insulator

CARRIER the main transmitted frequency upon which intelligence is impressed

CARD industry name for a small printed circuit board

CHARACTERISTIC IMPEDANCE a property of a transmission line

CHANNEL a circuit reserved for one form of intelligence

CHOKE an inductance used to impede alternating current

CLASS A one form of amplifier operation (also Class B & Class C)

CONDENSER another name for capacitor

COAXIAL CABLE a cable whose inner conductor is shielded by a metallic outer coating

COLLECTOR the output element of a transistor

CONDUCTANCE the opposite of resistance

CRYSTAL a substance such as quartz that may be ground to a specific frequency

CYCLE a complete move from zero to positive to zero to negative and back to zero

DARLINGTON a method of connecting two transistors

DEGENERATION a method of steadying the output of an amplifier

DEMODULATE to retrieve the intelligence within a carrier

DETECTION another name for demodulation

DIELECTRIC the insulator between two plates of a capacitor

DIELECTRIC CONSTANT a measure of value of a dielectric

DIODE a vacuum tube or semiconductor having only two elements

DIPOLE an antenna with two elements

DIRECTOR an element of a multi-element antenna

DOPE an impurity introduced into a semiconductor mix

DYNAMIC MICROPHONE a microphone activated by a moving coil

DECIBEL a unit for measuring increase or decrease of amplitude

DOLBY a circuit for reduction of noise

ELECTROLYTE a solution capable of carrying electric current

ELECTRON the portion of the atom that carries a negative charge

EMITTER one element of a transistor

EMITTER FOLLOWER one form of transistor circuit

ELECTROSTATIC FIELD the condition between two charged conductors

FET field effect transistor

FM frequency modulation

FEEDBACK a portion of the output returned to the input

FARAD the unit of measurement of the capacitance of capacitors

FLUORESCENCE light emitted when under electronic bombardment

FLIP FLOP a circuit in which the transistors alternate being on and off

FREQUENCY the number of times an action occurs within one second

FLUTTER unwanted changes in frequency caused by defective drive of records or tape

FILTER a circuit that passes or rejects designated frequencies

FULL WAVE a rectifier that uses both alternations of a cycle

GAIN ratio of the output to the input

GERMANIUM a semiconductor material used in the construction of transistors

GIGA prefix indicating multiplication by one billion

GROUND a common connecting point of zero potential

HALF WAVE a rectifier that uses only one alternation of each cycle

HARMONIC a full integer multiple of a frequency, such as second, third, and so forth

HEAT SINK a metal attached to a transistor for radiating away heat

HERTZ new name for cycle per second

HENRY unit of inductance

HETERODYNE interaction of two dissimilar frequencies

HYSTERESIS the lagging effect some materials have on an alternating current magnetic field

IMPEDANCE opposition offered to the flow of alternating current

INDUCTANCE a property of a wire or coil carrying current

INTEGRATED CIRCUIT a complete circuit built on a chip of semiconductor

INTERMEDIATE FREQUENCY the fixed frequency of the main amplifier in a superheterodyne receiver

ION a particle that has lost or added electrons

INTERMODULATION a form of internal distortion in a circuit

IONOSPHERE a layer of ionized particles above the earth

KILO multiplier prefix (by one thousand)

KLYSTRON a tube that maintains oscillations by cavity resonation

L symbol for inductance

LC symbol for an inductor-capacitor circuit

LIMITING the cutting down of amplitude to a constant level as in an FM receiver

MAGNETRON an oscillator tube used in radar transmitters

MEGA a multiplier prefix (by one million)

MILLI a divider prefix, one-thousandth

MICRO a divider prefix, one-millionth

MODULATE to impress intelligence on a higher frequency carrier

MIXER a circuit unit that combines two frequencies

MOTORBOATING a defect in amplifiers that sounds like boat exhaust

MULTIVIBRATOR a relaxation type oscillator

MULTIPLEX a system of impressing several channels on one carrier

NEUTRALIZATION a method of stabilizing an amplifier

NPN one of the two forms of transistors

NEGATIVE FEEDBACK a method of improving the linear response of an amplifier

OHM the unit of resistance

OP AMP operational amplifier

OSCILLATOR a circuit that generates a desired frequency

OSCILLOSCOPE an instrument that makes wave forms visible on a cathode ray tube

PARALLEL one form of connecting a number of units into a circuit

PIEZOELECTRIC the generation of voltage by stressing a crystal

POTENTIOMETER a variable resistance with three terminals

PARASITICS unwanted oscillations in a circuit

PHASE the time relationship between two wave forms

PICO a divider prefix, one-trillionth

PNP one of the two forms of transistor construction

POWER the rate of doing work

POWER FACTOR the ratio of actual power to the apparent power

PUSH PULL an amplifier circuit using two tubes or two transistors

Q a figure of merit for circuits and coils; also prefix for transistors

QUADROPHONY a system of reproduction using four speakers

R abbreviation for resistance

RADIATION RESISTANCE a parameter relating to antennas

RF abbreviation denoting radio frequency

REACTANCE opposition to the flow of alternating current

RECTIFIER device for changing alternating current to direct

REFLECTOR an element in a multi-element antenna

RESONANCE the state of being in tune

REVERBERATION the effect of multiply reflected sound

RUMBLE mechanical sound of the turntable carried to the speaker

RESISTANCE opposition to the flow of current

RIPPLE fluctuation in the output of a rectifier

RMS abbreviation for root-mean-square

SERIES a method of connecting units into a circuit

SHUNT a resistance in parallel

SOLENOID an electrical coil with a hollow center used as a magnet

STANDING WAVE condition formed when two waves travel a transmission line in opposite directions

SUPERHETERODYNE a circuit in which a received signal is amplified at several frequencies

SIGNAL TO NOISE RATIO ratio of wanted signal to unwanted noise in a circuit; a mark of merit

SQUELCH to cut out interstation noise when tuning

SUB-CARRIER a carrier impressed upon another carrier

SWEEP CIRCUIT the source of the voltages that carry the display over the cathode ray tube face

SCR silicon controlled rectifier

SILICON a semiconductor material

SOLDER a low melting point metal based on tin

TANK CIRCUIT a combination of elements that oscillates at a given frequency

THYRATRON a vacuum tube used to control the flow of current

TRANSIENT unwanted voltage or current pulse

TRIGGER to instigate action of a circuit

TWEETER a loudspeaker that is built to handle high frequencies

UHF ultra high frequency

UJT unijunction transistor

UNBALANCED LINE certain transmission lines, such as the coaxial

V abbreviation for volt

VARACTOR semiconductor diode with special features

VECTOR a line whose direction and length each denote given quantities

VOLT the unit for electrical pressure

VHF very high frequency

W abbreviation for watt

WATT unit of electrical power

WAVELENGTH distance from one positive peak to next positive peak

WOOFER loudspeaker designed to handle low frequencies

WOW slow changes in frequency caused by uneven turntable speed

X abbreviation for reactance

Z abbreviation for impedance

INDEX*

*Page numbers in italic (e.g., *39*) indicate that the reference is to the illustration or chart on the page indicated.

C

L

M

N

Q

R

S